INSECT CHEMICAL ECOLOGY

INSECT CHEMICAL ECOLOGY
AN EVOLUTIONARY APPROACH

EDITED BY
BERNARD D. ROITBERG &
MURRAY B. ISMAN

Chapman & Hall
New York   London

First published in 1992 by
Chapman and Hall
an imprint of
Routledge, Chapman & Hall, Inc.
29 West 35th Street
New York, NY 10001-2291

Published in Great Britain by
Chapman and Hall
2-6 Boundary Row
London SE1 8HN

Library of Congress Cataloging in Publication Data

Evolutionary perspectives on the chemical ecology of insects/edited
  by Bernard D. Roitberg and Murray B. Isman
     p.   cm.
  Includes bibliographical references and indexes.
  ISBN 0-412-01871-3—ISBN 0-412-01881-0 (pbk.)
  1. Insects—Ecophysiology. 2. Chemical ecology. 3. Insects—
  Evolution. 4. Insect-plant relationships. I. Roitberg, Bernard
D., 1953– . II. Isman, Murray B.
QL495.E93 1992
595.7052'22—dc20   92-12249
                                      CIP

British Library Cataloguing in Publication Data also available.

This book is dedicated to the memory of

Joshua Aaron Isman
(1991–1992)

and

Sylvie Marie Amar
(1991)

Contributors

May Berenbaum
Department of Entomology
320 Morrill Hall
University of Illinois
505 S. Goodwin Ave
Urbana, IL 61801
U.S.A.

M. Deane Bowers
Department of Environmental,
   Population and Organismic Biology
Campus Box 334
University of Colorado
Boulder, CO 80309
U.S.A.

Michael A. Caprio
Department of Entomology
University of California
Berkeley, CA 95616
U.S.A.

Marcel Dicke
Department of Entomology
Agricultural University Wageningen
P.O. Box 8031
6700 EH Wageningen
The Netherlands

Murray B. Isman
Department of Plant Science
University of British Columbia
Vancouver, BC V6T 1Z4
Canada

John Jaenike
Department of Biology
University of Rochester
Rochester, NY 14627
U.S.A.

Jeremy N. McNeil
Department de Biologie
Universite Laval
Sainte-Foy, P.Q. G1K 7P4
Canada

Daniel R. Papaj
Department of Ecology and
   Evolutionary Biology
University of Arizona
Tucson, AZ 85721
U.S.A.

P. Larry Phelan
Department of Entomology
Ohio Agricultural Research and
   Development Center
Wooster, OH 44691
U.S.A.

Mark D. Rausher
Department of Zoology
Duke University
Durham, NC 27706
U.S.A.

Bernard D. Roitberg
Behavioural Ecology Research Group
Department of Biology

Simon Fraser University
Burnaby, BC V5A 1S6
Canada

Maurice W. Sabelis
Department of Pure and Applied
  Ecology
University of Amsterdam
Kruislaan 302
1098 SM Amsterdam
The Netherlands

David Seigler
Department of Plant Biology
University of Illinois

Urbana, IL 61801
U.S.A.

Bruce E. Tabashnik
Department of Entomology
University of Hawaii
Honolulu, HI 96822
U.S.A.

Mark L. Winston
Department of Biological Sciences
Simon Fraser University
Burnaby, BC V5A 1S6
Canada

Contents

Preface

This volume is designed to provide a comprehensive treatment of insect chemical ecology from an evolutionary perspective. Great progress has been made toward answering the more causal aspects of chemical ecology, but ecological and functional questions have been addressed far less often. Here, we have attempted to address both causal and functional issues in insect chemical ecology in a balanced and complimentary manner.

We are pleased to have included contributions from both established scientists and some highly capable up-and-coming researchers in this volume. Unlike other volumes with similar titles, we have not asked the authors to review their own work or related research findings, but rather, we have posed problems for them to explore. By encouraging their speculations and novel approaches to important problems and issues in insect chemical ecology and evolutionary biology, we hope that this book will provide raw material for research ideas and discussion. We hope that both field- and theoretically-oriented scientists will find our approach stimulating.

The volume is divided into two parts. In the first part, problems of a more general nature are addressed. Insect chemical ecology is considered in the broad sense from the evolutionary, behavioral, physiological, and chemical viewpoints, with the overall goal of identifying boundary conditions for evolutionary processes. In the second part, specific problems are considered from the perspectives established in the first part. Here, the authors employ both functional and causal analyses to areas such as learning and sociality in insects.

We wish to thank all the authors for their efforts and perseverance, as well as the many peer reviewers whose comments and suggestions invariably led to the improvement of each chapter. We also thank our graduate students and research associates for stimulating discussions on several of the topics covered in this volume. Finally, we express our appreciation to the Natural Sciences and Engineering Research Council of Canada (NSERC) for their continuing support of our respective research programs throughout the period in which this book was prepared.

Bernard D. Roitberg
Murray B. Isman
June 28, 1991
Vancouver, Canada

Part I

General Considerations

Section Overview

Many, if not most, inter- and intraspecific interactions among insects and between insects and other organisms are mediated to at least some degree by chemicals acting as signals or messengers. The often-breathtaking specificity of these chemical mediators in extant ecological interactions suggests a long history of the uses of such chemicals by insects, molded and tuned by evolutionary processes. In this volume, we have chosen to view insect chemical ecology from an evolutionary perspective; our underlying purpose is to seek an understanding of the processes that shape these chemically-mediated interactions in insects, with the ultimate goal of predicting the directionality of evolution in this regard.

In attempting to understand the processes responsible for the evolution of chemically-mediated systems in insects, one must understand both the basic principles of evolution (i.e., how natural selection acts on interacting organisms) and possible constraints on evolution arising from the chemistry of natural products themselves (i.e., their physical and biological properties).

This first part of the book, comprising Chapters 1–5, is intended to provide a strong foundation upon which questions of a more specific nature are addressed in Chapters 6–11.

The first chapter seeks to provide a rationale for adopting an evolutionary perspective in conducting studies of chemical ecology. Roitberg makes a plea to consider the fundamental "four questions" raised by Tinbergen, in particular, the how and why questions corresponding to proximate mechanisms resulting in a phenomenon and the ultimate (or evolutionary) reasons underlying the phenomenon. To emphasize the importance of

1

considering all aspects of a problem in chemical ecology (lest invalid conclusions be drawn), he refers to his own area of expertise, marking pheromones, and in particular to the problem of "self-marking" by parasitoids. A simple model built around such a system illustrates the utility of this approach.

Chapter 2 presents a comprehensive treatment on the causes of evolutionary change in organisms and identifies some important unresolved evolutionary issues in chemical ecology. Rausher begins by providing working definitions of evolution and its two main processes, genetic drift and natural selection. He then proceeds to discuss ways of estimating the magnitude of variation and selection and how evolutionary change may be predicted. Although much of the ground covered relates to evolution in the broadest sense, Rausher has taken considerable care to make his definitions and mathematical treatments valuable for those investigating phenomena in chemical ecology in particular by detailing some of the common problems encountered in applying genetic methods and analyses to such studies. In the latter half of the chapter, Rausher addresses a long-standing controversy, namely the role of plant secondary compounds as defenses against herbivory and discusses the possible evolution of plant defenses against herbivory. The concept of coevolution, often used inappropriately, is defined by Rausher, and a plea is made for empirical evidence. Finally, the evolution of diet breadth (i.e., to specialize or generalize) in insects is discussed with reference to the author's previous studies addressing that question.

In Chapter 3, Berenbaum and Seigler examine plant natural products as mediators of plant-insect relationships, initially focusing on the availability of elements and simple molecules for the production of allelochemicals, then turning to biological factors (plant phylogeny, energetics, and toxicology) which constrain the synthesis of such substances. The authors discuss both the positive selecting agents (pollinators) and negative ones (herbivores) in relation to plant chemistry. Coevolution and plant chemistry are again explored, but in this case, on the macroevolutionary scale; the authors present three possible scenarios to account for the "explosion" of phytochemical diversity in angiosperms.

Both the proximate and ultimate factors underlying chemical information conveyance are discussed in Chapter 4. Dicke and Sabelis limit their perspective at the outset to those chemicals that elicit behavioral responses in arthropods, reviewing how chemical signals can be used "legitimately" between organisms and "illegitimately" when the signal is intercepted by an unintended party. These concepts and others become useful in their attempt to unravel the fascinating interaction between host plants, phytophagous mites, and predatory mites, which includes the possible recruitment of the natural enemies by attacked host plants. These authors are among the few who have analyzed the production of "infochemicals" from a cost/benefit point of view, and in that light, they discuss the potential impact of differences in plant fitness of an apparently small magnitude.

Chapter 5 returns the reader to an examination of chemical ecology from the insect's perspective. In reviewing the physiological bases for both sensory perception of semiochemicals, and thus behavioral response, and responses to physiological effectors per se (toxicants, nutrients), Isman identifies cases where selection has likely taken place and suggests neural and biochemical target sites for selection by xenobiotics in insects. In completing this section of the volume, Chapter 5 concludes with a call for detailed investigations that examine the genetic basis of apparent physiological differences in populations and taxa, as suggested by Rausher in Chapter 2.

1

Why an Evolutionary Perspective?

Bernard D. Roitberg
Simon Fraser University

1.1. Introduction

Nearly 30 years ago, the esteemed ethologist, Niko Tinbergen, posed "the four questions" that should be used to study behavior. These were questions of function (the ultimate or "why" question), causation (the proximate or "how" question), phylogeny (the lineage question), and ontogeny (the developmental question) (Tinbergen 1963). More important than the identification of these approaches was Tinbergen's plea that all four questions be given consideration when studying a biological phenomenon. The same plea should hold true for studies in the area of chemical ecology.

An analogy from civil engineering will help clarify this point. Suppose one has been asked to design and erect a building. There are several questions that need to be answered before proceeding. First, what is the *function* of the building? If the answer is, for example, "to provide space for the production of small electronic gadgets," then there are any number of ways that the building could be designed, yet it is unlikely that the client would be satisfied with many of those designs for reasons outlined below. Thus, the next question to be answered would be more *proximate*. For example, one would likely be constrained by the client's property size, building materials, and local building codes. Of course, building a structure with only the latter three aspects in mind would be as foolish as designing one purely from a functional perspective. Taken together, however, a consideration of both "how" and "why" criteria would lead to a much more satisfactory design. The next aspect that one might consider is one of *lineage*. Given that much of the machinery that will fill the rooms of the structure will have been derived from machinery employed at buildings already in place, the current design must take this constraint into account. On the other hand, designing a building strictly from the lineage perspective seems as foolish as the earlier one-dimensional approaches. Finally, there is the question of *ontogeny*. Suppose the client would like the building to be designed so that other products could be produced at some time in the future. Incorporating this flexibility into the design would ensure that the final form of the building will reflect the developmental changes that occur after initial occupation. One would not likely design a building with the sole criterion being developmental constraints. On the other hand, ignoring this criterion would be considered a major design flaw. In other words, integration of the four questions would lead to the development of a more satisfactory structure than would strict adherence to any one of the four aspects alone.

Now let us compare the building example to a problem in chemical ecology, sex pheromone communication in moths. First, simply identifying the *function* of a sex pheromone would not immediately lead one to identify the mechanisms employed (i.e., how the act is carried out) by the organisms of interest. On the other hand, simply knowing the chemical structure (i.e., the *causal* mechanisms) of the compounds released by a calling female without any knowledge of the species' mating system (i.e., the function of the sex pheromone) has a high risk of failure for use in a pheromone-based management program if, for example, different cues are employed under different conditions (see McNeil, this volume). Third, the sex pheromones of many insects share many components with related species (e.g., in leafroller moths). Thus, it is often useful to consider *phylogeny* when attempting to elucidate pheromone blends. Finally, the sex pheromone blend of some adult moths may be a function of larval diet. This demonstrates the importance of *ontogeny* in defining a specific problem in chemical ecology.

What these examples suggest is that there are two dangers in applying single-dimensional approaches to multi-dimensional problems. First, reductionism may not directly resolve specific questions, while detailed analyses of particular components can obscure the question at hand (the proverbial not seeing the forest for the trees). By contrast, a complementary approach that considers all four Tinbergen questions has proven to be particularly useful in answering questions by students of behavior (Charnov and Skinner, 1985).

Unfortunately, the complementary approach does not seem to have been appreciated by many of us engaged in studies of chemical ecology. For example, Jones (1988) recently surveyed papers published in the *Journal of Chemical Ecology* and lamented the lack of ecological perspective in a majority of those papers. He concluded that chemical ecologists would benefit by considering more interdisciplinary approaches to their systems under investigation.

My purpose in this opening chapter is neither to interpret the pattern that Jones observed nor to suggest that we all trade in our NMRs for dynamic programming software, but rather to repeat Tinbergen's plea for complementarity (see Sherman, 1988, and Armstrong, 1991, for further discussion on levels of analysis and "balanced approaches"). To do this, I will draw upon my experience with studies of host-marking and host discrimination in order to demonstrate that a consideration of both the functional and causative aspects of host-marking can give a better perspective on the overall problem than would either of these perspectives alone. My example considers parasitic insects, but my message is broad-ranging. The message is that evolutionary biology provides an excellent framework from which to study problems of chemical ecology because it integrates "the four questions."

1.2. The Problem of Self-Marking

In this section, I deal with a phenomenon that has been observed in many different species of insects: host discrimination, or the ability of parasites to recognize conspecific offspring within hosts. The functional interpretation of host discrimination suggests that such discrimination acts as a mechanism that permits individuals to reduce competition with conspecifics and among one's own offspring (Roitberg and Prokopy, 1987) while the causal approach considers the mechanisms by which recognition is achieved. Here, I follow the lead provided in the introduction to illustrate the need for a complementary approach. Little emphasis is given to phylogeny and ontogeny in the present example. The latter two perspectives would, however, require consideration if the general case were applied to a species-specific study.

Self-discrimination, (or the ability of parasitic insects to recognize the presence of their own offspring within hosts, is a special case of host discrimination that has been the focus of considerable discussion of late (Hubbard et al., 1987). It has been suggested that parasites should discriminate between hosts parasitized by themselves and those parasitized by conspecifics because, from the mother's perspective, the evolutionary payoff from the second larva outcompeting the first is much greater in the latter case. When the second larva is a sibling of the first, then the payoff from its success will be offset by the death of the sib because mothers are equally related to each of their offspring. Thus, from an evolutionary perspective, there is little benefit in superparasitizing a host harboring one's own offspring, whereas the potential benefits for superparasitizing a host already parasitized by a conspecific are higher.

Much of the discussion of self-discrimination has centered around the mechanism (host-specific chemical marks) by which recognition is achieved (the causal perspective) and the experimental protocols required to demonstrate such recognition. For example, van Alphen and Visser (1990) correctly pointed out that patch recognition must be distinguished from host recognition in order to demonstrate unique host marks. Accordingly, the following experimental protocol could demonstrate the presence of individual recognition. First, a parasite and a few conspecifics are allowed to oviposit in a population of hosts. Following this, the same individual is presented with a random mix of hosts parasitized by her and those by her conspecifics, and her acceptance or rejection of individual hosts is recorded and analyzed (Ofuya and Agele, 1989), Now, suppose that one is able to demonstrate that the female can discriminate between the two host types. Everyone will be satisfied that the "proper" protocol has been employed, and the demonstration will be deemed valid. Suppose, however, that the female does not discriminate between the two host types (van Dijken and Waage, 1987). What can one conclude? Simply, that the female does not possess her own unique marker. This might be an interesting conclusion at the purely proximate (causation) level; however, it does not address the real issue at hand—whether females can and do avoid self-superparasitism (a functional interpretation).

There are likely a number of means by which females might avoid self-superparasitism, and under such circumstances, self-marking may be unnecessarily redundant. I will contrast two scenarios in which self-marking may or may not be advantageous and employ evolutionary models to make my point. The details of the models are not important to understanding the logic underlying the two cases which will clarify the argument. In the first case, I will consider a population of parasites that forage randomly for hosts in an environment in which hosts are not spatially structured (i.e., they may move about continuously, such as host larvae within some food medium). None of the individual parasites in this population self-mark. I will then introduce a self-marking mutant into the population and compare its lifetime reproductive success with the non-self-markers, which I call the "normals." If mutants have higher success than normals, then I can conclude that self-markers do have an evolutionary advantage and that self-marking is not redundant.

The second scenario differs from the first in that hosts are spatially structured into clumps, and parasites remember how many eggs they have laid in a clump of such hosts as they exploit the patch. Each time they enter a patch, however, they possess no knowledge of the number of eggs they have laid in that patch previously. No individuals in the population can produce or recognize self-marks. Once again, I will introduce a mutant that can self-mark and self-recognize and ask how well it will fare. These two scenarios assess the likelihood that a population of non-recognizers could be "invaded" by self-markers (by invasion I mean that mutants will persist within the population and may increase in frequency).

Notice that the models described above are derived from both functional and causal perspectives. Thus, while the models are designed to ask *why* parasites might self-recognize, they also consider *how* this might be accomplished (i.e., chemical versus spatial recognition).

The details of how such evolutionary models are developed and employed can be found in the Appendix. Optimal decisions are derived for individuals harboring a given number of mature eggs and information obtained regarding host availability at particular times in their lives. Such models can predict the relative success of mutants (self-markers) under the two different scenarios described above under a variety of ecological and physiological conditions. Lack of mutant advantage should cause us to question conclusions derived from the experimental protocols described above. In other words, if we are unable to demonstrate mutant advantage in scenario two, then we might be compelled to reconsider the random presentation approach. This approach was, after all, a causal one designed to test for the presence of a self-marking pheromone, not necessarily avoidance of self-superparasitism.

Results from the two protocols are shown below. Let us first consider the scenario where hosts are *not* spatially structured. Figs. 1.1 and 1.2 show results from solving for the optimal behavior of mutants and normals that forage in habitats harboring different number of hosts that have already been parasitized by selfers and normals under several different ecological settings. (It is important to remember that, in these scenarios, both phenotypes make optimal decisions based upon the information that they have, but mutants appear to be at an advantage because they have access to more information). In both figures, fitness is shown for mutants relative to normals. Thus, anytime mutant fitness exceeds one, mutants have enhanced fitness.

Some obvious patterns emerge. First, as the probability of a superparasite winning at larval competition decreases, so does the fitness of the mutant. When such probabilities approach 5%, there is no mutant advantage (Fig. 1.1). Similarly, as encounter rate declines, so does mutant advantage (Fig. 1.2). The reason behind the latter result is that encounters with self-parasitized hosts rarely occur, and little advantage is accrued through recognition of such hosts. Under such conditions, mutants and normals behave in an almost identical fashion. Thus, it is only when the payoff for superparasitism is high that it would pay to self-mark. Such a conclusion is pleasingly intuitive. Remembering that there is no benefit from superparasitizing oneself (an extreme version occurs when females kill eggs present in the host and replace them with their own (Strand and Godfray, 1989), it will only pay to discriminate between oneself and others when the *differential* benefits are high. Similarly, when encounter rates are low, the benefits for self-avoidance will accumulate very slowly. In conclusion, when hosts are not spatially ordered, some but not all conditions favor self-marking. What this tells us is that particular ecological situations

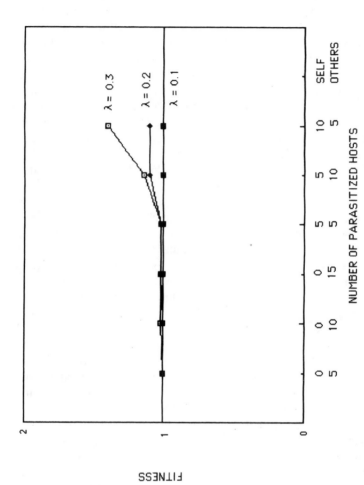

Figure 1.1. The fitness of self-markers when foraging in habitats harboring different numbers of already-parasitized hosts. The *x* axis indicates the number of hosts parasitized and marked by self-markers as well as those hosts parasitized by non-markers (others). The results are shown for three different host encounter rates (λ = 0.3, 0.2, and 0.1) in a habitat harboring 20 hosts.

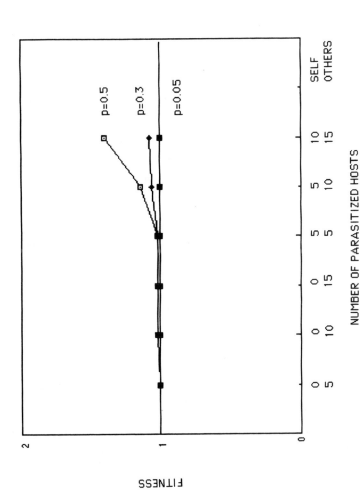

Figure 1.2. The fitness of self-markers when foraging in habitats harboring different numbers of already-parasitized hosts. The *x* axis indicates the number of hosts parasitized and marked by self-markers as well as those hosts parasitized by non-markers (others). The results are shown for three different probabilities of survival ($p = 0.5$, 0.3, and 0.05) for offspring that are deposited in an already-parasitized host in a habitat harboring 20 hosts.

do not favor self-recognition. We would not expect it to evolve even if the organisms were capable of producing unique marks.

Let us now consider a situation where hosts are clustered in groups of four within host-containing patches, and parasites can keep track of the number of eggs they have deposited from the moment that they enter the patch. The results produce a much more complicated pattern from which the following conclusions can be drawn. When parasites enter patches that have not been exploited previously, mutants will not outperform normals as long as normals keep track of the number of eggs they lay. When parasites enter patches that have been partially exploited, mutants will tend to outperform normals. However, the extent to which they do depends upon how many hosts are already parasitized and how many other parasites are present, in addition to the encounter rates and payoffs for superparasitism. The conclusion again is that there are a number of conditions that *do not* favor the evolution of self-marking. More importantly, it is difficult, if not impossible, to predict when this would be the case without complimentary studies on *how* (causal) and *why* (functional) the parasites search for hosts. For example, in a series of papers, I have shown that some tephritid fruit flies (e.g., *Rhagoletis pomonella*; see Roitberg, 1989) rapidly exit patches if they encounter marked hosts when first entering the patch. If this is the case, then these flies fit into the scenario in which simply keeping track of oviposition rates within a patch provides adequate information on the likelihood of self-superparasitism. Similarly, van Dijken and Waage (1987) suggested that upon encountering a parasitized host, females that had not previously laid any eggs must, by definition, be encountering a conspecific's eggs.

The models that I employed simplify host search to a considerable extent. One could incorporate more detail in the actual search process. This would allow us to indicate specific instances in which self-markers would or would not gain an advantage over normals (see Roitberg and Mangel, 1988, for an example of behavior-rich approaches). The general conclusions would not change, however: self-marking and recognition are not always advantageous in an evolutionary sense. Mackauer (1989) and Bai and Mackauer (1990) make similar economic arguments by a consideration of fecundity and the ability to resorb eggs. It is again interesting to note that their conclusions would not have arisen without complimentary approaches that emphasize *how* (egg absorption) and *why* (lifetime reproductive success) questions.

The complementary approach I present in this chapter differs in another fundamental way from others which simply ask if a physiological or behavioral trait is beneficial. Here, I consider the relative advantage of a trait when it is present at different frequencies in the population. It may often be the case that a trait, such as cheating, is advantageous when rare, but less so when common. For example, in my models, if self-markers are at an advantage and do become more common, then their behavior will have greater impact on egg dispersion patterns at the population level (i.e., eggs will be more evenly dispersed). As these patterns change, mutant advantage will shift as normals encounter conspecific-parasitized hosts at different rates, thus further influencing relative mutant advantage. Roitberg and Mangel (1988) and Hubbard et al. (1987) consider host-marking from a frequency-dependent perspective. Similarly, Sabelis and deJong (1988) consider frequency-dependent payoffs for release of semiochemicals by plants.

Finally, what is the danger in neglecting an evolutionary approach to self-marking? The danger is that related patterns of behavior may not be explainable or predictable. For example, less than one decade ago, all marking pheromones, self and non-self, were considered to be deterrents, and cases of non-deterrence were considered to be the result of errors by the parasite. A great deal of effort was devoted to elucidating the structure of the marking compounds, but it was still difficult to explain why great variability in responses to marks occurred in both the laboratory and the field (Klijnstra, 1985.). Recent considerations of superparasitism as an adaptive behavior (i.e., the pheromones are not deterrents per se, and the parasites do not err when they superparasitize) has led to the development of models of high predictive power (see van Alphen and Visser, 1989; Speirs et al., 1991).

1.3. Conclusions

Why should chemical ecologists care about conclusions derived from evolutionary models? Surely it will not make them better chemists or physiologists! Or will it? Perhaps the most important insight that should be gleaned from the foregoing example is that an evolutionary perspective can help us identify the critical issues (e.g., avoidance of superparasitism under particular conditions and not host-marking oneself). On the other hand, an evolutionary model that is devoid of reasonable biological assumptions can also lead us down the wrong path. Such approaches might, for example, predict that the parasites should remember all features of their environment and transfer energy between somatic and reproduction function with 100% efficiency, and thus disobey the second law of thermodynamics! Of course, this is nonsense, given the kinds of physiological limitations that natural selection must act upon. Thus, we return to Tinbergen's (1963) plea that we employ an integrated approach that considers "the four questions" of function, causation, phylogeny, and ontogeny. The remaining chapters in this volume employ such an integrated approach. Some of the authors are more explicit than others in their use of evolutionary theory, yet that is the foundation upon which all the chapters are built. This should behoove readers to scrutinize those areas in which they do not possess expertise in order to advance their understanding of processes in chemical ecology.

Appendix

To calculate the relative fitness that is, contribution to the gene pool (see Rausher, this volume), of these two behavior types in the two scenarios, I employ the tool of stochastic dynamic programming (see Mangel and Clark, 1988, for an excellent review of the topic). These types of models satisfy the criterion of complementarity in that they allow the use of explicit proximate parameters (e.g., egg load, information state; Mangel and Roitberg, 1989). The models discussed below are similar to those derived by Roitberg and Mangel (1988) for the evolution of host-marking pheromones.

To begin, it is important to state several assumptions. First, I assume that the parasites are pro-ovigenic, that is, they begin adult life with all their eggs mature and do not manufacture any more eggs during their life. Second, I assume that only one individual can complete development within each host (i.e., they are solitary parasites). Thus, the fitness a mother accrues from laying an egg in a healthy host always exceeds that from laying in a parasitized one because there is an increased probability of her offspring dying relative to its chances when alone. Third, I assume that all parasites can discriminate between parasitized and healthy (unparasitized) hosts.

Following is a detailed description of models for Scenarios 1 and 2.

Scenario 1

The scenario considered here is for parasites that search for hosts in an environment where hosts are not spatially structured and encounters with each host are essentially random events.

The first step toward solving for parasite fitness is to divide the parasite's life into T discrete time periods. The second step is to characterize individual parasites by their egg complement at the start of period t as $X(t)$. The state of the environment is determined by the number of hosts available. Thus, $N(t) = [N_h(t), N_p(t)]$ defines the number of healthy and parasitized hosts, respectively, that are available at the beginning of time period t. The $N_p(t)$ category can be further subdivided into N_{pn} and N_{pm} to differentiate between hosts parasitized by normal (N) and mutant (M) individuals. Finally, the fitness accrued from laying eggs in healthy, conspecific, and self-parasitized hosts is denoted by f_h, f_c, and f_p, respectively. Note that f_c always equals zero, since only one offspring can survive per host. Thus, the fitness accrued from having a second individual survive in a self-parasitized host is negated through the loss of the first individual.

Following Roitberg and Mangel (1988), I assume that the probability of encountering a host of type i during period t can be defined by:

$$\lambda_i = (1 - \exp^{-\varepsilon s}) N_i / \Sigma_j N_j$$

where e is a parameter that characterizes the search rate of the parasite and s is the total number of hosts in the habitat being searched.

Let $F_n(x, n, t, T)$ denote the maximum expected fitness of a normal individual at the start of period T, based on decisions made between periods t and T, given that $X(t) = x$ and $N(t) = n$. Further, I assume that no additional fitness is accrued after T such that $F_n(x, n, T, T) = 0$. Finally, I assume that individuals survive to period $t + 1$, given that they are alive in period t with some probability $\rho(t)$. All that is required to derive the models is to consider the mutually exclusive events that can occur during single periods of time. First, let us consider normal individuals.

1. An individual may not discover a host during period t, in which case, $X(t)$ still equals x and $N(t)$ still equals n.

2. An individual may discover a healthy host. I assume that such encounters will always lead to an oviposition, in which case, $X(t)$ now equals $x - 1$ and $N(t)$ now becomes $[n_h - 1, n_p + 1]$.

3. An individual may discover a parasitized host. If she chooses to reject that host, then $X(t)$ remains at x and $N(t)$ remains at n. If, however, she chooses to oviposit in that host, then she will receive fitness increment $(N - 1)/N f_p$. In other words, the frequency with which she is likely to encounter a parasitized host that was parasitized by a conspecific is $(N - 1)/N$. Furthermore, ovipositions that accrue no fitness (i.e., in a host already parasitized by herself) will occur with the frequency $1/N$. In both cases, $X(t)$ becomes $x - 1$ but $N(t)$ remains at n, since the host was already parasitized. (*Note*: in this scenario, hosts parasitized by mutant and normal parasites need not be distinguished. All that is important is that self-superparasitism accrues no fitness.)

The three cases are summarized in the following equation.

$$F_n(x, n, t, T) = (1 - \Sigma_i \lambda_i) \rho(t) F_n(x, n, t + 1, T)$$
$$+ \lambda_h \{ f_h + \rho(t) F_n(x - 1, n_p, t + 1, T) \}$$
$$+ \lambda_p \max \{ \rho(t) F_n(x, n, t + 1, T);$$
$$(1/N)(\rho(t) F_n(x - 1, n, t + 1, T)$$
$$+ (N - 1/N)(f_p + \rho(t) F_n(x - 1, n, t + 1, T) \}$$

Let the maximum expected mutant fitness be denoted by $F_m(x, n, t, T)$. One must now distinguish between hosts parasitized by self and conspecifics such that $N(t) = [n_h, n_{pn}, n_{pm}]$. Four mutually exclusive events must be considered. The first two are identical to those for normal parasites. The latter two follow:

3. An individual may discover a host parasitized by a conspecific. If she chooses to reject that host, then $X(t)$ remains at x and $N(t)$ remains at n. If, however, she chooses to oviposit in that host, then she will receive fitness increment f_p, $X(t) = x - 1$, and $N(t) = [n_h, n_{pn} - 1, n_{pm} + 1]$. The latter occurs because the parasite marks that host with one of her own marks.

4. An individual may discover a host parasitized by herself. If she chooses to reject that host, then $X(t)$ remains at x and $N(t)$ remains at n. If she chooses to oviposit, she will accrue no fitness, $X(t) = x - 1$ and $N(t) = n$.

The four cases are summarized in the following equation:

$$F_m(x, n, t, T) = (1 - \Sigma_i \lambda_i)\rho(t) F_m(x, n, t + 1, T)$$

$$+ \lambda_h \{f_h + \rho(t) F_m(x - 1, n_{pm}, t + 1, T)\}$$

$$+ \lambda_{pn} \max\{\rho(t) F_m(x, n, t + 1, T);$$

$$f_{pn} + (\rho(t) F_m(x - 1, n_{pm}, t + 1, T)\}$$

$$+ \lambda_{pm} \max\{\rho(t) F_m(x, n, t + 1, T);$$

$$(\rho(t) F_m(x - 1, n, t + 1, T)\}$$

The equations are solved "backwards in time," starting with $t = T - 1$ and going until $t = 1$. The maximization terms in the equations indicate that the decision (e.g., oviposit or reject) that gives the highest expected fitness is chosen (see Mangel and Clark, 1988, for more details).

Senario 2

In this scenario, the habitat is divided into cells, some of which harbor groups of four hosts. Thus, parasites' decisions occur at two levels. First, whether to continue searching within a cell, and second, whether to oviposit in a host when one is encountered. I assume that if the parasite leaves a cell, it randomly chooses to visit some other cell (there is no spatial structure at this level). Finally, I assume that as the parasite searches through a cell, it keeps track of how many eggs it has laid thus far in that cell. Each time it begins searching in a new cell, it carries no knowledge of whether it has visited that cell previously.

To denote parasite fitness in this scenario, one must consider two further variables. The first, an environmental variable $C(t)$, defines the availability of parasitized and unparasitized hosts in the current cell. There are six possible cell states: (1) no host present; and (2 through 6) hosts present with an increasingly larger proportion of parasitized hosts (i.e., 0 through 4). The second is a knowledge state variable $K(t)$, which contains information on how many eggs have been laid since arrival in the current cell.

The maximum expected fitness of a normal parasite at time t is denoted by:

$$F_n(c, x, k, t, T).$$

Four mutually exclusive events may occur during a period of time at two different levels. First, should the individual decide to remain in her present cell, the following can occur.

(1) No host is found. Variable values do not change.

(2) A healthy host is discovered. Thus, $C(t) = c + 1$, $X(t) = x - 1$, $K(t) = k + 1$.

(3) A parasitized host is discovered. If rejected, variable values do not change. If accepted, the parasite receives fitness f_p with frequency $\psi = (1 - (k/c)) - ((1 - (k/c))/N)$ and will receive fitness payoff 0 with frequency $(1 - \psi)$ and $X(t) = x - 1$ and $K(t) = k + 1$. In essence, the frequency of payoff ψ is determined by how many parasitized hosts are present and how many eggs the parasite knows she has laid, plus the change that she may have visited previously and oviposited.

Should the parasite choose to leave the cell, she will enter a new cell i with probability $c_i/\Sigma_j N_j$. In that case, $K(t) = 0$.

The four cases are summarized in the following equation.

$$F_n(x, c, k, t, T) = \max\big[(1 - \Sigma_i \lambda_i)\rho(t)F_n(x, n, t + 1, T)$$

$$+ \lambda_h\{f_h + \rho(t)F_n(x - 1, c + 1, k + 1, t + 1, T)\}$$

$$+ \lambda_{pn} \max\{\rho(t)F_n(x, c, k, t + 1, T);$$

$$\psi f_{pn} + (\rho(t)F_n(x - 1, c, k + 1t + 1, T)$$

$$+ (1 - \psi)(\rho(t)F_n(x - 1, c, k + 1, t + 1, T)\big];$$

$$[\Sigma_i \rho(t)F_n(x, c_i, k, t + 1, T)]$$

Mutant parasite fitness $F_m(x, c, k, t, T)$ is derived as above except that encounters with parasitized hosts must be distinguished between encounters with conspecific and self-parasitized hosts. The following equation summarizes the four cases.

$$F_m(x, c, k, t, T) = \max\Big[(1 - \Sigma_i \lambda_i)\rho(t)F_m(x, n, t + 1, T)$$

$$+\lambda_h\{f_h + \rho(t)F_n(x - 1, c + 1, k + 1, t + 1, T)\}$$

$$+\lambda_{pn}\max\{\rho(t)F_n(x, c, k, t + 1, T);$$

$$f_{pn} + (\rho(t)F_m(x - 1, c, k + 1, t - 1, T)\}$$

$$-\lambda_{pm}\max\{\rho(t)F_n(x, c, k, t + 1, T);$$

$$(\rho(t)F_m(x - 1, c, k, t + 1, T)\}\Big];$$

$$[\Sigma_i\rho(t)F_n(x, c_i, k, t + 1, T)]$$

As with the first scenario, the equations are solved backwards in time.

Acknowledgements

I thank Bob Lalonde and Rob McGregor for comments on the manuscript. This work was supported by a Natural Sciences and Engineering Research Council of Canada (NSERC) operating grant.

References

Alphen, J. J. M. van and M. Visser. 1990. Superparasitism as an adaptive strategy. *Annu. Rev. Entomol.* 35:59–79.

Alphen, J. J. M. van, M. J. van Dijken, and J. K. Waage. 1987. A functional approach to superparasitism: host discrimination needs not be learnt. *Neth. J. Zool.* 37:167–79.

Armstrong, D. P. 1991. Levels of cause and effect as organizing principles for research in animal behaviour. *Can. J. Zool.* 69:823–9.

Bai, B. and M. Mackauer. 1990. Host discrimination by the aphid parasitoid Aphelinus asychis (Hymenoptera: Aphelinidae): when superparasitism is not adaptive. *Can. Entomol.* 122:363–72.

Charnov, E. L. and S. W. Skinner. 1985. Complementary approaches to the understanding of parasitoid decisions. *Environ. Entomol.* 14:383–91.

Dijken, M. J. van and J. K. Waage. 1987. Self and conspecific superparasitism by the egg parasitoid *Trichogramma evanescens*. *Entomol. Exp. Appl.* 43:183–92.

Hubbard, S. F., G. Marris, A. Reynolds, and G. W. Rowe. 1987. Adaptive patterns in the avoidance of superparasitism by solitary wasps. *J. Anim. Ecol.* 56: 387–402.

Jones, C. G. 1988. What is chemical ecology? *J. Chem. Ecol.* 14:727–30.

Klijnstra, J. W. 1985. Oviposition behavior as influenced by the oviposition deterring pheromone in the large white butterfly, *Pieris brassicae*. Ph.D. dissertation, Agricultural University, Wageningen, The Netherlands.

Mackauer, M. 1989. Host discrimination and larval competition in solitary endoparasitoids. In *Critical Issues in Biological Control*, eds. M. Mackauer, L. Ehler, and J. Rolands, pp. 41–62. Intercept Press, Andover, U.K.

Mangel, M. and C. W. Clark. 1988. *Dynamic Modeling in Behavioural Ecology*. Princeton Univ. Press, Princeton.

Ofuya, T. and S. Agele. 1989. Ability of ovipositing *Callosobruchus maculatus* to discriminate between seeds bearing their own eggs and those bearing eggs of other females. *Ecol. Entomol.* 14:243–6.

Roitberg, B. and R. Prokopy. 1987. Insects that mark host plants. *BioScience* 37:400–6.

Roitberg, B. D. and M. Mangel. 1988. On the evolutionary ecology of marking pheromones. *Evol. Ecol.* 2:289–315.

Roitberg, B. D. 1989. Variation in the behaviour of parasitic insects: bane or boon. In *Critical Issues in Biological Control*, eds. M. Mackauer, L. Ehler, and J. Rolands, pp. 25–39, 41–62. Intercept Press, Andover, U.K.

Sabelis, M. W. and C. M. de Jong. 1988. Should all plants recruit bodyguards? Conditions for a polymorphic ESS of synomone production in plants. *Oikos* 53:247–52.

Sherman, P. W. 1988. The levels of analysis. *Anim. Behav.* 36:616–9.

Speirs, D., T. Sherratt, and S. F. Hubbard. 1991. Parasitoid diets: does superparasitism pay? *Trends in Ecol. Evol.* 6:22–5.

Strand, M. R. and H. C. J. Godfray. 1989. Superparasitism and ovicide in parasitic Hymenoptera: theory and a case study of the ectoparasitoid *Bracon hebetor*. *Behav. Ecol. Sociobiol.* 24:421–32.

Tinbergen, N. 1963. On the aims and methods of ethology. *Z. Tierpsychol.* 20:410–33.

2

Natural Selection and the Evolution of Plant-Insect Interactions

Mark D. Rausher

Duke University

2.1. The Causes of Evolutionary Change

For more than a third of a century, researchers have debated the functions of plant characters, such as secondary chemicals, sclerophylly, trichomes, and leaf shape. Since the early work of Fraenkel (1956, 1959, 1969), Dethier (1947, 1954), and Ehrlich and Raven (1964), many biologists have interpreted the often-detrimental effects of these characters on herbivores as indicating that their primary function is to defend plants against attack by their natural enemies. On the other hand, a number of biologists have suggested that these characters serve other ecological and physiological functions and that the effects of these characters on herbivores are fortuitous and of no real ecological importance to plants. At the heart of this debate lies an evolutionary question, since the issue of what ecological function these characters perform is equivalent to asking what evolutionary forces caused those characters to evolve and be maintained in plant populations.

Although similar evolutionary questions arise in attempts to understand chemical interactions among other organisms, they are seldom addressed directly by researchers. One reason may be the extreme difficulty of examining and quantifying evolutionary forces in nature. Consequently, biologists are often forced to infer the nature of those forces from indirect evidence. Another reason, however, is undoubtedly that chemical ecologists are often not trained in more than the rudiments of evolutionary biology and population genetics, and hence, often are not equipped to bring the tools and types of reasoning associated with these disciplines to bear on the study of chemically-mediated interactions.

In this chapter, I hope to illustrate the power of these disciplines to address fundamental and long-standing questions about interactions among plants and herbivores. In particular, I outline some basic principles of evolution and natural selection and then show how these principles may be used to address issues that have proved intractable to examination by other approaches.

2.1.1. Defining Evolution

Although several definitions of evolution exist (see Endler, 1986), I will adopt the one favored by many population geneticists: evolution is any change in the genetic constitution of a population from one generation to the next. Since, to a first approximation, the genetic makeup of a population can be characterized by a set of allele frequencies for all loci in the genome, evolution can be equivalently defined as a change in allele frequencies at one or more loci from one generation to the next. Often, such a change will be accompanied by changes in the average value of one or more morphological, physiological, or behavioral characters, since such characters are often influenced by an individual's genetic makeup. However, changes in allele frequencies do not necessarily result in obvious alterations of an organism's phenotype. For example, changes in the relative frequencies of redundant base-pair-codon alleles at a particular locus will not alter the gene product emanating from that locus and hence, may have little, if any, effect on the average phenotype of individuals in a population. Similarly, allele frequencies at different loci affecting a continuous, or *quantitative*, trait may change without affecting the overall mean of the trait, if changes at some loci are compensated for by changes at other loci. Nevertheless, these types of changes must be explained by any comprehensive theory of evolution that purports to account for changes in mean phenotypes over time, since the precise genetic make-up of an individual (i.e., the sequence of base pairs in each of the chromosomes) can be considered to be an aspect of an organism's phenotype as legitimately as chemical defense profile, pheromonal composition, degree of host preference, tail length, or any other "normal," complex character. It thus seems appropriate to adopt a definition of evolution that focuses on genetic changes in populations.

One question that immediately arises concerning this definition is whether it is sufficiently comprehensive to encompass long-term, "macroevolutionary" phenomena, such as the evolution of insects from annelid-like ancestors or the adaptive radiation of troidine swallowtail butterflies onto host plants in the family Aristolochiaceae. While there has been much recent disagreement over whether known "microevolutionary" processes can account for macroevolutionary patterns (Gould and Eldridge, 1977; Stanley, 1979, 1982; Gould, 1982; Wright, 1982; Oster and Alberch, 1982, Charlesworth et al., 1982; Schopf, 1982), few would disagree with the assertion that long-term evolutionary change results from a series of simultaneous and/or sequential changes in allele frequencies. The primary debate is over the process responsible for those changes (i.e., the relative importance of individual, population, and species selection) and the rates at which those changes occur (gradually and consistently versus alternating bursts of rapid allele-frequency change and relative stasis). Short- or long-term evolution is thus almost universally agreed to consist of change in the genetic makeup of a population or species. The main task of evolutionary biologists is to determine how and why such genetic changes occur.

2.1.2. Genetic Drift

Evolutionary biologists have identified two major processes that alter allele frequencies and that are thus responsible for most evolutionary change: *genetic drift* and *natural selection*. Genetic drift is a change in allele frequencies at a particular genetic locus that arises because each generation constitutes a finite sample of the alleles (and genotypes) present in the previous generation. Consider, for example, a population of leaf-feeding beetles in which the frequencies of alleles A and a in the eggs (zygotes) of generation 1 are 0.6 and 0.4, respectively, and in which an individual's genetic makeup at this locus has no influence on its probability of surviving or reproducing. Some individuals die before reproducing because they are eaten by predators, others because they fail to obtain enough food to avoid starvation, and still others from accidental causes, such as drowning during heavy rains. One can think of the process that determines which alleles will be carried by the surviving individuals as being determined by the toss of a biased coin weighted to come up "heads" 60% of the time (corresponding to the A allele) and "tails" 40% of the time (corresponding to the a allele). Just as a fair coin is not expected to produce exactly 50% heads and 50% tails in a finite number of flips, so the frequency of the two alleles in the surviving adult beetles of generation 1, as determined by the biased coin, is not expected to be exactly 0.6 and 0.4. A change in allele frequencies will thus occur within a generation due to the random sampling of survivors from among the original zygotes of that generation. The formation of gametes and the fusion of those gametes to form generation 2 zygotes involves a second sampling episode that may also alter allele frequencies. The combined change in allele frequencies from these two sampling events constitutes a change in genetic composition of the population from generation 1 to generation 2 and hence constitutes evolution by genetic drift.

As has been pointed out by numerous authors (Wright, 1969; Roughgarden, 1979), evolution by genetic drift has several important characteristics.

1. *Genetic drift causes greater changes in allele frequencies in small populations than in large populations.* When tossing a fair coin, the propor-

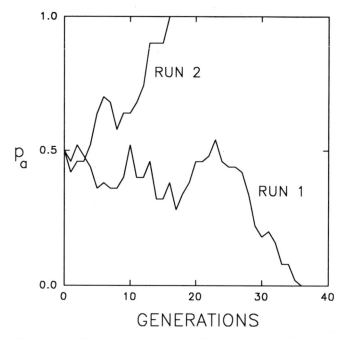

Figure 2.1. Simulated changes in allele frequencies due to genetic drift at a single locus for a population size of 20. p_a is the frequency of allele *a*. Allele frequency at each generation was determined by sampling 40 alleles with replacement from the collection of alleles present in the previous generation.

tion of heads is expected to deviate less and less from 50% the more times the coin is tossed. Similarly, the more adult survivors there are in a population (i.e., the greater the population size), the less the allele frequencies are expected to differ from those of the previous generation due to sampling effects.

2. *Given sufficient time, genetic drift will eliminate all but one allele at a locus.* During any given generation, the frequency of a particular allele is as likely to increase as decrease (Fig. 2.1), and the magnitude of the change may be large or small. Eventually, the frequency of one allele will, by chance, become 1.0 (i.e., becomes "fixed"). Although it cannot be predicted with certainty which allele will eventually become fixed, the probability of fixation for a particular allele is equal to the initial frequency of that allele in the population. This property of drift implies that although the possibility of novel mutations, and the characters associated with them, eventually coming to predominate in populations simply by drift is small (because the initial frequency of that allele is small), this will

occur for *some* mutations because the number of mutations that occur over evolutionary time is large.

3. *Genetic drift does not cause adaptive evolution.* Adaptive evolution may be loosely defined as evolutionary change that increases the mean fitness of individuals in a population. Because drift acts randomly with respect to the biological properties of alleles, it contributes equally to fixation of detrimental and beneficial alleles and hence, on average, will not improve a population's mean fitness. Moreover, drift is probably most important for characters that have little effect on fitness, since natural selection will not override changes in allele frequencies caused by drift. Alteration by drift of allele frequencies for such characters is clearly evolution, but it is not adaptive evolution because the ability of individuals to function in a given environment, as judged by their fitness, is not changed.

4. *Genetic drift may constrain future adaptive evolution.* Genetic variants affect an individual's phenotype by altering developmental or biochemical pathways. The range of novel phenotypes that may evolve in a population is thus limited by the ways by which developmental pathways characteristic of that population may be changed. The ways they may be changed are, in turn, constrained by the current state of those pathways, i.e., by past evolutionary events that have produced the current developmental system. For example, loss of a functional biochemical pathway through drift (see Section 2.2.3, The Evolution of Defenses) may preclude the possibility of subsequent elaboration of that pathway for defense against herbivores.

2.1.3. Natural Selection

Natural selection is the second important process that causes evolutionary change. As with evolution, biologists disagree over the detailed definition of what constitutes natural selection (Endler, 1986). Nevertheless, it may be loosely defined as the differential average survival and/or reproductive success of individuals with different genotypes. It arises because genetic differences among individuals cause corresponding morphological, physiological, or behavioral differences that affect an individual's ability to pass genes on to the next generation. As will be seen below, such differential transmission of genetic material may result in a change in allele frequencies and, hence, in evolution.

As has been pointed out repeatedly (see references on p. 4 of Endler, 1986), there are three requisites for the evolution of a character by natural selection: (1) the character must be variable; (2) at least part of the variation in the trait must be caused by underlying genetic variation; and (3) there must be a correlation between the trait and fitness. In the following sections, I briefly discuss how one may determine whether these requisites apply in a concrete situation. I also consider when a demonstration that these three requisites are satisfied is sufficient to allow an inference that evolution by natural selection occurs.

2.1.3.1. Variation

Virtually all complex characters are variable. Consequently, requisite (1) for evolution by natural selection will almost always be satisfied for any study organism. In understanding how natural selection acts on variation, however, it is necessary to characterize variation quantitatively. The manner in which this is done depends on whether a character is *qualitative* or *quantitative*. A qualitative character is one that can be placed into discrete categories. An example is provided by the cyanide polymorphism exhibited by several plant species. Within some populations of *Lotus*, *Trifolium*, and *Pteridium*, some plants release cyanide when their leaves are damaged, while others fail to do so (Jones, 1972; Cooper-Driver and Swain, 1976; Harper, 1977). Cyanide production is thus a qualitative character with two possible values. Qualitative characters are often, though not always, controlled by one or a few genes with large effects on the character.

Quantitative characters take on a more or less continuous range of values. An example is the proportion that α-pinene constitutes of all monoterpenes in ponderosa pine xylem resin, which varies continuously among individuals from 0.0 to 0.29 in populations from the western U.S. (Smith, 1964; Sturgeon, 1979). Quantitative characters are often thought to be controlled by a large number of loci, each with a small, additive effect on the character (Falconer, 1981), though several genetic analyses suggest they may often be controlled by a small number of genes (Spickett, 1963; Spicket and Thoday, 1966; Wehrhan and Allard, 1965) or by genes with non-additive effects (Fasoulas and Allard, 1962; Russell and Eberhart, 1970; Edwards et al., 1987).

2.1.3.2. Heritability of Variation

Variation is a character arises from two main sources: genes and the environment. Environmental variation occurs because different individuals experience different environments during development. These environmental differences affect the course of development and, hence, the phenotype, which is the final product of development. Alary polymorphism in aphids provide an illustrative example. Within an aphid clone, some individuals ("alates") possess wings and well-developed flight musculature, while other individuals ("apterae") lack wings and have poorly developed musculature. Because aphids reproduce parthenogenetically, all individuals in a clone are genetically identical (Blackman, 1979; Suolmalainen, 1950; Suolmainen et al., 1980). Thus, the differences in wing morphology cannot be caused by underlying genetic differences and must be due to direct effects of the environment on aphid development. In fact, it is known for many species that whether an individual becomes alate or apterous is determined by a variety of environmental factors, including degree of crowding on a host plant, host quality, and photoperiod (Johnson and Birks, 1960; Lees, 1966; Sutherland, 1969a, b).

Characters also vary because individuals differ genetically. Different gene products produced by different individuals (or, in the case of some regulatory genes, the difference in the DNA structure itself) channel development in different directions to produce different phenotypes. Cyanide polymorphisms again provide a cogent example (Nass, 1972; Conn, 1979). In *Lotus* and *Trifolium*, failure to produce cyanide arises because an individual is homozygous for a mutant allele at one of two loci. One allele produces a defective enzyme that catalyzes a step in the synthesis of cyanogenic glycosides, a chemical that breaks down to release cyanide when leaves are damaged. When this allele is homozygous, cyanide cannot be released because its chemical precursor is not produced. The second allele produces a defective enzyme that normally catalyzes the breakdown of cyanogenic glycosides. Homozygotes produce the needed chemical precursor but are incapable of releasing cyanide from it. Genetic makeup at two loci thus determines almost completely an individual's cyanide phenotype.

The way in which quantitative traits are influenced by genes and environment is conceptually straightforward. Consider a trait such as rostrum (proboscis) length in aphids. Some evidence suggests that longer rostra facilitate feeding on plants with long trichomes (Moran, 1986). This character varies continuously among individuals within an aphid clone. Since all individuals within a clone are genetically identical, this variation must be due to environmental differences experienced by different individuals within the clone. As is characteristic of much environmentally-induced variation in quantitative characters, however, little is known about the particular environmental factors that affect rostrum length. A comparison of mean rostrum length among clones would also presumably reveal that cones differ in this trait. As long as the clones were grown under the same environmental conditions, such a result would indicate a genetic contribution to overall variation in rostrum length.

2.1.3.3. Selection

The third requisite for evolution by natural selection is selection. As defined by many evolutionary biologists (Haldane, 1954; Lande and Arnold, 1983; Futuyma, 1986), natural selection is simply differential fitness among phenotypes, that is, among individuals that differ in one or more traits.

Defining fitness precisely is complicated by the need to consider factors such as the age structure of a population and whether generations overlap (Charlesworth, 1980). However, in the simplest case of an organism with discrete, non-overlapping generations, fitness may be defined as the number of offspring an individual produces in its lifetime. This measure of fitness incorporates a survival component, since individuals that fail to survive and therefore, fail to reproduce, leave no offspring and hence have a fitness of zero.

Two main types of selection are commonly recognized: directional and stabilizing/disruptive. Directional selective occurs when a correlation exists between the measure of a trait in an individual and the fitness of that individual. For example, Berenbaum et al. (1986) report correlations in wild parsnip between seed number, their measure of fitness in the wild parsnip, and the concentrations of several furanocoumarins in the leaves and seeds. In particular, the concentration of bergapten in the leaves is negatively correlated with seed number, indicating selection for lower levels of this compound. As this example suggests, directional selection tends to alter the mean of a trait in a population.

Stabilizing (disruptive) selection occurs when an intermediate phenotype has the highest (lowest) fitness. An example of stabilizing selection is provided by Rausher and Simms (1989), who examined levels of resistance in the morning glory, *Ipomoea purpurea*, to damage by several types of insects. Phenotypes exhibiting intermediate levels of corn-earworm damage to developing pods produced significantly more seeds than phenotypes exhibiting more or less damage. This type of selection tends primarily to decrease the variance about the mean phenotype in the population without changing the mean. Disruptive selection, by contrast, tends to increase the variance about the mean.

2.1.3.4. *Estimating the Magnitude of Variation and Selection*

Because most traits examined by chemical ecologists will not be inherited in a simple Mendelian fashion, the remainder of this chapter concentrates on the evolution of quantitative traits. Investigators who are fortunate enough to be studying single-locus characters may consult any number of excellent accounts of the evolution of such traits (Wright, 1969; Crow and Kimura, 1970; Cook, 1971; Roughgarden, 1979).

As shown below, quantitative estimates of the three requisites of evolution by natural selection can be used to predict the magnitude of evolutionary change in one or more characters, It is thus worth considering briefly how each of these requisites may be quantified.

Variability in a character x may be characterized simply by the phenotypic variance, V_p, that is, the variance of that character among individuals using the standard statistical formula

$$V_p(x) = \frac{\Sigma f(x_i)(x_i - \bar{x})^2}{\Sigma f(x_i)} \tag{2.1}$$

where $f(x_i)$ is the number of individuals with a value of the character equal to x_i, $\bar{x}$ is the mean value of the character in the population, and the sums extend over all i. The phenotypic variance may be estimated for a population simply by collecting a random sample of individuals and measuring the value of the character on each individual.

As described below, the evolutionarily relevant measure of genetic variation for a character is its additive genetic variance, V_a. This variance can be defined in terms of another quantity, the breeding value, BV, of an individual for the trait. Ideally, an individual's breeding value could be obtained by mating that individual with a large number of other individuals selected randomly from the population. The average value of the trait measured in one offspring from each mating, expressed as a deviation from the population mean for that trait, is then one-half that individual's breeding value. Numerically, the breeding value of individual j, BV_j may be expressed as:

$$BV_j = 2\left[\frac{\Sigma x_{ji}}{n} - \bar{x}\right] \tag{2.2}$$

where x_{ji} is the value of the character in the offspring of the ith mating of individual j, and $\bar{x}$ is the mean value of the trait among the parental generation. The additive genetic variance for a trait is then simply the variance of the breeding values in the population, i.e.,

$$V_a = \frac{\Sigma f(BV_j)\left(BV_j - \overline{BV}\right)^2}{\Sigma f(BV_j)} \tag{2.3}$$

where $\overline{BV}$ is the mean breeding value in the population, $f(BV_j)$ is the number of individuals with breeding value BV_j, and the sums are taken over all j. Because breeding values are determined from measurements of offspring, the additive genetic variance estimates the amount of variation in the population that can be passed from one generation to the next, or heritable variation.

In practice, one usually does not measure additive genetic variance in this manner, since it is usually not possible to mate an individual with a large number of others. To overcome this problem, geneticists have devised a number of breeding designs and associated statistical techniques to estimate V_a. Details of these approaches, which include parent-offspring regressions, sib analyses, and factorial and diallel crosses, may be found in several excellent sources (Bulmer, 1980; Falconer, 1981; Becker, 1984). Knowledge of these details is not necessary, however, for a conceptual understanding of the evolution of quantitative traits.

The intensity of directional selection on a quantitative trait is often measured by the selection differential, s. In general, the selection differential is defined as the covariance between fitness and the value of the trait, i.e.,

$$s = \frac{\Sigma f(w_i, x_j)(w_i - \overline{w})(x_j - \overline{x})}{\Sigma f(w_i, x_j)} \qquad (2.4)$$

where w is relative fitness, x is the value of the trait, and $f(w_i, x_j)$ is the number of individuals with fitness equal to w_i and a trait value of x_j, the summations being over all i and j. When selection operates exclusively by means of differential survival, this formula reduces to the difference between the mean value of the trait in the surviving individuals and the mean value of the trait in all individuals before selective mortality occurs.

2.1.3.5. Predicting Evolutionary Change

The three requisites for evolution by natural selection are linked together quantitatively by the standard equation for evolutionary change in a quantitative character (Falconer, 1981):

$$R = \left[\frac{V_a}{V_p}\right]s \quad \text{or} \quad R = h^2 s \qquad (2.5)$$

Here, h^2 is the heritability of the trait, which is equal to the ratio of the additive genetic variance to the phenotypic variance. The quantity R, known as the response to selection, is the change in the mean value of the trait in the population from one generation to the next. This basic equation thus says that the expected magnitude of evolutionary change over a single generation is equal to the product of the heritability and the selection differential. For the expected change to be non-zero, both of these quantities (and hence V_a and V_p) must also be non-zero.

A great number of studies, as well as practical experience of animal and plant breeders, have shown that Equation (2.5) predicts evolutionary change in single characters reasonably accurately under relatively simplified laboratory or agricultural conditions, in which selection is imposed on a single trait. (Falconer, 1953; Clayton et al., 1957). One might therefore assume that this equation could also be used to predict or infer evolutionary change in natural populations. However, this type of assumption can be misleading for two important reasons (see also Endler, 1986): (1) the three requisites for evolution by natural selection described above and embodied in Equation (2.5) are necessary, but not sufficient, conditions for evolutionary change, and (2) correlations among characters may produce indirect selection on a trait that counteracts the direct selection embodied in Equation (2.5).

2.1.3.6. Insufficiency of the Three Conditions

It is not difficult to image how a trait might fail to evolve, even though the three requisites for evolution by natural selection were satisfied. Consider a hypothetical alkaloid and suppose its concentration in a plant's leaves is determined by both the plant's genotype and by soil nitrogen level. Suppose further that variation in the concentration of this alkaloid is selectively neutral, that is, genotypes that differ in average concentration do not differ predictably in average fitness. There can then be no evolutionary change in mean concentration due to selection. (Technically, this conclusion follows from Crow and Nagylaki's [1976] demonstration that directional evolutionary change occurs if and only if there is a correlation between the breeding values of fitness and a trait, i.e., if and only if there is a genetic correlation between fitness and the trait. See also Price, 1970; Uyenoyama, 1986.)

Imagine, however, that soil nitrogen content limits plant growth and varies greatly within the field in which the plant grows. If one examined plants of just one genotype, one would find some growing in relatively nitrogen-rich soil and others growing in relatively nitrogen-poor soil. The former plants would not only produce higher concentrations of the compound, but might also grow larger and produce more seeds, that is, have higher fitness. There would thus be a positive environmental correlation between fitness and alkaloid concentration. Other genotypes would exhibit a similar environmental correlation, resulting in an overall phenotypic correlation between fitness and alkaloid level and hence a positive value for the selection differential, s. Under these circumstances, h^2 and s are both positive. Equation (2.5) would thus predict a positive evolutionary response. As noted above, however, this response would not occur because genotypes do not differ predictably in fitness.

This hypothetical example illustrates a general problem with using the selection differential, s, as a measure of selection. The selection differential reflects a correlation between fitness and a trait, a correlation that is, in turn, determined by both environmental and genetic correlations. Numerically,

$$s = \text{cov}_a + \text{cov}_e \qquad (2.6)$$

(Falconer, 1981), where cov_a and cov_e are the additive genetic and environmental covariances, respectively, between fitness and the trait. (As a reminder, a correlation, r, between two variables x and y is related to the covariance between those two variables by the formula $cov(x, y) = r(x, y)$ $[\sqrt{Var(x) Var(y)}]$.) Yet, as indicated above, genetic change in the trait depends only on the genetic correlation (or covariance). Consequently, to the extent that the absolute value of the environmental correlation is greater than that of the genetic correlation (more strictly, to the extent that $|cov_e|$ is greater than $|cov_a|$), the selection differential will reflect the environmental correlation rather than the genetic correlation between fitness and the trait and will give an inaccurate picture of selection.

It is not known how frequently the selection differential may be a biased indicator of selection for the reason described above. Many examples exist of genetic correlations differing in sign from corresponding environmental correlations between two traits (Falconer, 1981). There is no reason to doubt that similar differences often exist when one of the characters measured is fitness, but there have been few attempts to estimate this type of bias in the selection differential. One way of avoiding this potential problem is to measure selection acting directly on genetic variation for a trait by calculating the genetic covariance between the trait and fitness (Rausher and Simms, 1989), but this approach may not be feasible for organisms whose breeding behavior is unsuitable for generating offspring from complex mating designs, For such organisms, one must be very cautious about inferring or predicting the direction or magnitude of evolutionary change in a trait from measurements of the selection differential associated with that trait.

2.1.3.7. Selection and Evolutionary Change in Correlated Characters

The validity of Equation (2.5) depends critically on the assumption that selection acts on a single trait or, more precisely, that selection does not also act on correlated traits. In laboratory selection experiments, the validity of this assumption can often be assured because selection is imposed by the investigator. In nature, however, complex phenotypes of numerous traits are exposed to a wide variety of potential selective agents. One can therefore expect that quite often, several traits will be exposed to selection and evolve simultaneously.

In this situation, the predictions of Equation (2.5) are no longer valid because this equation does not account for indirect selection on a character due to selection on correlated characters. To see how such selection might arise, consider an example from Berenbaum et al. (1986), who report a negative correlation between first flowering date and the proportion that bergapten constitutes of all furanocoumarins in the seeds of the wild parsnip, *Pastinaca sativa* (see Fig. 2.2). Suppose that in a particular

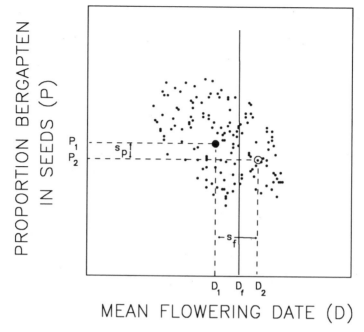

Figure 2.2. The effect of selection on correlated characters. The cluster diagram portrays a negative correlation between the mean flowering date and the proportion of begapten in seeds of wild parsnip (see Berenbaum et al., 1986). D_f and the solid line indicate the date of a hypothetical late killing frost. The filled circle represents the mean of the population before selection (D_1, P_1). All individuals to the right of the solid line survive the late frost and set seed. The open circle indicates the mean of these surviving individuals (D_2, P_2). s_{fd} is the selection differential for the flowering date. Because of the correlation between the proportion of bergapten and the flowering date, the selected individuals also have a lower proportion of bergapten, i.e., $P_2 < P_1$, giving rise indirectly to the selection differential, s_p, on the proportion of bergapten.

year, the seeds of early-flowering plants are killed by a late frost, while late-flowering plants escape frost damage. This type of selection would be truncation selection, in which only individuals with values of a trait above a certain threshold are able to reproduce. As shown in Fig. 2.2, the selection differential on flowering date, s_{fd} is simply the difference between the mean flowering date of all individuals not affected by frost (selected individuals) and the mean flowering date of all individuals in the population. Because the proportion of bergapten is correlated with the flowering date, however, the proportion of bergapten in the seeds of selected individuals (to the right of the solid line in Fig. 2.2) is lower than in the seeds of unselected individuals. Consequently, there is a measureable negative selection differential for the proportion of bergapten, s_p. It is important to note, however, that this selection is indirect in that it arises from direct selection on a correlated character, flowering date.

Both genetic and phenotypic correlations among characters can render the predictions of Equation (2.5) invalid, albeit in different ways. Phenotypic correlations render the selection differential, s, an inaccurate indicator of direct selection on a character, while genetic correlations cause the mean of a character to change because of selection on other characters. These effects can be visualized by considering a generalization of Equation (2.5) for predicting evolutionary change in a set of correlated characters (Lande, 1979; Lande and Arnold, 1983):

$$\Delta \bar{z} = GP^{-1}s \tag{2.7}$$

Here $\bar{z}$ is a vector $(\bar{z}_1, \bar{z}_2, \ldots \bar{z}_n)$ of values for characters $1, 2, \ldots n$, G is the genetic variance-covariance matrix (a matrix whose diagonal elements, g_{ii} consist of the genetic variances of the traits, V_{a_i}, and whose off-diagonal elements, g_{ij}, are the genetic covariances between the characters, $cov_{a_{ij}}$), P is the phenotypic variance-covariance matrix (analogous to genetic variance-covariance matrix, but with phenotypic variances, p_{ii}, and covariances, p_{ij} as elements), and s is a vector composed of the selection differentials for the characters ($s = s_1, s_2, \ldots s_n$).

An alternate form of Equation (2.7) is

$$\Delta z = G\beta \tag{2.8}$$

where $\beta = P^{-1}s$ is known as the vector of selection gradients. Each element, β_i, represents the strength and magnitude of selection acting directly on a character. It may be thought of as being the slope of a regression of the log of relative fitness on the character, with the values of all the other characters held constant (Lande and Arnold, 1983). Graphically, selection gradients may be portrayed for two characters on a diagram of the adaptive landscape, which is the relationship between the logarithm of mean fitness in a population and the mean values of a set of traits. In Fig. 2.3, I present a hypothetical adaptive landscape for two traits. The contours represent combinations of trait means that have the same mean fitness. At any point on the contour surface, the vector of selection gradients points in the direction of steepest ascent up the

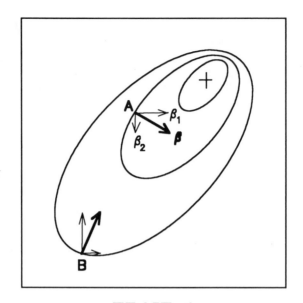

TRAIT 1

Figure 2.3. Hypothetical adaptive landscape for two traits. Contours represent combinations of values of the two traits that have equal fitness. The + indicates the combination with highest fitness, i.e., the adaptive peak or optimum. The vector of selection gradients, β, for points A and B are shown as thick arrows on main diagram and inset. This vector points in the direction of steepest assent up the contours. The thin horizontal (β_1) and vertical (β_2) arrows are the orthogonal components of β and represent the selection gradients for traits 1 and 2, respectively. Note that the vectors of selection gradients do not necessarily point at the optimum.

contour surface (arrows in Fig. 2.3) and, in a loose sense, indicates the direction in which natural selection is pushing the population.

To demonstrate the influence of both phenotypic and genetic correlations (or covariances) on evolutionary change, consider two characters that are correlated with each other, but not with any other characters. Written out in full, Equations (2.7) and (2.8) then become

$$\Delta \bar{z}_1 = V_{a_1}\beta_1 + \text{cov}_a \beta_2 \qquad (2.9A)$$

$$\Delta \bar{z}_2 = \text{cov}_a \beta_1 + V_{a_2}\beta_2 \qquad (2.9B.)$$

First, consider the influence of genetic correlations between the characters. In the absence of a genetic correlation, $\text{cov}_a = 0$. Equations 2.9A and 2.9B reduce to

$$\Delta \bar{z}_1 = V_{a_1} \beta_1 \tag{2.10A}$$

$$\Delta \bar{z}_2 = V_{\alpha_2} \beta_2 \tag{2.10B.}$$

These equations say that the change in the mean of each character is determined by the amount of additive genetic variance for that character and by the strength of selection acting directly on that character, which is equivalent to the prediction for a single character (Equation 2.5). If, however, the genetic correlation between the characters is not zero, then the terms involving cov_a in Equations 2.9A and 2.9B also influence evolutionary change. In particular, strong selection for an increase in trait 1 ($\beta_1 \gg 0$) may cause a decline in trait 2, even if direct selection favors an increase in trait 2 ($\beta_2 > 0$). Genetic correlations thus allow direct selection on one trait to influence genetic change in a second.

By contrast, phenotypic correlations among characters affect the interpretation of the selection differential. Consider once more the bergapten example. When there are no phenotypic correlations among characters, then

$$\beta_1 = \frac{s_1}{V_{p_1}} \quad \text{and} \quad \beta_2 = \frac{s_2}{V_{p_2}} \tag{2.11}$$

(This may be shown by calculating P^{-1} and forming the product $P^{-1}s$.) Equations 2.9A and 2.9B then become

$$\Delta \bar{z}_1 = \left[\frac{V_{a_1}}{V_{p_1}} s_1 \right] + \text{cov}_a \, \beta_2 \tag{2.12A}$$

$$\Delta \bar{z}_2 = \text{cov}_a \, \beta_1 + \left[\frac{V_{a_2}}{V_{p_2}} s_2 \right] \tag{2.12B.}$$

The terms in brackets represent evolutionary change caused by direct selection on each character and are the same as the response to selection predicted for a single character by Equation (2.5). In other words, when there are no phenotypic correlations between characters, the selection differential and Equation (2.5) provide an accurate prediction of the evolutionary change caused by direct selection on a character.

When phenotypic correlations exist, however, the relationship between the selection differential and the selection gradients becomes more complex. In particular, for two correlated traits,

$$\beta_1 = \frac{s_1}{\left(V_{p_1} - b_{12}\right)} - \frac{s_2 b_{12}}{\left(V_{p_1} - b_{12}\right)} \qquad (2.13A)$$

$$\beta_2 = \frac{s_2}{\left(V_{p_2} - b_{21}\right)} - \frac{s_1 b_{21}}{\left(V_{p_2} - b_{21}\right)} \qquad (2.13B.)$$

where b_{ij} is the slope of the regression of character i on character j. Direct selection on trait 1 thus involves a combination of the selection differentials for both traits (s_1 and s_2), and hence is conceptually different from the type of direct selection embodied in the selection differential in Equation (2.5).

My own work in attempting to understand how natural selection molds resistance to herbivores in the morning glory, *Ipomoea purpurea*, provides an example of how failure to consider phenotypic correlations among characters may lead to very inaccurate pictures of how selection acts. Ellen Simms and I (Rausher and Simms, 1989; Simms and Rausher, 1987, 1989) have shown that local populations of morning glories harbor genetic variation for resistance to four different types of insect herbivores, with resistance being assayed as the inverse of the proportion of damage to leaves or developing pods (1 − Proportion Damage). We have also used the techniques of Lande and Arnold (1983) to estimate selection gradients for these indices of resistance. If each type of resistance is examined separately, there is apparently significant directional selection favoring increased resistance to corn earworms and decreased resistance to tortoise beetles, but not detectable directional selection acting on resistance to flea beetles or to generalist insects (Table 2.1). Stabilizing selection favoring an intermediate level of resistance to each type of herbivore is also apparent.

When correlations among resistances and total plant area are taken into account by multivariate analysis, however, the pattern of selection revealed is very different. Resistances to all four types of insects experienced directional selection (Table 2.1). Only resistance to corn earworms was subject to significant stabilizing selection, while resistance to generalist insects experienced disruptive selection. Had we been particularly interested in resistance to generalist insects, for example, and examined that character by itself, we would have been totally misled by the selection differentials for this trait.

It should be clear that biologists interested in quantifying selection on ecologically interesting characters should strive to measure selection gradi-

Table 2.1. Comparison of directional (β) and stabilizing/disruptive (γ) selection gradients for the proportion of damage to *Ipomoea purpurea* caused by four types of herbivores. Negative values of γ indicate stabilizing selection, positive values indicate disruptive selection. (See Rausher and Simms, 1989, for details.) Univariate selection gradients, calculated by analyzing each trait separately, do not adjust gradients for phenotypic correlations among traits. Multivariate gradients, calculated in a single multivariate analysis including damage by all types of herbivores as well as total leaf area, adjust gradients for correlations among characters. Note that in the univariate analyses, all characters exhibited an optimum phenotype within the range of possible phenotypes. This property also holds for corn earworm resistance in the multivariate analysis when all other characters are held at their mean values.

N.Resistance to	Multivariate Selection Gradients		Univariate Selection Gradients	
	β	γ	β	γ
Corn earworms	-0.119**	-0.358**	-0.072*	-0.332***
Flea beetles	0.238**	0.077	-0.046	-0.467***
Tortoise beetles	-0.169**	0.345	0.318***	-1.117***
Generalist insects	-0.285**	0.061**	0.014	-0.616*

ents rather than selection differentials, as well as the genetic correlations (covariances) among traits. Fortunately, simple statistical methods have been developed for estimating selection gradients (Lande and Arnold, 1983, Arnold and Wade, 1984a, b). Even these methods have drawbacks, however. Environmental correlations between characters and fitness can bias estimates of selection gradients, just as they can bias estimates of selection differentials (see above; Price et al., 1988, Rausher and Simms, 1989). Moreover, failure to incorporate important correlated traits in an analysis can affect estimates of selection gradients (Lande and Arnold, 1983; Endler, 1986). Despite these problems, the measurement of selection using these techniques can suggest how fundamental evolutionary processes shape interactions between organisms. Follow-up, manipulative experiments can then confirm the validity of inferences drawn from such measurements.

2.1.4. *Constraints on Evolutionary Change*

The existence of selection on a trait does not guarantee that trait will evolve. If it did, all organisms would live forever and produce large numbers of offspring each breeding season, since, all else being equal, a decrease in mortality and an increase in reproductive rate confers an increase in fitness. A complete understanding of evolutionary change thus requires explanations of processes that constrain, as well as promote, evolution. Constraints on the evolution of individual traits may be grouped into two major categories: (1) lack of appropriate genetic variation for that trait, and (2) trade-offs involving two or more traits.

2.1.4.1. Lack of Variation

As is clear from Equations 2.5 and 2.8, a character cannot evolve if there is no genetic variation affecting that trait. There are many reasons why such variation might be lacking in a population: (1) a bottleneck (drastic reduction in population size) may have led to fixation at many loci by genetic drift (Nei et al., 1975; but see Goodnight, 1988, on how a bottleneck may actually increase additive genetic variance); (2) recent intense directional selection may have depleted additive genetic variation, as commonly occurs in artificial selection experiments (Falconer, 1955; Clayton, et al., 1957; Yoo, 1980); (3) long periods of stabilizing selection may have eliminated most additive variation (Lewontin, 1964; Felsenstein, 1979); and (4) inbreeding (mating with self or with close relatives) may be associated with reduced genetic variation (Hamrick et al., 1979). Unfortunately, little is known about how commonly evolution in natural populations is retarded by lack of genetic variation.

2.1.4.2. Trade-offs and Genetic Correlations

Trade-offs are often believed to be a second major constraint on evolutionary change. An evolutionary trade-off exists between two beneficial traits whenever evolutionary change that enhances one of the traits causes a correlated diminishment of the other. As can be seen from Equations 2.9A and 2.9B, such a correlated response to selection will occur only when there is a negative genetic correlation between the two characters. Consequently, negative genetic correlations are often interpreted as evolutionary constraints (Cheverud, 1984; Berenbaum et al., 1986).

Lerner (1958) has suggested how trade-offs might arise. Consider two beneficial traits that are each influenced by a set of common loci, as well as by loci uniquely affecting each trait. The common loci may be categorized into two types. At one type, a particular allele, designated $+/+$ or $-/-$, respectively enhances or diminishes both traits, as might occur if a locus produces a protein catalyzing the production of a common biochemical precursor for two plant secondary compounds. At the other type, a particular allele, designated $+/-$ or $-/+$, enhances one trait and diminishes the other, as might occur with a gene that influences the proportions of that common precursor that are converted into the two secondary compounds. The genetic correlation between these two traits then depends, to a first approximation, on the relative number of the two types of loci. If loci with $+/+$ and $-/-$ alleles are more common, the genetic correlation will be positive, while if loci with $+/-$ and $-/+$ alleles predominate, the genetic correlation will be negative.

Lerner's argument postulates that $+/+$ alleles should be fixed rapidly by selection because they enhance both beneficial traits and thus increase fitness greatly. By contrast, at loci with $+/-$ and $-/+$ alleles, the net force of selection may often be very small, since for each allele the beneficial effect of one trait is compensated for by the adverse effect of the other trait. Consequently, these loci will tend to remain genetically variable after fixation occurs at loci with $+/+$ and $-/-$ alleles, yielding a negative correlation between the two traits.

The way negative genetic correlations may constrain evolution can be demonstrated with the aid of Equations 2.9A and 2.9B. The simplest case involves two traits with equal variances and an adaptive landscape in which fitness declines symmetrically as the population mean moves away from an optimal value (Fig. 2.4A). Consider a population in which the mean for each trait is less than the optimum. If both traits are heritable and there are no trade-offs between the two traits, i.e., the genetic correlation is zero, then selection will quickly move the population to the optimum (Fig. 2.4A). By contrast, if there is a perfect trade-off (the genetic correlation equals -1), selection will move the population along a line reflecting this genetic constraint but will not move the population toward the optimum (Fig. 2.4B).

The reason for this response to selection is straightforward. Consider a population with a mean for character 1 slightly below its optimum and a mean for character 2 considerably below its optimum ("X" in Fig. 2.4B). The selection gradient for trait 1 is then small compared to the gradient for trait 2 (Fig. 2.4B). With a genetic correlation of -1, the covariance between the two traits is equal in absolute value to the additive genetic variance of each trait. The covariance term of Equation (2.9A) will then be negative and larger in absolute value than the variance term, and hence character 1 will actually decline. On the other hand, for trait 2, the variance term in Equation (2.9B) dominates because the selection gradient for that trait is greater. Trait 2 will thus increase. The combined responses of the two traits yields the trajectory depicted in Fig. 2.4B. When the population eventually reaches point on the main diagonal (connecting origin and peak), the two selection gradients are equal, and there is no further net change in either mean, even though there remains additive genetic variance for both traits, and both traits are subject to selection.

This analysis demonstrates that a perfect trade-off can constrain two traits from evolving to their optimal values. When the trade-off is not

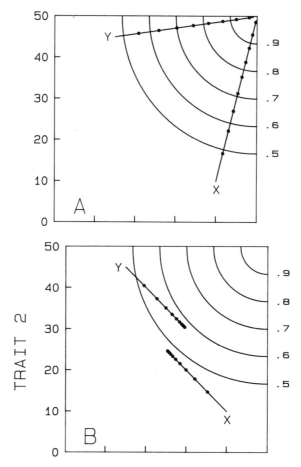

Figure 2.4. Evolutionary trajectories for two correlated traits. Optimal combination of traits is in upper
right corner. Contours of adaptive landscape are
symmetrical about the optimum. For convenience,
only one quadrant of the landscape is depicted. Solid
lines represent evolutionary trajectories from initial
points X and Y. Segments between arrowheads represent 50 generations. In all figures, $V_{a_1} = V_{a_2} = 1$. (A)
Trajectories for $cov_a = 0$. All trajectories converge
toward optimum, showing absence of constraint. (B)
Trajectories for $cov_a = 1$. Trajectories converge to a
point on the main diagonal then stop. (C) Trajectories
for $0 < cov_a < -1$. Trajectories eventually converge
on optimum, but do so slowly for cov_a close to -1.

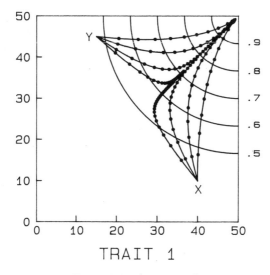

Figure 2.4. (Continued).

perfect (the genetic correlation is less than 0 but greater than −1), the population eventually reaches the optimum (Fig. 2.4C). However, the rate at which the optimum is attained is reduced compared to when the correlation is zero. Moreover, the more negative the correlation, the slower the population approaches the optimum. When correlations approach −1 significant movement toward the optimum requires thousands of generations, by which time the optimum may be expected to have changed anyway. From this analysis, it seems that trade-offs may act as a constraint on the evolution of two traits toward a joint optimum, but that the genetic correlations underlying the trade-offs must be close to −1 to be effective as constraints. (For an interpretation of constraints involving three or more correlated characters, see Charlesworth, 1990.)

One caveat must be added to this conclusion. The trajectories depicted in Fig. 2.4 were determined by iterating Equations 2.9A and 2.9B and assuming that the genetic variances and covariances (the G matrix in Equations 2.9A and 2.9B) remain constant. Although Lande (1975, 1977, 1980) has provided theoretical arguments to justify this assumption, Turelli (1984, 1985, 1988) has described alternative analyses that imply that genetic variances and covariances will change as populations evolve. One likely change for a population evolving toward an optimum, as in Fig. 2.4C, is that the covariances between traits will become more negative as progress toward the optimum fixes alleles that affect both traits in the same direction (the +/+ alleles from the discussion above). Evolution toward the optimum would then gradually increase the efficacy of a trade-off in preventing further change and could prevent a population starting with a weak trade-off from ever attaining the optimum. Evolution would then be constrained by the absence of the particular type of genetic variation needed to allow further progress (Antonovics, 1976). This is, in fact, Lerner's argument.

2.2.   Unresolved Evolutionary Issues in Chemical Ecology

Although the field of chemical ecology has blossomed in the last 25 years, there remain many outstanding problems that investigators have not been able to resolve. In the remainder of this chapter, I review several of these problems and suggest how the tools of evolutionary biology and population genetics might be used to address them. Because of my own experience and interests, I concentrate primarily on problems involving interactions between plants and herbivores. Nevertheless, the general approach and outlook I espouse could be and in fact, is being, applied to questions involving plant-pollinator interactions, animal behavior, and community ecology.

2.2.1.   Definitions

A discussion of the evolution of plant-herbivore interactions requires that potentially ambiguous terms be defined precisely. For my purposes, I adopt the following definitions.

Resistance is the degree to which a plant (or plant part) avoids damage by herbivores. Resistance is a relative rather than absolute quantity, so that the resistance level of an individual plant may be quantified only with reference to that of some other plant used as a standard. Operationally, resistance is often measured as the inverse of the amount or proportion of foliage (or other plant tissue) that is damaged by herbivores (Painter, 1958; Beck, 1965; Carter, 1973; Horber, 1980; Rausher and Simms, 1989). Defined this way, resistance level is determined both by genetic makeup and by environmental factors (Fig. 2.5). Genetic constitution influences resistance via its effect on particular plant traits that deter oviposition and feeding by herbivores. Environmental variables influence the level of resistance in two ways: by influencing the development and expression of plant traits that deter herbivory and by affecting the local abundance and behavior of herbivores.

A *resistance trait* is any plant character that influences the amount of damage a plant suffers (and, therefore, influences the level of resistance) under a specific environmental regime. For example, cyanogenesis in bracken and in *Trifolium* is a resistance character because production of

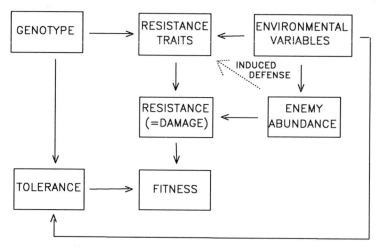

Figure 2.5. Path diagram indicating causal influences of genotype and environmental variables on resistance, tolerance, and fitness.

cyanide reduces damage caused by deer, slugs, and snails (Cooper-Driver and Swain, 1976; Dirzo and Harper, 1982). Sclerophylly in *Aristolochia reticulata* (Aristolochiaceae) is a resistance character because it reduces feeding damage by larvae of the pipevine swallowtail butterfly, *Battus philenor* (Rausher, 1981).

A *defense* or *defensive trait* is any resistance trait that has evolved or is maintained in a plant population because of selection exerted by herbivores or other natural enemies. The term "defense" thus implies something about the evolutionary *raison d'etre* for the trait, whereas the term "resistance" does not.

Tolerance, sometimes also known as compensation, is the ability of plants to experience damage without a reduction in fitness (Crawley, 1983; Paige and Whitham, 1987). It is thus the difference between the fitness of a plant experiencing damage and the fitness that plant would have had if it had suffered no damage. Since a plant either experiences damage or it does not, the tolerance of an individual plant cannot be assessed. Rather, one can measure only the average tolerance of a group of plants subjected to a particular type and amount of damage by comparing the averaging fitness of those plants to the average fitness of a group of similar control plants that are not damaged. Nevertheless, such a comparison can reveal whether plant populations are genetically variable for tolerance if the average tolerances of several genotypes of families are compared. Moreover, additive genetic covariances between tolerance and fitness can be used to measure selection on degree of tolerance. As with resistance, the tolerance of an individual is determined both by the individual's genetic makeup and by environmental factors, including herbivores and pathogens (Fig. 2.5).

2.2.2. The Raison d'Etre of Plant Secondary Compounds and Other Resistance Factors

Plants possess a diverse array of characters that appear to provide protection from the deprivations of insects, slugs, vertebrate grazers, and other herbivores. Many plants seem to be veritable chemical arsenals, bristling with toxins and growth inhibitors. Many have spines, trichomes, egg mimics, and tough, nutrient-poor leaves that deter herbivory. Ever since Stahl (1888), some biologists have interpreted the primary function of these types of resistance characters to be defense (Fraenkel, 1956, 1959, 1969; Ehrlich and Raven, 1964; Janzen, 1969, 1973a; Whittaker and Feeny, 1971; McKey, 1974; Feeny, 1975, 1976; Rhoades and Cates, 1976; Levin, 1976; Rausher, 1981; Rhoades, 1983). By contrast, other biologists have argued that these traits serve other important ecological or physiological functions and that the effects of these characters on resistance are fortuitous (Del Moral, 1972; Robinson, 1974; Jermy, 1976, 1984; Siegler and Price, 1976; Siegler, 1977; Swain 1977). This debate is central to understanding the evolution of plant-herbivore interactions because it is essentially a debate about what types of selection pressures direct the evolution of a large suite of plant characters.

Proponents of the defensive function of resistance characters have offered two primary types of evidence in support of their position. First, they have cited examples of herbivores reducing the viability or reproductive success of the plants on which they feed. Such examples are not uncommon (Cantlon, 1969; Morrow and Lamarche, 1978; Rausher and Feeny, 1980; Parker and Root, 1981; Louda, 1983; Marquis, 1984) and suggest that herbivores commonly have the potential to exert intense selection on plants. Examples of genetic variation for resistance to herbivores in natural plant populations are also accumulating rapidly (Jones et al., 1978; Moran, 1981; Dirzo and Harper, 1982; Kinsman, 1982; Fritz et al., 1986; Simms and Rausher, 1987). However, as Jermy (1984) points out, demonstrating the potential for selection is not the same as demonstrating that selection occurs; none of these studies has attempted to determine whether fitness differs among genetic variants for resistance. There has thus been no direct evidence that herbivores impose natural selection on plant traits. Moreover, Jermy (1976, 1984) argues that the above examples are exceptional cases and that most herbivores, particularly insects, are normally too rare to impose significant selection on plants.

The second type of evidence offered in support of the defensive function of resistance characters is the fact that they are resistance characters. Demonstrating that secondary plant compounds are toxic, repellant, or growth-inhibiting to herbivores has become a minor industry supporting professional ecologists. Sclerophyllous leaves, leaf shape, pubescence, trichomes, modified egg-mimic stipules and low leaf-water content are all known to affect, or strongly suspected of affecting, some herbivores adversely and reduce damage to some plants (Gilbert, 1971, 1975; Rausher, 1978; Williams and Gilbert, 1981; Rausher, 1981; Scriber and Slansky, 1981).

The effect of these traits on resistance is exactly what is expected if resistance traits have evolved as defenses. As Jermy and other have stressed, though, adverse effects on herbivores and an attendant reduction in herbivory is also expected if resistance is a pleiotropic effect of traits that have some other primary ecological or physiological function. Moreover, evidence exists for the functioning of plant secondary compounds as regulators of growth and biosynthesis, energy and nutrient storage compounds, shields against ultraviolet radiation, and facilitators of frost tolerance (Chew and Rodman, 1979). Leaf shape and pubescence greatly affect heat load on plant tissues and may therefore function primarily in thermoregulation (Jeffree, 1986). Sclerophylly and low foliar water content are often correlated with water- or nutrient-poor habitats and primarily may reflect adaptations to conserve water or minerals (Mooney and Dunn, 1970; Alberdi et al., 1974; Grubb, 1986). Such potential primary functions of resistance traits are ubiquitous and indicate that a simple demonstration of toxicity or deterrence is not sufficient to allow one to infer that a trait functions primarily as a defense.

If these types of evidence are sufficient to determine whether a particular trait has evolved to defend a plant from herbivores, what types are needed? Occasionally, one may be fortunate enough to observe natural evolutionary experiments. For example, while leaves of most plants in the genus *Acacia* are cyanogenic, some species have evolved mutualistic relationships with ants in the genus *Pseudomyrmex*. These species produce swollen, hollow thorns used by the ants as domiciles and proteinaceious "Beltian bodies" on the tips of the leaflets that are harvested by the ants as food. Janzen (1966) has demonstrated convincingly that the ants inhabiting these "ant Acacias" remove herbivorous insects and deter browsing by mammals, and thus providing a sort of biological resistance to the plants. The ants also prune away vines and other neighboring plant species, thus presumably reducing the intensity of competition experienced by the *Acacias*. Rehr et al. (1973) have shown that *Acacia* species that have evolved this type of mutualistic relationship with ants have tended to lose the capacity for cyanogenesis. *Pseudomyrmex* thus replaces the primary function of cyanogenesis in *Acacia*. Since there is no evidence that cyanogenesis functions in plants as an allelopathic chemical reducing competition, while cyanide is toxic to many, if not most, herbivores, it seems very likely that the function replaced was resistance to herbivores. This work by Janzen and collaborators provides a very strong case for a resistance trait (cyanogenesis) have evolved primarily as a defense.

Natural evolutionary experiments of this type are too rare to provide more than anecdotal evidence bearing on the question of whether, in general, plant secondary compounds and other resistance traits have evolved as defenses. An alternative approach is direct measurement of selection imposed by herbivores on plants. This approach assumes the general applicability of the uniformitarianism principle that the selection pressures experienced on plants now are not fundamentally different from those experienced in the past or those that will be imposed in the future. Consequently, if in the past, selection pressures imposed by herbivores were common and were responsible for the evolution of apparently defensive traits, similar pressures should be common and detectable today. By examining a number of representative (randomly chosen?) plant-herbivore associations, it should be possible to determine how frequently the plants in such systems are subject to selection imposed by the herbivores.

Adopting this approach requires addressing three fundamental questions.

1. *Is there genetic variation for resistance?* As noted earlier, selection cannot lead to evolutionary change without genetic variation on which to act. Moreover, because environmental correlations may bias estimates of selection based on phenotypic correlations between resistance and fitness, selection is best measured on genetic variation (see above). The presence of genetic variation is thus a prerequisite for measuring the intensity and direction of selection. Ideally, one would like to be able to identify genetic variation for a particular resistance trait. However, it is also possible to ask whether there is genetic variation for resistance itself, measured as the inverse of damage (Simms and Rausher, 1987, 1989). The existence of such variation implies that there are one or more unknown genetically variable characters that confer resistance.

2. *Is genetic variation for a resistance trait (or for resistance itself) subject to selection?* If a plant is completely tolerant to herbivore damage, genetic variation in a resistance trait may be selectively neutral. Genetic drift may then be the primary process governing the evolution of that trait. Ruling out this possibility by demonstrating that the trait is subject to selection will often be the next logical step in determining the trait's primary function, since measuring whether selection occurs will often be easier than identifying the agents of selection.

3. *Is selection imposed by herbivores?* Convincing demonstration that herbivores impose selection on a resistance trait involves measuring selec-

tion on the trait both in the presence and in the absence of herbivores. If the measurements differ, herbivores exert a significant, though not necessarily the only, evolutionary force on the trait.

Two recent studies illustrate the power of this approach. Berenbaum et al. (1986) examined the selective forces directing the evolution of furanocoumarins in the wild parsnip, *Pastinaca sativa*. Furanocoumarins are toxic to a wide variety of herbivores (Berenbaum, 1978, 1983; Berenbaum and Feeny, 1981) and are thus potentially resistance factors. Moreover, their concentrations and proportions are also genetically variable. Concentrations of some furanocoumarins and the proportion of unblets damaged by the parsnip webworm are significantly correlated, evidence that these compounds confer resistance to this herbivore. Finally, Berenbaum et al. (1986) also demonstrate significant phenotypic and genetic covariances between the concentrations of several furanocoumarins and seed production, a result indicating that some furanocoumarins are subject to selection. This study also provides some evidence that the selection detected is in part imposed by parsnip webworms: the correlations between furanocoumarin content and secondary ray number (an estimate of potential seed production) differ for plants grown in the field (webworm present) and in the greenhouse (webworm absent). This evidence must be interpreted with caution, however, since the presence of parsnip webworms is not the only environmental factor (and potential selective agent) that differs between field and greenhouse.

Rausher and Simms (1989) and Simms and Rausher (1987, 1989) found that a population of the annual morning glory *Ipomoea purpurea* is genetically variable for resistance to each of four types of insect herbivores. For the three types of folivores studied, selection on variation for resistance was not detected. If the selection gradients measured during this study are typical of other years and other localities, this result suggests that whatever traits are responsible for variation in resistance, they are evolving primarily by genetic drift. The evolution of resistance would then be a passive evolutionary process with no inherent direction.

By contrast, for resistance to the pod-feeding corn earworm, *Heliothis zea*, resistant genotypes had higher fitness than susceptible genotypes. Moreover, removal of insects from some plants by spraying with insecticide eliminated selection on resistance to corn earworms. This result provides direct evidence that herbivorous insects, presumably the corn earworm itself, impose selection on resistance (and hence on whatever unidentified traits are responsible for resistance).

One additional interesting result of this study was that removal of insects by spraying eliminated all additive genetic variance for fitness. Fisher's fundamental theorem of natural selection (Fisher, 1958) states that the rate of increase of mean fitness in a population is proportional to the amount of additive genetic variation in fitness. The magnitude of the additive variance for fitness is thus an index of the amount of selection on all characters. Since resistance to corn earworms was not perfectly correlated genetically with fitness, some other characters affected by insects must have been contributing to variance in fitness. A likely possibility is that the *Ipomoea* population examined was genetically variable for tolerance as well as for resistance. If so, the presence of insects would generate fitness variation, but this variation would disappear with the removal of insects because all plants would exhibit the same base level fitness.

These recent studies illustrate that it is possible to measure selection on resistance and resistance traits. They thus indicate that we may one day know through direct measurements of selection pressures whether feeding by herbivores is a common and significant evolutionary force guiding the evolution of plant secondary compounds and other purported defenses.

2.2.3. The Evolution of Defenses

Although the jury is still out on the ecological function of most resistance traits, it is probably safe to assume that at least some traits function primarily as a defense against herbivores. For such defenses, a question that has engendered much speculation is how natural selection determines the level of resistance that evolves in a population. Several workers have argued that the mean value of quantitative resistance traits will often evolve toward an intermediate optimum determined by a trade-off between costs and benefits of resistance (Janzen, 1973a, b; McKey, 1974; Feeny, 1976; Rhoads, 1979, 1983; Krischik and Denno, 1983; Coley et al., 1985).

The logic of this argument may be understood by considering how it might apply to the evolution of the foliar concentration of some defensive secondary compound, say a particular type of alkaloid (see Fig. 2.6A). The production of this alkaloid is envisioned to entail a cost because energy and materials used to manufacture it are diverted away from other structures and physiological processes that enhance fitness. The relationship between concentration and cost is depicted as linear in Fig. 2.6A for convenience, but linearity is not essential to the argument.

Alkaloid production is also envisioned to benefit a plant by reducing herbivory, thereby reducing the detrimental effects of herbivory on plant fitness. Consequently, a second set of assumptions integral to the argument is that herbivory reduces plant fitness and that an increase in the concentration of foliar alkaloid reduces damage. If both assumptions are valid, the benefit should increase monotonically with alkaloid concentration. This relationship is portrayed as plateauing in Fig. 2.6A because at alkaloid concentrations greater than a certain value, all herbivores will be eliminated. A further increase in alkaloid levels can then provide no additional benefit to the plant.

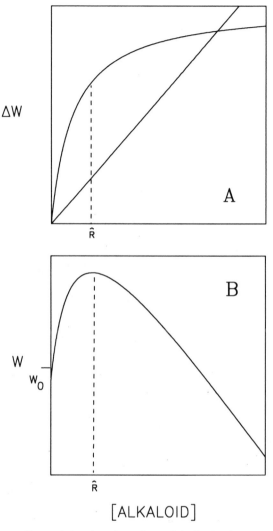

[ALKALOID]

Figure 2.6. Cost-benefit model of selection on alkaloid concentration. Points on curve represent additive genetic values for costs, benefits, and fitness as functions of additive genetic value of alkaloid concentration. (A) costs (straight line) and benefit (curve) as function of alkaloid concentration. Costs and benefits are measured as incremental changes in fitness from a base fitness, $W(0)$. See text for explanation. (B) Relationship between fitness and alkaloid concentration, determined by Equation 2.14. $\hat{R}$ is equilibrium alkaloid concentration.

Under these assumptions, the expected fitness, $W(R)$, for an individual with additive genetic value for alkaloid concentration equal to R is given by

$$W(R) = W(0) + B(R) - C(R) \qquad (2.14)$$

where $W(0)$ is a base fitness equal to that of a genotype that fails to produce alkaloids, and $B(R)$ and $C(R)$ are the benefit and cost for the same individual, as derived from the curves in Fig. 2.6A (Simms and Rausher, 1987). The genotype with the highest fitness is the one for which the difference between cost and benefit is greatest. As depicted in Fig. 2.6B, this genotype produces an intermediate concentration of alkaloid, $\hat{R}$. Alkaloid concentration is thus subject to stabilizing selection.

The validity of this argument can be examined in two ways: (1) by determining whether resistance traits are commonly subject to stabilizing selection, and (2) by determining whether the argument's underlying assumptions apply to such traits. The only study of which I am aware that adopts the first approach is my own investigation with Ellen Simms of selection on resistance to insects in morning glories (Rausher and Simms, 1989). We explicitly attempted to detect stabilizing selection on resistance to three types of folivore and the pod-feeding corn earworm. Our measure of stabilizing selection was the stabilizing selection gradient, γ, which is analogous to the selection gradient, β, for directional selection (see above). Only for resistance to corn earworms was there any indication of stabilizing selection, and the evidence was equivocal. When we examined selection on phenotypic variation, we found it to be significantly stabilizing; but when we examined selection acting on genetic variation, we found that it was, if anything, disruptive. In general, then, the expectation that level of resistance is maintained at some intermediate optimum was not upheld for morning glories.

The assumptions underlying the cost-benefit argument have been examined by a number of authors. Attempts to determine whether resistance is costly fall into a number of categories. In one set of studies, investigators have attempted to infer the existence of costs from the observation that defenses are often reduced or absent in populations (or species) in areas where herbivores are absent or where plants are afforded protection from herbivores by some other mechanism (Jones, 1972; Rehr et al., 1973; Janzen, 1973c, 1975; Feeny, 1975; Cates, 1975; Ellis et al., 1977, Compton et al., 1983). For example, it has been argued (Feeny, 1975) that the loss of the capacity for cyanogenesis by Central American ant-acacias (Rehr et al., 1973; also see above) is evidence for their being a cost associated with this type of defense. However, this loss can also be explained by the operation of genetic drift: when the defensive function of cyanogenesis was taken over by ants, mutations disrupting the machinery of cyanide production could accumulate by drift and render populations and species acyanogenic. The same argument applies to any situation in which defenses are absent when herbivores are also absent. Consequently, a correlation between resistance level and herbivore abundance cannot be taken as evidence for costs of resistance.

A slightly different problem is associated with demonstrations of negative correlations across habitats between resistance and fitness. For example, Cates (1975) showed that populations of wild ginger, *Asarum caudatum* in western Washington are polymorphic for resistance to the slug *Ariolimax columbianus*, as well as for growth rate and seed production. In a common garden, plants from microhabitats in which slugs are present tend to be resistant and have low growth rate and seed production, while plants from microhabitats lacking slugs tend to be non-resistant, grow slowly, and produce many seeds.

While Cates and others (e.g. Krischik and Denno, 1983) have interpreted the negative correlation between resistance and seed number to indicate a cost of resistance, other interpretations are just as plausible. One is that plants in different microhabitats constitute different local populations, between which there is limited gene flow. If so, consistent differences in environmental conditions between slug-infested and slug-free habitats may have selected for differences in *Asarum* life-history traits, including growth rate and seed number. The observed negative correlation between resistance and fitness would then not reflect a necessary trade-off due to pleiotropic effects of resistance genes on seed production; instead, it would reflect an across-habitat correlation between the selection pressures acting on two genetically independent traits. The plausibility of this interpretation is enhanced by the existence of clear differences in soil moisture and shade between sites with and without slugs (Cates, 1975).

The best evidence about costs of resistance is supplied by studies that attempt to measure such costs directly by assessing whether the fitness of non-resistant genotypes exceeds that of resistant genotypes when herbivores are absent. The rationale underlying this comparison can be understood by referring to Equation (2.14). In the absence of herbivores, the benefit term, $B(R)$, is zero regardless of level of resistance, since in the absence of herbivores herbivory cannot be reduced. Any difference in fitness among resistance genotypes must therefore reflect differences in costs, since the cost term is the only remaining term in the equation that varies with resistance.

Windle and Franz (1979) examined the relative competitive abilities of two cultivars of barley that were isogenic except for a locus controlling tolerance to the greenbug, *Schizaphis graminum*. In mixtures, the susceptible cultivar was a stronger competitor than the tolerant cultivar. Although lifetime fitness of the plants was not measured, this result suggests that a cost, reflected as reduced growth in the presence of a competitor, is associated with tolerance.

Dirzo and Harper (1982) found that acyanogenic morphs outperformed cyanogenic morphs of *Trifolium repens* under natural conditions, despite the fact that slugs grazed preferentially on the acyanogenic morph. Cyanogenic genotypes were less tolerant of frost than acyanogenic genotypes. In addition, cyanogenic morphs were more susceptible to fungal infection. Each of these differences can be interpreted as a cost of slug resistance contributing to the greater fitness of the susceptible, acyanogenic genotype.

This interpretation is somewhat complicated by the fact that Dirzo and Harper did not examine whether grazing by slugs adversely affected growth and reproduction in *T. repens*. Reports of compensation in response to feeding by herbivores (Crawley, 1983) indicate that it may be dangerous to assume automatically that herbivory is detrimental to plants. If it is beneficial for *Trifolium*, then the higher fitness of the acyanogenic morph in Dirzo and Harper's experiments may simply reflect differential feeding by slugs and not a cost associated with redirecting resources away from some other character contributing to fitness. As Dirzo and Harper point out, this potential problem could be alleviated by demonstrating that slug grazing reduces fitness and that in the absence of slugs, the acyanogenic genotype still has higher fitness than the cyanogenic genotype.

Berenbaum et al. (1986) report reductions in umbel production in genotypes of the wild parsnip that produce high levels of several types of furanocoumarins conferring resistance to the parsnip webworm. While the magnitude of these apparent costs is large, their interpretation is obscured because they were measured in an artificial, greenhouse environment. Experiments with *Drosophila* have shown that when organisms are brought into novel environments, patterns of genetic correlations can differ from those exhibited in the organisms' natural environment (Service and Rose, 1985; this is a special case of the phenomenon of genotype-environment interactions; see Parsons, 1977, Murphy et al., 1983, Luckinbill and Clare, 1985). Consequently, unless costs are assessed under natural environmental conditions, their demonstration may be somewhat suspect.

Simms and Rausher (1987, 1989) report two attempts to detect costs of resistance in the morning glory, *Ipomoea purpurea*. In both studies, resistance of different genotypes (half-sib families) was measured in the natural environment with herbivores present. In addition, fitness was assessed for the same genotypes in the same environment, but with insects removed by spraying with insecticides. (In an additional experiment, no effect of the insecticide was found on fitness. See Simms, 1992.) The absence of a negative genetic correlation between level of resistance and fitness for any of the insects examined suggests that resistance to insect herbivores is not costly in this plant. Two caveats must be added to such an interpretation, however. First, although Simms and Rausher failed to detect costs, they cannot rule out the possibility that they existed but were too small to measure. Second, the apparent lack of costs applies only to the resistance characters associated with extant genetic variation in the plant population examined. In particular, there may have been resistance traits in the *Ipomoea* population examined that were genetically fixed, perhaps because of strong selection exerted in the past by herbivores. The costs of such traits cannot be assessed by the methods used by Simms and Rausher, since those methods rely on analysis of genetic variation for resistance.

The results of these attempts to detect and measure costs of resistance indicate that we are a long way from knowing how commonly such costs impose significant selection pressures on plant resistance traits. Several of the studies are suggestive of the existence of costs, but each has problems that prevent it from being considered definitive. Moreover, Simms and Rausher's results suggest that costs of resistance may not be universal. Additional studies that avoid the problems of previous investigations are needed.

The second major assumption underlying the cost-benefit model of the evolution of resistance is that reducing herbivore damage benefits plants. Although much evidence indicates that this assumption is often valid (for reviews see Harper, 1977; Crawley, 1983; Dirzo, 1984; Coley et al., 1985), several authors have argued that herbivory may benefit some plants under some circumstances (Reardon et al., 1972; Dyer, 1975; Dyer and Bokhari, 1976; Owen and Wiegert, 1976, 1982; Simberloff et al., 1978; Owen, 1980). In a review of the relevant literature, however, Belsky (1986) concluded that a beneficial impact of herbivory has been demonstrated only for plants grown in artificial situations—in greenhouses and under cultivation. Moreover, Janzen (1979) argues cogently that crop plants can be particularly deceptive when assessing the impact of herbivory, since defoliation often leads to branching and the production of more bushlike plants that prosper in crop systems in which plants are well-spaced, but that would unlikely do so in more intensely competitive natural environments.

More recently, however, Paige and Whitham (1987) have provided convincing evidence of overcompensation (an increase in fitness caused by herbivory) in natural environments. They showed that browsing by mammals stimulates the type of branching described by Janzen, and that increased branching leads to an increase in the number of seeds produced per plant. Exact compensation or tolerance, i.e., absence of an effect of herbivory on fitness, has also been shown to occur in natural plant populations (Crawley, 1983; Maschinsky and Whitham, 1989). There is thus no justification for assuming automaically that resistance necessarily benefits a plant by reducing herbivory.

These considerations suggest that a cost-benefit approach to understanding the evolution of resistance will not be universally applicable, since for many plants two key assumptions of that approach may not be appropriate. For such plants, some other model is needed. One possibility that does not assume that resistance is costly is that the level of defense that evolves is determined by a combination of natural selection and genetic drift. Under this scenario, the intensity of selection on a resistance trait depends on the mean level of that trait in the plant population. When average resistance is low, herbivore populations are abundant and impose strong selection to increase mean resistance. As the mean resistance level increases, however, herbivore populations may decline, and hence impose weaker and weaker selection over time. Eventually, when resistance has evolved to a threshold level that protects plants almost completely and herbivore populations have been reduced greatly, there will effectively be no further selection to increase mean resistance. Nothing, however, constrains mean resistance from increasing above this threshold by genetic drift. Level of resistance may thus fluctuate over time, always being prevented by selection from falling below the threshold.

Distinguishing among these and other models of the evolution of resistance requires reliable estimates of the selection pressures acting on genetically variable resistance traits, as well as the costs and benefits of resistance and how they vary over time. In addition, it requires a determination of how changes in mean level of resistance affect the sizes of herbivore populations. To date, such information does not exist for any plant-herbivore system. Obtaining it should be a major goal of researchers in chemical ecology.

2.2.4. The Nature of Coevolution

The term "coevolution" was coined by Ehrlich and Raven (1964) to describe a particular type of evolutionary interaction between plants and herbivores. They posited that this kind of interaction is common in nature and described how it could produce patterns of association observed between specific insect and plant taxa. Subsequently, the term "coevolution" has been applied to several different types of evolutionary interactions. Evidence about the commonness of any of these kinds of interactions is virtually nonexistent, in part because devising experimental procedures that can distinguish among these processes is exceedingly difficult. However, the different types of coevolution differ in the types of genetic correlations among resistances that allow them to operate. It may thus be possible to infer something about the relative commonness of these processes by examining the genetic correlation structure of resistances to different herbivores in a variety of natural plant populations. In this section, I describe the logic underlying this approach and evaluate what empirical data exists. I begin by describing in more detail three different coevolutionary processes.

2.2.4.1. Non-Reciprocal Coevolution

Coevolution as envisioned by Ehrlich and Raven (1964) consists of a series of evolutionary events that begin with a plant population that is fed on by a variety of herbivore species that collectively depress plant fitness. The plant population is assumed to be genetically variable for a trait that reduces herbivory by many, if not all of these herbivores. Selection imposed by the herbivores then leads to evolutionary elaboration of this trait, which in turn greatly reduces herbivory and may even eliminate local populations of specialist herbivores. Freed of a major drain on their resource budgets, the resistant plants become better competitors and are able to colonize new habitats. The entry into novel habitats occasionally is accompanied by reduced gene flow between the new and old habitats, allowing speciation to occur as the descendent plant populations diverge in response to selection pressures in different habitats. Repeated speciation of this type produces an "adaptive radiation" and results in a new plant taxon characterized by the novel resistance trait.

This new plant taxon constitutes a relatively "open niche" for an herbivore that could break through the resistance barrier characteristic of the taxon. Ehrlich and Raven suggest that occasionally herbivores that feed on other plants will put together the requisite genetic variation that confers the necessary counterresistance. Such genotypes thereby effectively escape into a new "adaptive zone" that is relatively free of competitors and predators. The resulting selective advantage over other genotypes causes the herbivore population to incorporate plants of the novel taxon into its diet. Again, this colonization of the novel plant taxon will normally occur within a single local herbivore population and will involve one or a few of the plants in that taxon. However, subsequent dispersal into new regions containing related plants with the same type of resistance may lead to speciation and adaptive radiation of the insect. Over time, this type of evolutionary colonization of a host plant taxon may be repeated by a variety of herbivores, so that eventually the number of insects associated with the taxon builds up and the cumulative amount of herbivory increases. The original resistance trait then becomes less and less effective.

Eventually, however, in one plant species of the taxon there will arise a new resistance trait and the coevolutionary cycle will begin again, but with a new set of herbivores.

One characteristic of coevolution as described by Erhlich and Raven is that different sets of herbivores impose selection on a given plant species in different coevolutionary cycles. It may thus be termed non-reciprocal coevolution to refer to the fact that evolutionary changes in the plants impose selection on a different set of herbivores from those that selected for those changes.

Evidence suggesting the operation of non-reciprocal coevolution consists primarily of associations between particular insect taxa and particular plant taxa on which they feed. For example, virtually all species of the tribe Troidini of swallowtail butterflies feed as larvae in plants of the genus Aristolochiaceae, and most are restricted to the genus *Aristolochia* (Scriber, 1973). Similarly, within the Pierid butterflies, most species in the subfamily Pierinae feed on plants in the families Cruciferae, while legumes are the main host plants for most species in the subfamily Coliadinae (Ehrlich and Raven, 1964).

The Pierid example illustrates a second pattern expected from coevolution of this type: related herbivore taxa are not necessarily associated with related plant taxa. Another example is provided by swallowtail butterflies in the genus *Papilio*. Although most species feed on plants in the Rutaceae, Lauraceae, and related families, members of the *Papilo machaon* group feed almost exclusively on plants in the unrelated family Umbelliferae (Feeny et al., 1983). Dethier (1947, 1954) pointed out, however, that umbellifers share several essential oils in common with rutaceous plants and showed that larvae of the umbellifer-feeding black swallowtail, *Papilio polyxenes*, were attracted by the odors of these essential oils. He and others (Feeny, 1975) suggested that these shared secondary compounds may have acted as a "chemical bridge" that facilitated a behavioral shift by an ancestral member of the *machaon* complex from the Rutaceae to the Umbelliferae. Such shifts would also be facilitated by the ability of insects to detoxify or otherwise nullify the adverse effects of the shared compounds. Thus, while related insect taxa often feed on taxonomically unrelated plants, those plants are often chemically similar, a pattern expected of the type of coevolution described by Ehrlich and Raven.

While these patterns are consistent with the operation of nonreciprocal coevolution involving some plants and their associated herbivores, they are not universal. Ehrlich and Raven (1964), Gilbert (1979), and Futuyma (1983) describe several examples of insects within families or subfamilies that feed on taxonomically, and presumably chemically, unrelated plants. Nonreciprocal coevolution is thus not the only process that influences patterns of association among plants and their herbivores.

From the above account, it should be evident that in nonreciprocal coevolution, selection imposed by herbivores arises because a novel trait confers protection from attack by a suite of different herbivores. In other words, because the same trait confers resistance to a variety of herbivores, plant genotypes that are resistant to one herbivore should be resistant to others, i.e., resistances to different herbivores are expected to exhibit positive genetic correlations.

2.2.4.2. *Reciprocal Coevolution*

Since Erhlich and Raven's seminal paper, ideas about the nature of evolutionary interactions among plants and herbivores have evolved and multiplied. By 1979, Gilbert recognized that the term coevolution "is widely used to refer to any situation in which a pair of organisms are simultaneously or alternatively acting as selective agents for one another." A year later, Janzen (1980) defined coevolution as "an evolutionary change in a trait of the individuals in one population in response to a trait of the individuals of a second population, followed by an evolutionary change in the second population to the change in the first." At about the same time, Fox (1981) defined coevolution as occurring "between two interacting species when each exerts selective pressures affecting each other's gene pool ... coevolution implies ongoing stepwise evolution within both populations, with the properties of first one and then the other continuing to evolve in response to specific changes in the properties of the other."

This process, which may be termed reciprocal coevolution, is clearly different from that described by Ehrlich and Raven because it postulates that the same species are involved in successive cycles of coevolutionary change. In addition, nonreciprocal coevolution involves sequential adaptive radiations, and hence diversification of interacting plant and herbivore species, as an integral aspect of the process. By contrast, reciprocal coevolution need involve no speciation; rather, it has been likened to an evolutionary arms race between two (or a few) interacting populations (Feeny, 1975; Berenbaum, 1983; Thompson, 1986; Fox, 1988). Finally, because different pairs (or groups) of interacting plant and herbivore species may coevolve in different directions, reciprocal coevolution tends to cause divergence between the chemical and other resistance characteristics of the host plants of related herbivore species. In other words, reciprocal coevolution tends to disrupt the pattern of related herbivore taxa feeding on chemically similar plant taxa, that is, the pattern generated by nonreciprocal coevolution.

Direct evidence for the operation of reciprocal coevolution is very difficult to obtain because, like most evolutionary processes, it occurs over many generations. The most important indirect evidence is provided by

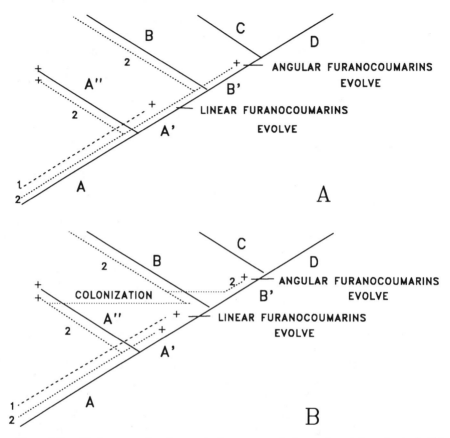

Figure 2.7. Cladograms showing evolutionary associations of evolving plants (solid lines) and herbivores (broken lines). (A) Reciprocal coevolution between plant lineage AB' and herbivore species 2. (B) Non-reciprocal coevolution between plant lineages B and B' and herbivore species 2. See text for explanation.

detailed analyses of apparent reciprocal adaptations, in particular, plant-herbivore associations. As suggested by others (Futuyma 1983a, b), perhaps the best apparent example is provided by the work of Berenbaum (1983) (Fig. 2.7A). Linear furanocoumarins, which are produced by most umbellifers and which are toxic to most herbivores, are presumed to have evolved as defenses in ancestral unbellifers. The presence of these compounds is then presumed to have selected for detoxification mechanisms in some herbivores, such as the black swallowtail, *Papilio polyxenes* (represented by herbivore species 2 in Fig. 2.7A), but to have caused the extinction of other herbivores that lacked genetic variation to respond to such selection (represented by herbivore species 1 in Fig. 2.7A). At some later date, herbivores tolerant of linear furanocoumarins are assumed to have selected for the production of angular furanocoumarins (in umbellifer species B'). These herbivores then either became extinct on or evolved to avoid umbellifers containing angular furanocoumarins, against which they could not evolve detoxification mechanisms.

We are able to deduce the existence of two coevolving lineages (delimited by the bracketed lines in Fig. 2.7A) only because between the times when linear and angular furanocoumarins evolved, the plant species split into two independently evolving lineages (B and B' in Fig. 2.7A) and the tolerant herbivore species remained associated with each. It is thus present today, and hence available to demonstrate the toxicity of angular furanocoumarins to herbivores tolerant of linear furanocoumarins, because there are modern unbellifer lineages that have no evolved angular furanocoumarins.

Similar interpretations can be placed on traits found in other plant-insect associations. Heliconiine butterflies have apparently evolved the ability to tolerate cyanogenic glycosides in their *Passiflora* host plants (Spencer, 1988). In turn, some species of *Passiflora* have evolved hooked trichomes and egg-mimicking stipules that kill larvae and deter oviposition respectively (Gilbert, 1971; Williams and Gilbert, 1981). Since it is difficult to imagine other functions for these traits, it seems reasonable to conclude that they evolved in response to selection imposed by butterfly herbivory. Similarly, butterflies in the genus *Battus* (Papilionidae: Triodini) have evolved mechanisms for detoxifying aristolochic acids in their host plants, compounds that are extremely toxic to most insects (Rausher, 1979). In east Texas, *Aristolochia reticulata* have evolved tough, nutrient-poor mature leaves that reduce herbivore damage by *Battus philenor* larvae. That *B. philenor* was a primary selective agent for this trait is suggested by the strong detrimental impact of this species on survival and reproduction in *A. reticulata*, even though its mature leaves are resistant to feeding (Rausher and Feeny, 1980).

Three problems exist with interpretations of this type. First, as pointed out by Miller (1987), in none of these examples has an appropriate phylogenetic analysis on both insects and plants been performed to determine whether the two phylogenies exhibit parallel cladogenesis, as is required if reciprocal coevolution has occurred. Moreover, in the few cases in which such analyses have been performed (Mitter and Brooks, 1983; Miller, 1987), insect and host phylogenies exhibit a poor match, suggesting that reciprocal coevolution may be less common than previously believed.

A second problem is that even if detailed phylogenies of an insect taxon and its associated host plants appear to match perfectly, one cannot necessarily infer that the match was produced by reciprocal coevolution.

Consider, for example, another possible scenario for the evolution of furanocoumarins portrayed in Fig. 2.7B. Initially, herbivore species 1 and 2 are associated with umbellifer species A. At some point, lineage A splits into two, A' and A'', and species 2 remains associated with each until it becomes extinct for ecological reasons on lineage A'. At some later date, herbivore 1 selects for the production of linear furanocoumarins in lineage A', is unable to adapt to them, and becomes extinct. Somewhat later still, host A'' becomes extinct for ecological reasons. Herbivore 2 is then left without host plants, but is able to persist by colonizing host lineage B and evolving to tolerate linear furanocoumarins. It is now preadapted to colonize host lineage B' and soon does so. Thereafter, it exerts sufficient selection pressure on B' to cause this host to evolve angular furano-coumarins, to which the insect fails to adapt. The insect thus becomes extinct on lineage B', but persists to the present on lineage B.

Under this scenario, herbivore 1 imposed selection on umbellifer A', but not vice versa; umbellifer B imposed selection on herbivore 2, but not vice versa; and herbivore 2 imposed selection on umbellifer B', but not vice versa. In other words, there is no reciprocal coevolution under this scenario. In fact, the events portrayed in Fig. 2.7B actually represent nonreciprocal coevolution on a very fine scale. Yet this scenario cannot be distinguished from that in Fig. 2.7A because both present the same information to the investigator. The difference corresponds to colonization and extinction events that leave no historical trace. Thus, the operation of reciprocal coevolution can be disproved, but not proved, by phylogenetic analysis.

The third problem with inferring the operation of reciprocal coevolution from patterns such as those described above is that there is normally no evidence that the herbivores have imposed significant selection on their host plants. As discussed in a previous section, the lack of such evidence precludes ruling out the possibility that the purported chemical defenses evolved for reasons other than to protect the plants from herbivory. In other words, these patterns are fully consistent with the operation of sequential evolution (*sensu* Jermy, 1976, 1984) rather than coevolution of any type.

2.2.4.3. Diffuse Coevolution

A reaction to the emphasis on pairwise, reciprocal coevolution was the development, in the early 1980s, of the concept of diffuse coevolution (Janzen, 1980; Fox, 1981; Futuyma, 1983). Fox and Futuyma each suggest that reciprocal coevolution is most likely to occur when a plant population is attacked by only a few species of herbivores, since there will be few constraints in the form of trade-offs between resistance to one herbivore and resistance to another. By contrast, diffuse coevolution, as defined by these authors, is a process that occurs when a plant population hosts many different herbivorous enemies. In such a situation, the selection pressures exerted by any particular herbivore are postulated to be relatively weak. Moreover, different herbivores are assumed frequently to exert conflicting selection pressures on resistance traits, preventing the plant from undergoing extensive reciprocal coevolution with any of its associated herbivores. Consequently, the only types of resistance factors that are likely to be favored by herbivore-generated selection are generalized defenses that protect the plant from most of its herbivore enemies. Finally, Fox (1981, 1988) suggests that unlike the constantly escalating "arms race" analogy often associated with reciprocal coevolution, diffuse coevolution results in continuous but minor and fluctuating adjustments in the plant's resistance characteristics, i.e., steady fine-tuning without long-term, major evolutionary change.

The impression one gets from these accounts is that coevolution is in some way constrained by the complex web of interactions in species-rich plant-herbivore associations and, in particular, by opposing selection pressures generated by different herbivores on particular resistance traits. The most obvious genetic interpretation of such constraints is that for many pairs of herbivores, resistance to one herbivore is negatively genetically correlated with resistance to the other. As discussed earlier, such correlations tend to retard, or even prevent, the evolution of resistance to both herbivores (see Fig. 2.4).

Nevertheless, diffuse coevolution is not a necessary theoretical consequence of a plant population being attacked by many herbivores. Reciprocal coevolution between the plant and each herbivore is also possible under some circumstances. In particular, if resistances to the different herbivores are pairwise uncorrelated genetically, then the plant's evolutionary response to selection imposed by each herbivore may occur independently of selection imposed by the other herbivores (see Equations 2.10A and 2.10B). Whether reciprocal or diffuse coevolution predominates in species-rich plant-herbivore associations is thus an empirical question and can not be settled by recourse to theoretical arguments.

2.2.4.4. Empirical Evidence

The preceding discussion suggests that examination of the genetic correlation structure of resistances in natural plant populations may provide some indication of the relative importance of nonreciprocal, reciprocal, and diffuse coevolution in molding the evolution of plant-herbivore interactions. In particular, lack of correlations among resistances to different insects is necessary for reciprocal coevolution, whereas extensive

Table 2. Numbers of negative, positive, and zero pairwise genetic correlations for resistance to herbivorous insects in several plant species.

Plant Species	Number of Herbivores	Sign of Correlation −	Sign of Correlation 0	Sign of Correlation +	Reference
Solidago altissima	17	9	100	27	Maddox and Root (1990)
Ipomoea purpurea	4	0	6	0	Simms and Rausher (1989)
Salix lasiolepis	4	0	3	3	Fritz and Price (1988)
Salix sericea	3	1	1	1	Fritz (1990)

strongly negative correlations are conducive to diffuse coevolution. Extensive positive correlations indicate that a single resistance trait (or a group of traits controlled by the same genes) is effective against most herbivores, as is expected under nonreciprocal coevolution.

In an attempt to evaluate the relative commonness of these three types of correlation structure, I have compiled the results of several studies that have estimated the genetic correlations among resistances to herbivores attacking the same plant species (Table 2.2). The number of such studies is small, largely because I have tabulated only studies in which genetic correlations among resistances were calculated for a single, interbreeding, natural population. Correlations determined by comparing different populations (Hare and Futuyma, 1978) are omitted because between-population correlations do not necessarily reflect the within-population correlations that determine the course of evolution (Rausher 1984a). Similar data available from agricultural studies have also not been included because they are often obtained by crossing lines of unknown, geographically distant origin. Thus, the correlations measured often reflect between-population differences. In addition, many correlations reported in or derivable from the agricultural literature are based on comparisons of a small number of lines, and thus may reflect chance associations of resistance genes affecting different herbivores rather than pleiotropic effects of the same set of genes (Rausher, 1983).

It is dangerous to draw definitive conclusions from so few studies. Nevertheless, the available evidence provides little obvious support for the importance of diffuse coevolution. The negative correlations expected under diffuse coevolution seem to be quite rare. Those that occur tend to be weak (mean for all negative correlations listed in Table 2.1 = 0.50) and hence would not be expected to constrain the evolution of resistance significantly (see above and Fig. 2.4). Similarly, positive correlations are relatively uncommon, though they seem to occur somewhat more frequently than negative correlations, at least in the studies by Maddox and Root (1990) and Fritz and Price (1988). By far, the dominant pattern seems to be a lack of correlation between resistances. It thus seems that resistance to particular insects in the plant species examined is free to evolve independently of resistance to most of the other species attacking those plants. In other words, the genetic correlation structures revealed by these investigations suggest that reciprocal coevolution may be the dominant mode of evolutionary interaction between plants and their insect herbivores.

Two caveats need be added to this conclusion, however. One is that the zero correlations reported in Table 2.1 are actually non-significant correlations. It is thus not possible to say that they represent a true absence of genetic correlation between resistances to the corresponding herbivores. However, because all of these non-significant correlations are small, they would probably not constrain evolution significantly. They are thus unlikely to contribute to diffuse coevolution.

A second caveat is that independence of the genes and associated traits that control resistance to different herbivores does not guarantee that evolution of resistance to one herbivore will be independent of the evolution of resistance to others. In particular, interactions among herbivores may either reduce or augment selection pressures that might otherwise drive reciprocal coevolution. For example, consider two competing herbivore species. Genotypes that selectively eliminate one herbivore may not experience enhanced fitness if the abundance of, and hence damage caused by, the other herbivore increases because of competitive release. In this situation, the presence of the second herbivore influences the magnitude of the selection pressure exerted by the first herbivore on genetic variation in resistance. In the absence of the second herbivore, reciprocal coevolution between the plant and the first herbivore might be possible; the presence of the second herbivore might prevent such coevolution. Unfortunately, very little is known about how commonly such competitive interactions occur among herbivores in nature or about how they affect selection pressures herbivores exert on plants.

Before the nature of coevolution can be better understood, several important questions will need to be addressed for a variety of plant-herbivore associations. These include:

1. *What types of selection pressures do herbivores impose on plants?* As described in a previous section, there exists very little evidence indicating whether herbivores are important selective agents in the lives of plants. Until we have a better idea of whether they are, we will not have a very good feeling even for the potential for coevolution of any kind. Consequently, many more studies are needed that attempt to determine under what ecological conditions herbivores impose selection on the plants they

eat. In addition, we need studies that reveal the nature of that selection—its magnitude, whether it is directional or stabilizing, whether it is constant or intermittant, and whether it varies spatially from population to population.

2. *What selection pressures do plants exert on herbivores?* In general, biologists have found it easier to assume that herbivores evolve in response to specific characteristics of their host plants than vice versa. Nevertheless, I am aware of no investigation that directly demonstrates such selection on a natural herbivore population. Moreover, a crucial question bearing on the commonness of reciprocal coevolution in nature is, exactly what herbivore characters evolve in response to changes in host-plant quality. As several authors have pointed out (Janzen, 1980; Futuyma, 1983; Castillo-Chavez et al., 1988; Rausher, 1990), herbivores faced with a recently evolved plant toxin have three possible evolutionary responses: (1) they may evolve detoxification mechanisms that, in effect, enhance the quality of the foliage containing the toxin; (2) they may evolve behavioral avoidance of the plant containing the toxin; or (3) they may not evolve at all and risk extinction. These alternatives have different consequences for the continuation of reciprocal coevolution. Evolution of avoidance and extinction clearly break up this type of coevolution, while evolution of detoxification capabilities allows it to continue. While there have been preliminary attempts to construct models that predict when either behavioral avoidance or physiological counteradaptation will evolve (Gould 1984, 1988; Castillo-Chavez et al., 1988; Rausher, 1992), more general models are needed to address this question. One of the prime goals of these models should be the identification of crucial parameters that need to be measured in natural populations in order to allow at least a reasonable guess as to the evolutionary counterresponse of herbivores to evolution in their hosts. It will then be the task of the empirical biologists, by measuring those parameters in a number of apparently coevolving systems, to determine the number of cycles of adaptation and counteradaptation through which reciprocal coevolution procedes.

3. *How common is diffuse coevolution?* Another way of asking the same question is, how commonly do trade-offs in resistance to different herbivores limit the evolution of resistance to particular herbivores? As a first step in addressing this question, additional attempts to measure the genetic correlation structure among resistances to different insects (Table 2.2) are needed. Beyond that, however, we need some assessment of the actual evolutionary response to selection pressures exerted by a large community of herbivores. As a first approximation, Equation (2.7) could be used to provide such an assessment. By measuring the genetic correlation structure (G matrix), the phenotypic correlation structure (P matrix),

and the selection differentials (s vector) for a set of resistance traits, one could calculate the expected magnitude and direction of evolutionary change. One could thus determine whether resistance traits were locked in evolutionary stalemates (*sensu* Berenbaum et al. 1986; Berenbaum and Zangerl, 1988) because of opposing selection pressures. By repeating this exercise for a number of years on the same population, one could also obtain at least an idea of whether these matrices and vectors remain approximately constant, and hence whether long-term secular coevolution was slowed by temporal variation in selection pressures and the genetic parameters of the population. Such studies will strain the physical and financial resources of investigators, but it is unlikely that any real advance in our understanding of coevolution will occur unless they are undertaken.

4. *Is the genetic variation normally extant in populations the real raw material for coevolution?* Studies, such as those just suggested, that attempt to measure coevolutionary forces in natural populations assume the uniformitarianism principle validly applies, that is, that genetic variation for resistance normally extant in plant populations, as well as selection that acts on that variation, are representative of that involved in major coevolutionary change. One reason this principle might not apply is that reciprocal evolution is expected to fix newly arisen variation for effective defense rapidly, assuming that the selection pressures generated by herbivores are reasonably intense. If such variation arises only rarely, then most attempts to examine genetic variation for defense will not detect it because they will be undertaken between the time variation for the last defensive trait was fixed and the time variation for the next trait arises. In this interval, there may exist only genetic variation for resistance that arises due to pleiotropic effects of variation in traits that have other important ecological and physiological functions. Measuring selection on this variation, as well as determining whether resistance is costly and whether resistances to different herbivores are correlated, might then give a distorted picture of the evolutionary forces and constraints that mold coevolution. Since much of our ability to apply the tools of evolutionary genetics to understanding coevolution hinges on the applicability of the uniformitarianism principle, it is essential that its validity be somehow assessed.

2.2.5. The Evolution of Diet Breadth

2.2.5.1. Mechanistic versus Evolutionary Explanations

A central question in the study of plant-herbivore interactions is what determines the degree of diet specificity in herbivorous insects. Phrased in this manner, however, the question is ambiguous because it allows for both mechanistic and evolutionary answers (Rausher, 1985). It is thus necessary to distinguish between two possible problems posed by this question: (1) how do properties of individual insects and the plants available to them determine diet breadth?; and (2) what evolutionary processes have molded those properties to produce the degree of specificity observed for any particular insect?

The first problem is one of behavior, physiology, and phytochemistry. Ever since the early work of Verschaffelt (1910), Brues (1920, 1924), Detheir (1947), and Fraenkel (1956, 1959), it has been evident that host-plant chemistry, as well as physical factors such as leaf roughness, pilosity, and shape are important cues used by insects when selecting plants on which to feed or lay eggs. Subsequently, sufficient evidence has accumulated to indicate that in many, if not most, specialist insect species, a highly restricted diet is largely accomplished by having a nervous system that responds positively to a unique chemical profile characteristic of one or a few plant species (David and Gardiner, 1962, 1966; Reese, 1969; Ichinose and Honda, 1978; Feeny et al., 1983). By contrast, for generalist insects, the mechanistic determinants of host breadth are less well understood, though the consensus seems to be that acceptance of a plant depends on the presence of nonspecific chemical stimuli such as sugars and water, as well as on the absence of deterrent compounds (Fraenkel, 1969; Futuyma 1983, Isman, this volume).

Although further attempts to examine the interactions between insect nervous systems and plant chemical properties will doubtless enhance our understanding of insect behavior, the basic mechanistic cause of diet specialization (or lack thereof) in insects is fairly clear. The same cannot be said of the evolutionary processes influencing diet breadth.

Evolutionary explanations of diet specialization attempt to explain why a particular insect has evolved to use particular plants with particular chemical cues for feeding or oviposition. Evolutionary biologists tend to view such cues as token stimuli, i.e., indicators of the identity of plants that are either suitable or unsuitable for insect survival, growth, and reproduction. Implicit in this view is the recognition that there is no necessary connection between the physiological and pharmacological properties of the token stimuli and the suitability of the plant parts that contain them. Evolutionary biologists are thus comfortable with the explanation that ancestral pierid butterflies may originally have evolved to use glucosinolates as oviposition cues (Renwick, 1988) because the crucifers containing them for some reason harbored few predatory insects, and, therefore, pierid larval survival was higher on crucifers than on other plants. In this explanation, the evolutionary pressure favoring specialization is generated by a difference in predator-caused mortality between crucifers and non-crucifers, a difference that has nothing to do with glucosinolates. The same selection pressures could just as easily have caused the ancestral pierid to evolve to select only plants with small, yellow, four-petalled flowers, an oviposition cue that in most habitats would also cause egg laying to be restricted to crucifers. The use of particular plant chemicals as feeding or oviposition cues may thus often be a historical artifact determined by (1) the cues available that are uniquely associated with the plant species on which feeding or oviposition is favored, and (2) the amounts of genetic variation for such cues.

Two observations have stimulated much of the thinking on the evolution of diet breadth in herbivorous insects. One is that specialist insects are much more common than generalists (Strong et al., 1984; Bernays and Graham, 1988). The second is the claim that there has been a historical trend toward increasing specialization (Dethier, 1954; Feeny, 1975). Although these observations have often prompted a search for reasons why specialization might be advantageous, it must be remembered that many herbivorous insects have very broad host ranges (Scriber, 1973, 1988), which are presumably as adaptive as the narrow host ranges of specialist species. Consequently, the diet breadth of any particular species is most profitably viewed as the outcome of a compromise between selection pressures promoting specialization and those promoting generalization.

2.2.5.2. *Factors Favoring Specialization*

Factors promoting a narrow host range have been classified into two categories. Several authors have suggested that diversity among plant species in chemical constitution, as well as in other intrinsic plant characters (phenology, thermal load on leaves, physical resistance), generates costs of adaptation that promote specialization (Dethier, 1954; Krieger, et al., 1971; Feeny, 1975; Rausher, 1983; Futuyma, 1983). On the other hand, others have argued that the primary factor promoting a narrow host range is variation among plants in "ecological suitability," that is, suitability for survival, growth, and reproduction as determined by extrinsic factors, such as predator load and intensity of competition from other herbivores (Gilbert, 1979; Futuyma, 1983; Bernays and Graham, 1988).

2.2.5.2.1. Specialization and Costs of Adaptation

Dethier (1954) was the first to suggest that many insects may be specialized because a generalized diet incurs substantial metabolic costs. Costs are presumed to arise from having to maintain a battery of detoxification mechanisms directed at the different toxic secondary compounds produced by the different plants on which a generalist insect might feed.

One expected result of such costs would be that the growth rates, survival, and/or fecundities of generalist insects are not as great as for specialist insects when the two types of insects utilize the hosts of the specialist species. Averaged over the host plants actually used, then, the average fitness of specialists is expected to be greater, favoring specialization.

Early evidence in support of one aspect of this argument was provided by Krieger et al. (1971), who found that the activity of microsomal oxidase enzymes in larval miguts was an order of magnitude greater in polyphagous than in monophagous lepidoptera. This result was not substantiated in a similar study, however (Rose 1985), and Gould (1984) raise other problems with the conclusions drawn by Krieger et al.

Several attempts to detect metabolic costs associated with polyphagy have involved comparisons of growth performance and survival of related generalist and specialist species (Waldbauer, 1968; Schroeder, 1976, 1977; Smiley, 1978; Scriber and Feeny, 1979; Fox and Morrow, 1981; Futuyma and Wasserman, 1982), but these have yielded no consistent or easily interpretable pattern. For example, Smiley (1978) found that the family generalist *Heliconius cydno* grew more slowly than the specialist *H. erato* on the latter's host, *Passiflora biflora*. By contrast, Futuyma and Wasserman (1982) detected no differences in feeding efficiency between the generalist moth *Malacosoma disstria* and the specialist *M. americanum* where both were grown on leaves on *Prunus serotina*, a host of the specialist.

One possible explanation for these inconsistencies is that these early studies involved among, rather than within, species comparisons. Costs may not be evident in comparisons among species because species often differ in many traits affecting growth, survival, and reproduction, and these differences can mask differences due to costs of polyphagy. To circumvent this problem, investigators have turned to genetic analyses in attempts to detect costs within populations of herbivorous insects.

The genetic manifestation of costs of polyphagy is a negative genetic correlation between fitness of an insect on one host and fitness on another host. This equivalence can be seen from the following argument: suppose a specialist insect, highly adapted to feeding on a single host species, is reared in alternate generations on this host species and on a second species containing a moderately toxic secondary compound. If appropriate genetic variation exists, the ability to detoxify this compound will evolve in the laboratory population. As a consequence, fitness on the second plant species will improve greatly, and the insect will have become a physiological generalist. There will be a cost associated with this broadened diet only if in selecting for improved fitness on the second host, fitness on the original host species has declined, which will occur only if there is a negative genetic correlation between fitness on the two hosts (see Equations 2.9A and 2.9B).

In the last ten years, several attempts have been made to detect costs of polyphagy by measuring genetic correlations among fitnesses of insects reared on different hosts (Gould, 1979, 1988; Rausher, 1984; Via, 1984; Weber, 1985; Hare and Kennedy, 1986; Futuyma and Philippi, 1987; Fry 1990). Although a few negative correlations were detected in this collection of studies, in a large majority of cases, either no correlations existed or correlations were positive. These results certainly suggest that costs associated with polyphagy are less common than previously suspected and that other explanations for the evolution of specialist diets should be sought (Futuyma and Philippi, 1987; Bernays and Graham, 1988). However, there are several reasons for not yet abandoning the search for costs. Most of the studies that have attempted to detect genetic trade-offs in fitness on different hosts suffer from one or more of the following limitations (Rausher, 1988): (1) low sensitivity to detect even moderately strong correlations $(0.5 < r < 0.25)$ because of small sample sizes; (2) failure to measure more than a few fitness components; (3) correlations may be positively biased because of the expression of general "vigor" genes in artificial laboratory environments that would not normally be expressed under field conditions (see Service and Rose, 1985); and (4) there is a possibility that, for the species examined, natural selection has favored modifier genes that eliminated trade-offs that once existed.

Because of these limitations, discarding the hypothesis that metabolic costs often contribute to the evolution of a narrow host range seems premature. Instead, additional information is needed in three crucial areas. First, we need studies that are sensitive enough to detect weak trade-offs in fitness on different hosts. These studies need to be carried out under field conditions and need to assess all major insect fitness components, admittedly a daunting task. Second, to avoid limitation (4) we need to determine whether trade-offs exist when insects are challenged by novel host plants and secondary compounds. The work of Gould (1979, 1988) and Fry (1990) is an excellent step in this direction. Along the same lines, we need information on how easily trade-offs can be modified by selection and how common the necessary genetic variation occurs in natural populations. Such information might be obtained by selection experiments designed specifically to break negative genetic correlations in fitness across hosts. Finally, we desperately need the development of a formal theory for the joint evolution of preference and performance in herbivorous insects. Among other things, such a theory would indicate under what conditions small trade-offs in fitness can be expected to favor the evolution of behavioral and physiological specialization. If theory were to tell us that very strong trade-offs are necessary, we could be more confident that failure to detect costs is not an artifact of necessarily limited sample sizes..

Metabolic costs associated with detoxification are not the only types of trade-offs in adaptation to different hosts that might tend to favor the evolution of specialization. Other possibilities include costs associated with coordinating life-history events with host plant phenology, with tolerance of microclimatic conditions, and with escape from predators via crypsis. For example, in many tree-feeding insects, a mismatch between the timing of egg hatch and leaf production by host plants is a major mortality factor (Varley et al., 1974; Feeny, 1976, Schneider, 1980). Larvae that hatch before buds have broken or after leaves toughen often die from starvation. Mitter et al. (1979) and Schneider (1980) have demonstrated that clones of the parthenogenetic moth *Alsophila pometaria* associated with different hosts have diverged at the time of hatching, and this divergence is in the direction expected if clones have been selected to match their normal hosts phenologically. Since timing of egg hatch (or of adult eclosion, which in turn determines when eggs hatch) in insects is often controlled by environmental cues not associated with the host plant (temperature, photoperiod; Tauber et al., 1986), evolution of a closer match to the phenology of one host presumably often entails less of a match with the phenology of a second host, and hence a cost. More information is needed on how commonly these types of non-metabolic trade-offs exist in herbivorous insects before their importance in fostering the evolution of specialization can be evaluated.

2.2.5.2.2. Specialization and Ecological Factors

The suitability of plant species for growth and survival of particular insects may vary greatly for a variety of reasons other than differences in foliage quality. In particular, plants may differ in such "ecological factors" as predator, pathogen, or competitor load and abundance.

It has been argued that the potential for parasites and predators to favor specialization in insect herbivores is apparent from the many demonstrations that mortality due to predators and parasites varies among host species (Price et al., 1980, Futuyma, 1983, Bernays and Graham, 1988). However, this argument has two sides. The use of several host species in spite of observed variation in enemy-induced mortality could be taken to indicate that such variation is not an effective force favoring specialization.

Perhaps the clearest example of enemy-induced specialization is provided by the Japanese pierids (Sato and Ohsaki, 1987; Ohsaki and Sato, 1990, 1991). *Pieris napi* oviposits almost exclusively on *Arabis gemmifera*, even though the foliage of this species is much less suitable for larval growth than that of other native crucifers. However, larvae escape parasitism when they feed on *A. gemmifera* because these host plants grow below dense vegetation which prevents the larvae from being discovered by parasitoids, while *P. napi* larvae grown on other native crucifers are heavily parasitized. Parasitoid pressure thus seems to favor specialization on *A. gemmifera*, even though it is a nutritionally inferior host. By contrast, *Pieris melete* is not constrained by parasitism pressure because it is able to encapsulate parasitoid eggs. As might be expected, this species uses many more native crucifers than *P. napi* and avoids the nutritionally poor *A. gemmifera*.

It has also been argued that incorporation of plant characteristics as defenses against parasites and predators generates selection for a restricted diet in insect herbivores (Price et al., 1986; Bernays and Graham, 1988). This argument is based on two examples. Mushroom-breeding *Drosophila* are much more tolerant of the toxin amanitin than related species that do not breed in mushrooms, and parasitism of flies by nematodes was much lower in amanitin-containing than in amanitin-lacking mushrooms (Jaenike, 1985; Jaenike et al., 1983). Courtney (1988) has suggested that amanitin tolerance evolved because tolerant flies could, to some degree, escape parasite-induced mortality by breeding in mushrooms producing amanitin. It is not legitimate to conclude from these results, however, that parasitoids have caused specialization in host use by amanitin-tolerant *Drosophila*, since the diet of these species includes many fungi that do not produce amanitin (Jaenike, 1978b, 1985). In other words, these flies are specialized on fungi, and there is no evidence that, on average, parasitism is less on muchrooms than on non-mushroom hosts of related species.

The second example involves oviposition site choice by lycaenid butterflies involved in mutualisms with ants. The ants provide protection to the butterfly larvae from predators and parasitoids, and several species have evolved to oviposit only on plants that harbor ant mutualists (Atsatt 1981a, b). However, it is not clear whether this restriction of diet has been caused by increased survival on those plants due to reduced parasitism and predation or because larvae are able to feed on ant brood or farmed homopteran colonies or gain other advantages from association with ants (Atsatt, 1981b).

The above considerations indicate that the evidence that variation in parasite and predator load commonly fosters the evolution of specialization is scanty. Even less evidence exists for the importance of competition. Although several authors have argued that herbivorous insects seldom compete because they are regulated at low densities by predators and parasites (Hairston et al., 1960; Lawton and Strong, 1981; Price, 1983; Strong et al., 1984), documented cases of such competition are increasing (see Fritz, 1990 for review). Nevertheless, I know of only one study that provides any evidence that competition restricts diet breadth (Joern and Lawlor, 1980).

While the evidence supporting the argument that ecological factors such as predation and competition favor specialization is minimal, this argument remains plausible. Indeed, examples of extreme specialization on a single host even when other plants with foliage of equal suitability are available (Wiklund, 1975; Smiley, 1978), strongly suggest that some ecological factor favors specialization. Additional information is needed on the identity of those factors and how they operate.

2.2.5.3. Factors Favoring a Broad Host Range

Opposed to the factors promoting specialization listed above may be any of several factors that presumably tend to promote the evolution of a broad diet. Since very little is known about these factors in natural populations of herbivorous insects, I discuss them only briefly.

2.2.5.3.1. Search Costs

It has been frequently suggested that specialization may entail a cost to a searching herbivore (Levins and MacArthur, 1969; Jaenike, 1978a; Courtney, 1982, 1984, 1986; Rausher, 1985, 1989). If preferred host plants are rare, then an insect genotype that limits feeding or oviposition to those plants may die before locating them or lay fewer eggs in a lifetime than a genotype that is less specialized behaviorally. If this fitness differential is great enough to counteract costs of specialization, the evolution of a broad diet would be expected.

Attempts to measure search costs have been limited almost exclusively to determining whether female insects searching for oviposition sites ever lack mature eggs. If they do, they incur no search costs, since oviposition rate is limited by the rate at which eggs mature rather than the rate at which suitable oviposition sites are found. Although these attempts suggest that in some insects search costs may be large, while in others, small or absent (Courtney, 1982, 1984, 1986), interpretation of these results is not clear-cut. For species in which searching females always have a large load of mature eggs, it is not certain that the females will not be able to lay most or all of those eggs before dying. If they can, the apparent costs are not real. By contrast, for species in which females commonly have few mature eggs while searching, it is difficult to know whether a narrowing of the diet would cause egg load to increase.

The ideal measure of search costs is the difference in the average number of eggs laid by genotypes that differ in diet breadth. In principle, it could be measured by releasing a group of females that are genetically variable for diet breadth into a well-delimited habitat and collecting a random sample of the eggs laid. If suitable marker genes are available and particular marker alleles can be engineered to be associated with particular diet-breadth genotypes, one could determine the proportion of eggs laid by females of different genotypes. A regression of this proportion on host-breadth would yield a selection gradient that estimated the search costs associated with a broad host range. Insects that are highly localized because of patchy host-plant distribution or because of intrinsic barriers to dispersal (Ehrlich, 1961; Gilbert and Singer, 1973) would be admirable candidates for a study of this type.

2.2.5.3.2. Unpredictability of Host Plants

If individual plant species cannot be located predictably because of spatial or temporal variation in abundance or distribution, the survival and reproductive success of specialist genotypes will be much more variable than that of generalist genotypes that can use whatever plants are locally and presently available. All else being equal, the long-term fitness of the generalist genotype will be greater because it is less variable (Hartl and Cook, 1973). This inference can be understood most easily by realizing that a specialist genotype has a greater chance of completely failing to locate a suitable host, and hence of failing to reproduce. Unpredictability will thus often tend to favor a broad diet.

Futuyma (1976) assessed the validity of this hypothesis for folivorous Lepidoptera that use relatively early successional forbs (a relatively unpredictable resource) and for those that use woody plants (a predictable resource). He found a pattern opposite from that expected, which he interpreted to be a result of more extensive coevolution between plants and insects in floristically diverse, successional habitats. His results do not necessarily indicate, however, that unpredictability failed to generate selection for a broad host range. Such selection may have occurred, but been counteracted by costs associated with a broad diet. Moreover, since Futuyma did not employ any objective measure of predictability, it is not clear whether his categorization of plants as predictable or unpredictable was appropriate. For insects like Lepidoptera that have high dispersal capabilities and sophisticated sensory apparatus for detecting host plants at a distance, early successional forbs may be located just as predictably as woody shrubs and trees. It is thus not clear whether Futuyma's study provided a valid test of whether insects using unpredictable host plants have broader diets.

Similar attempts to interpret the role of host-plant predictability in molding diet breadth (Jaenike, 1978b) suffer from similar limitations. Until an objective method of measuring predictability from the searching herbivore's perpsective is developed, attempts to examine the influence of predictability on host range are probably destined to fail.

2.2.5.3.3. Mode of population regulation

Most models of the evolution of herbivore diet breadth (Levins' and MacArthur, 1969; Jaenike, 1978a; Courtney, 1983, 1984; Futuyma, 1983b; Rausher, 1985) implicitly assume that a population of herbivores composed of subpopulations of individuals using different host species is regulated at the level of the entire population ("hard selection" sensu Wallace, 1968). Recently, however, I have examined the effect of independent regulation of subpopulations (Wallace's "soft selection") on diet breadth (Rausher, 1984b, 1985, 1990; Rausher and Englander, 1987).

The important difference between hard and soft selection is the relationship between the number of individuals entering a subpopulation and the number leaving. Under hard selection, the number leaving is determined by the number entering and the quality of the habitat occupied by that subpopulation. Under soft selection, subpopulations have their own carrying capacities, so that the number of individuals surviving in a subpopulation is more or less independent of the number entering.

As an example, an herbivorous insect population that oviposited on several host plants would experience hard selection if the population were regulated by parasitoids at the pupal stage and all individuals had an equal likelihood of being parasitized regardless of the host plant on which they grew as larvae. The insect would experience soft selection, however, if the number of individuals successfully pupating was regulated independently on each host, as might occur if different species for parasitoids were associated with different host plants, or if strong intraspecific competition occurred on individual host plants (Whitham, 1978, 1980).

The most striking result of the models I have analyzed is that independent regulation of subpopulations selects for a broad diet, whereas under hard selection, either a broad or a narrow diet may be favored. The reason for this expectation under soft selection is fairly easy to understand intuitively. Consider a population that consists of one specialist genotype, which oviposits on only one of the plant species available. Larvae on this host will experience crowding and density-dependent mortality. Now consider a mutant generalist genotype that oviposits on several additional plant species as well as the host of the specialist genotype. Because only a fraction of the offspring of the generalist genotype will end up on the specialist's host, its offspring will on average experience much less crowding and density dependent mortality. As a result, the mean number of offspring surviving to adulthood will be greater for the generalist genotype than for the specialist. The generalist will increase in frequency until the risk of density-dependent mortality is roughly equalized in the different habitats. It is interesting to note that escape from parasitism and predation under soft selection tends to favor a broad host range rather than specialization, in contrast to the common expectation (Price et al., 1986; Bernays and Graham, 1988), which implicitly assumes hard selection.

While the different predictions about diet breadth under hard and soft selection are intriguing, we unfortunately know nothing about whether differences in mode of population regulation contribute to variation in diet breadth among herbivores. This lack of knowledge is not surprising, given the general difficulty of determining how populations are regulated.

2.2.5.4. Balance Between Costs and Benefits

It seems an obvious truism to say that the diet breadth of any particular insect herbivore is ultimately determined by whether costs associated with a broad diet are greater or less than the cost of specialization. The main importance of this truism is that it focuses attention on the evolutionary pressures that vary among species that differ in diet breadth. To understand why one species is more specialized than another, one must determine whether search costs are greater for the generalist, whether trade-offs in fitness across hosts are stronger for the specialist, etc., in other words, why the balance between costs and benefits of specialization tips one way for one species, the other way for the second.

This approach to understanding diversity of diet breadths will require several advances. First, methods must be developed for accurately measuring costs. I have tried to outline some possible ways of achieving this goal. Second, appropriate models must be developed that will allow us to compare, say, search costs with metabolic costs associated with maintaining the digestive capabilities of processing different types of foliage. These models will allow us to say whether one type of cost outweighs another. Finally, we will eventually need to determine how the balance between costs and benefits correlates with ecological variables. Only then will we have any real hope of understanding why most herbivorous insects are specialists.

Acknowledgements

The thoughts expressed here have been greatly clarified by comments from and discussions with Robert Fritz, Fred Gould, John Jaenike, and two anonymous reviewers. I am especially grateful to Anne Lacey for sacrificing her time and energy to help bring this manuscript to completion. Supported by N.S.F. grant BSR-8817899.

References

Alberdi, M., P. Weinberger, M. Olivia, and M. Romero. 1974. Ein beitrag zur chemischen Kennzeichnung des Skleromorphie-Grades von Blättern immergrü Gehölze. *Beitrage zur Biologie der Pflanzen* 50:305–320.

Antonovics, J. 1976. The nature of limits to natural selection. *Ann. Missouri Bot. Gard.* 63:224–247.

Arnold, S. J. and M. J. Wade. 1984a. On the measurement of natural and sexual selection: theory. *Evolution* 38:709–719.

Arnold, S. J. and M. J. Wade. 1984b. On the measurement of natural and sexual selection: applications. *Evolution* 38:720–734.

Atsatt, P. R. 1981a. Ant-dependent food-plant selection by the mistletoe butterfly *Ogyris amaryllis* (Lycaenidae). *Oecologia* 48:60–63.

Atsatt, P. R. 1981b. Lycaenid butterflies and ants: selection for enemy-free space. *Am. Nat.* 118:638–654.

Beck, S. D. 1965. Resistance of plants to insects. *Annu. Rev. Entomol.* 10:207–232.

Becker, W. A. 1984. *Manual of Quantitative Genetics*. Academic Enterprises, Pullman, WA.

Belsky, A. J. 1986. Does herbivory benefit plants? A review of the evidence. *Am. Nat.* 127:870–892.

Berenbaum, M. R. 1978. Toxicity of a furanocoumarin to armyworms: a case of biosynthetic escape from insect herbivores. *Science* 201:532–534.

Berenbaum, M. R. 1983. Coumarins and caterpillars: a case for coevolution. *Evolution* 37:163–179.

Berenbaum, M. R. and P. Feeny. 1981. Toxicity of angular furanocoumarins to swallowtails: escalation in the coevolutionary arms race. *Science* 212:927–929.

Berenbaum, M. R. and A. R. Zangerl. 1988. Stalemates in the coevolutionary arms race: syntheses synergisms, and sundry other sins. In *Chemical Mediation of Coevolution*, ed. K.C. Spencer, pp. 113–132. Academic Press, New York.

Berenbaum, M. R., A. R. Zangerl, and J. K. Nitao. 1986. Constraints on chemical coevolution: wild parsnips and the parsnip webworm. *Evolution* 40:1215–1228.

Bernays, E. and M. Graham. 1988. On the evolution of host specificity in phytophagous insects. *Ecology* 69:886–892.

Blackman, R. L. 1979. Stability and variation in aphid clonal lineages. *Biol. J. Linn. Soc.* 11:259–277.

Brues, C. T. 1920. The selection of food-plants by insects with special reference to lepidopterous larvae. *Am. Nat.* 54:313–332.

Brues, C. T. 1924. The specificity of food-plants in the evolution of phytophagous insects. *Am. Nat.* 58:127–144.

Bulmer, M. G. 1980. *The Mathematical Theory of Quantitative Genetics*. Clarendon Press, Oxford, England.

Cantlon, J. E. 1969. The stability of natural populations and their sensitivity to technology. In *Diversity and Stability in Ecological Systems. Brookhaven Symposia in Biology* 22:197–203.

Carter, W. 1973. *Insects in Relation to Plant Disease.* John Wiley, New York.

Castillo-Chavez, C., S. A. Levin, and F. Gould. 1988. Physiological and behavioral adaptation to varying environments: a mathematical model. *Evolution* 42:986–994.

Cates, R. G. 1975. The interface between slugs and wild ginger: some evolutionary aspects. *Ecology* 56:391–400.

Charlesworth, B. 1980. *Evolution in Age-Structured Populations.* Cambridge Univ. Press, Cambridge, England.

Charlesworth, B. 1990. Optimization models, quantitative genetics, and mutation. 44:520–538.

Charlesworth, B., R. Lande, and M. Slatkin. 1982. A neo-Darwinian commentary on macroevolution. *Evolution* 36:474–498.

Chew, F. S. and J. E. Rodman. 1979. Plant resources for chemical defense. In *Herbivores. Their Interaction with Secondary Plant Metabolites*, eds. G. A. Rosenthal and D. H. Janzen, pp. 271–301. Academic Press, New York.

Clayton, G. A., J. A. Morris, and A. Rotertson. 1957. An experimental check on quantitative genetical theory. I. Short-term responses to selection. *J. Genet.* 55:131–151.

Coley, P. D., J. P. Bryant, and F. S. Chapin, III. 1985. Resource availability and plant antiherbivore defense. *Science* 230:895–899.

Compton, S. G., D. Newsome, and D. A. Jones. 1983. Selection for cyanogenesis in the leaves and petals of *Lotus corniculatus* L. at high latitudes. *Oecologia* 60:353–358.

Conn, E. E. 1979. Cyanide and cyanogenic glycosides. In *Herbivores. Their Interaction with Secondary Plant Metabolites*, eds. G. A. Rosenthal and D. H. Janzen, pp. 387–412. Academic Press, New York.

Cook, L. M. 1971. *Coefficients of Natural Selection.* Hutchinson Univ. Library, London.

Cooper-Driver, G. A. and T. Swain. 1976. Cyanogenic polymorphism in bracken in relation to herbivore predation. *Nature* 260:604.

Courtney, S. P. 1982. Coevolution of Pierid butterflies and their cruciferous foodplants. IV. Hostplant apparency and *Anthocharis cardamines* oviposition. *Oecologia* 52:317–321.

Courtney, S. P. 1983. Models of hostplant location by butterflies: the effect of search images and search efficiency. *Oecologia* 59:317–321.

Courtney, S. P. 1984. The evolution of egg clustering by butterflies and other insects. *Am. Nat.* 123:276–281.

Courtney, S. P. 1986. The ecology of Pierid butterflies: dynamics and interactions. *Adv. Ecol. Res.* 15:51–131.

Crawley, M. J. 1983. *Herbivory. The Dynamics of Animal-Plant Interactions.* Univ. of California Press, Berkeley, CA.

Crow, J. F. and M. Kimura. 1970. *An Introduction to Population Genetics Theory.* Harper & Row, New York.

Crow, J. F. and T. Nagylaki. 1976. The rate of change of a character correlated with fitness. *Am. Nat.* 110:207–213.

David, W. A. L. and B. O. C. Gardiner. 1962. Oviposition and the hatching of the eggs of *Pieris brassicae* (L.) in a laboratory culture. *Bull. Entomol. Res.* 53:91–109.

David, W. A. L. and B. O. C. Gardiner. 1966. Mustard oil glucosides as feeding stimulants for *Pieris brassicae* larvae in a semi-synthetic diet. *Entomol. Exp. Appl.* 9:247–255.

Del Moral, R. 1972. On the variability of chlorogenic acid concentration. *Oecologia* 9:289–300.

Dethier, V. G. 1947. *Chemical Insect Attractants and Repellents.* Blakiston, Philadelphia.

Dethier, V. G. 1954. Evolution of feeding preferences in phytophagous insects. *Evolution* 8:33–54.

Dirzo, R. 1984. Herbivory: a phytocentric overview. In *Perspectives on Plant Population Ecology*, eds. R. Dirzo and J. Sarukhan, pp. 141–165. Sinauer, Sunderland, MA.

Dirzo, R. and J. L. Harper. 1982. Experimental studies on slug-plant interactions. IV. The performance of cyanogenic and acyanogenic morphs of *Trifolium repens* in the field. *J. Ecol.* 70:119–138.

Dyer, M. I. 1975. The effects of red-winged blackbirds (*Agelaius phoenicius* L.) on biomass production of corn grains (*Zea mays* L.). *J. Appl. Ecol.* 12:719–726.

Dyer, M. I. and U. G. Bokhari. 1976. Plant-animal interactions: studies of the effects of grasshopper grazing on blue grama grass. *Ecology* 57:762–772.

Edwards, M. D., C. W. Stuber, and J. F. Wendel. 1987. Molecular-marker-facilitated investigations of quantitative-trait loci in maize. I. Numbers, genomic distribution, and types of gene action. *Genetics* 116:113–125.

Ehrlich, P. R. 1961. Intrinsic barriers to dispersal in checkerspot butterfly. *Science* 134:108–109.

Ehrlich, P. R. and P. H. Raven. 1964. Butterflies and plants: a study in coevolution. *Evolution* 18:586–608.

Ellis, W. M., R. J. Keymer, and D. A. Jones. 1977. On the polymorphism of cyanogenesis in *Lotus corniculatus* L. VIII. Ecological studies in Anglesey. *Heredity* 39:45–66.

Endler, J. A. 1986. *Natural Selection in the Wild.* Princeton Univ. Press, Princeton, NJ.

Falconer, D. S. 1953. Selection for large and small size in mice. *J. Genet.* 51:470–501.

Falconer, D. S. 1955. Patterns of response in selection experiments with mice. *Cold Spring Harbor Symposium in Quantitative Biology* 20:178–196.

Falconer, D. S. 1981. *Introduction to Quantitative Genetics*. 2nd. ed. Longman, London.

Fasoulas, A. C. and R. W. Allard. 1962. Non-allelic gene interactions in the inheritance of quantitative characters in barley. *Genetics* 47:899–907.

Feeny, P. 1975. Biochemical coevolution between plants and their insect herbivores. In *Coevolution of Animals and Plants*, eds. L.E. Gilbert and P.H. Raven, pp. 3–19. Univ. of Texas Press, Austin.

Feeny, P. 1976. Plant apparency and chemical defense. *Rec. Adv. Phytochem.* 10:1–40.

Feeny, P., L. Rosenberry, and M. Carter. 1983. Chemical aspects of oviposition behavior in butterflies. In *Herbivorous Insects: Host-Seeking Behavior and Mechanisms*, ed. S. Ahmad, pp. 27–76. Academic Press, New York.

Felsenstein, 1979. Excursions along the interface between disruptive and stabilizing selection. *Genetics* 93:773–795.

Fisher, R. A. 1958. *The Genetical Theory of Natural Selection*. Dover, New York.

Fox, L. R. 1981. Defense and dynamics in plant-herbivore systems. *Am. Zool.* 21:853–864.

Fox, L. R. 1988. Diffuse coevolution within complex communities. *Ecology* 69:906–907.

Fox, L. R., and P. A. Morrow. 1981. Specialization: species property or local phenomenon. *Science* 221:887–893.

Fraenkel, G. 1956. Insects and plant biochemistry. The specificity of food plants for insects. *Proc. 14th Int. Cong. Zool, Copenhagen.* pp. 383–387.

Fraenkel, G. 1959. The raison d'etre of secondary plant substances. *Science* 129:1466–1470.

Fraenkel, G. 1969. Evaluation of our thoughts on secondary plant substances. *Entomol. Exp. Appl.* 12:473–486.

Fritz, R. S. 1990. Community structure and species interactions of phytophagous insects on resistant and susceptible host plants. In *Ecology and Evolution of Plant Resistance.* eds. R. S. Fritz and E. L. Simms. Univ. of Chicago Press, Chicago (in press).

Fritz, R. S. and P. W. Price. 1988. Genetic variation among plants and insect community structure: willows and sawflies. *Ecology* 69:845–856.

Fritz, R. S., C. F. Sacchi, and P. W. Price. 1986. Competition versus host-plant phenotype in species composition: willow sawflies. *Ecology* 67:1608–1618.

Fry, J. D. 1990. Trade-offs in fitness on different hosts: evidence from a selection experiment with a phytophagous mite. *Am. Nat.* 136: (in press).

Futuyma, D. J. 1976. Food plant specialization and environmental predictability in Lepidoptera. *Am. Nat.* 110:285–292.

Futuyma, D. J. 1983a. Evolutionary interactions among herbivorous insects and plants. In *Coevolution*, eds. D. J. Futuyma and M. Slatkin, pp. 207–231. Sinauer, Sunderland, MA.

Futuyma, D. J. 1983b. Selective factors in the evolution of host choice by phytophagous insects. In *Herbivorous Insects: Host-Seeking Behavior and Mechanisms*, ed. S. Ahmad, pp. 227–244. Academic Press, New York.

Futuyma, D. J. 1986. *Evolutionary Biology*. 2nd ed. Sinauer, Sunderland, MA.

Futuyma, D. J. and T. E. Philippi. 1987. Genetic variation and covariation in responses to host plants by *Alsophila pometaria* (Lepidoptera: Geometridae). *Evolution* 41:269–279.

Futuyma, D. J. and S. S. Wasserman. 1982. Food plant specialization and feeding efficiency in the tent caterpillars *Malacosoma disstria* Hubner and *M. americanum* (Fabricus). *Entomol. Exp. Appl.* 39:106–110.

Gilbert, L. E. 1971. Butterfly-plant coevolution: has *Passiflora adenopoda* won the selection race with heliconiine butterflies? *Science* 172:585–586.

Gilbert, L. E. 1975. Ecological consequences of a coevolved mutualism between butterflies and plants. In *Coevolution of Animals and Plants*, eds. L. E. Gilbert and P. H. Raven, pp. 210–240. Univ. of Texas Press, Austin, TX.

Gilbert, L. E. 1979. Development of theory in the analysis of insect-plant interactions. In *Analysis of Ecological Systems*, eds. D. Horn, R. Mitchell, and G. Stairs, pp. 117–154. Ohio State Univ. Press, Columbus, Ohio.

Gilbert, L. E. and M. C. Singer. 1973. Dispersal and gene flow in a butterfly species. *Am. Nat.* 107:58–72.

Goodnight, C. J. 1988. Epistasis and the effect of founder events on the additive genetic variance. *Evolution* 42:441–454.

Gould, F. 1979. Rapid host range evolution in a population of the phytophagous mite *Tetranychus urticae* Koch. *Evolution* 33:241–250.

Gould, F. 1984a. Role of behavior in the evolution of insect adaptation to insecticides and resistant host plants. *Bull. Entomol. Soc. Amer.* 30:34–41.

Gould, F. 1984b. Mixed function oxidases and herbivore polyphagy: the devil's advocate position. *Ecol. Entomol.* 9:29–34.

Gould, F. 1988. Genetics of pairwise and multispecies plant-herbivore coevolution. *Chemical Mediation of Coevolution*, ed. K. C. Spencer, pp. 13–55. Academic Press, New York.

Gould, S. J. 1982. Darwinism and the expansion of evolutionary theory. *Science* 216:380–387.

Gould, S. J. and N. Eldredge. 1977. Punctuated equilibria: the tempo and mode of evolution reconsidered. *Paleobiology* 3:115–151.

Grubb, P. J. 1986. Sclerophylls, pachyphylls and pycnophylls: the nature and significance of hard leaf surfaces. In *Insects and the Plant Surface*, ed. B. Juniper and R. Southwood, pp. 137–150. Edward Arnold, London.

Hairston, N. G., F. E. Smith, and L. B. Slobodkin. 1960. Community structure, population control, and competition. *Am. Nat.* 94:421–425.

Haldane, J. B. S. 1954. The measurement of natural selection. Proc. IX Int. Congr. *Genetics* (Caryologica supplement) 1:480–487.

Hamrick, J. L., Y. B. Linhart, and J. B. Mitton. 1979. Relationships between life history characteristics and electrophoretically detectable genetic variation in plants. *Annu. Rev. of Ecol. & Systemat.* 10:173–200.

Hare, J. D. and D. J. Futuyma. 1978. Different effects of variation in *Xanthium strumarium* L. (Compositae) on two insect seed predators. *Oecologia* 39:109–120.

Hare, J. D. and G. G. Kennedy. 1986. Genetic variation in plant-insect associations: survival of *Leptinotarsa decemlineata* populations on *Solanum carolinense*. *Evolution* 40:1031–1043.

Harper, J. L. 1977. *Population Biology of Plants*. Academic Press, London.

Hartl, D. L. and R. D. Cook. 1973. Balanced polymorphism of quasineutral alleles. *Theor. Pop. Biol.* 4:163–172.

Horber, E. 1980. Types and classification of resistance. In *Breeding Plants Resistant to Insects*, eds. F. G. Maxwell and P. R. Jennings, pp. 15–21. John Wiley, New York.

Ichinose, T. and H. Honda. 1978. Ovipositional behavior of *Papilio protenor demetrius* Cramer and the factors involved in its host plants. *Appl. Entomol. Zool.* 13:103–114.

Jaenike, J. 1978a. On optimal oviposition behavior in phytophagous insects. *Theor. Pop. Biol.* 14:350–356.

Jaenike, J. 1978b. Resource predictability and niche breadth in the *Drosophila quinaria* species group. *Evolution* 32:676–678.

Jaenike, J. 1985. Parasite pressure and the evolution of amanitin tolerance in *Drosophila*. *Evolution* 39:1295–1301.

Jaenike, J., D. A. Grimaldi, A. E. Sluder, and A. L. Greenleaf. 1983. α-Amanitin tolerance in microphagous *Drosophila*. *Science* 221:165–167.

Janzen, D. H. 1966. Coevolution of mutualism between ants and acacias in Central America. *Evolution* 20:249–275.

Janzen, D. H. 1969. Seed-eaters versus seed size, number, dispersal and toxicity. *Evolution* 23:1–27.

Janzen, D. H. 1973a. Community structure of secondary compounds in plants. *Pure Appl. Chem.* 34:529–538.

Janzen, D. H. 1973b. Comments on host-specificity of tropical herbivores and its relevance to species richness. In *Taxonomy and Ecology. Systematics Association Special*, Vol. 5, ed. V.H. Heywood, pp. 201–211. Academic Press, London.

Janzen, D. H. 1973c. Dissolution of mutualism between *Cecropia* and its *Azteca* ants. *Biotropica* 5:15–28.

Janzen, D. H. 1975. Behavior of *Hymenaea courbaril* when its predispersal seed predator is absent. *Science* 189:145–147.

Janzen, D. H. 1979. New horizons in the biology of plant defenses. *Herbivores: Their Interaction with Secondary Plant Metabolites*, eds. G. Rosenthal and D. H. Janzen, pp. 331–350. Academic Press, London.

Janzen, D. H. 1980. When it is coevolution? *Evolution* 34:611–612.

Jeffree, C. E. 1986. The cuticle, epicuticular waxes and trichomes of plants, with reference to their structure, functions and evolution. In *Insects and the Plant Surface*, eds. B. Juniper and R. Southward, pp. 23–64. Edward Arnold, London.

Jermy, T. 1976. Insect-host plant relationship: co-evolution or sequential evolution? *Symposia Biologica Hungarica* 16:109–113.

Jermy, T. 1984. Evolution of insect/host plant relationships. *Am. Nat.* 124:609–630.

Joern. A. and L. R. Lawlor. 1980. Food and microhabitat utilization by grasshoppers from arid grasslands: comparisons with neural models. *Ecology* 61:591–599.

Johnson, B. and P. R. Birks. 1960. Studies on wing polymorphism in aphids. I. The developmental process involved in the production of different forms. *Entoml. Exp. Appl.* 3:327–339.

Jones, D. A. 1972. Cyanogenic glycosides and their function. In *Phytochemical Ecology*, ed. J.B. Harborne, pp. 103–124. Academic Press, New York.

Jones, D. A., R. J. Keymer, and W. M. Ellis. 1978. Cyanogenesis in plants and animal feeding. In *Biochemical Aspects of Plant-Animal Coevolution*, ed. J.B. Harborne, pp. 21–24. Academic Pres, New York.

Kinsman, S. 1982. Herbivore responses to *Oenothera biennis* (Onagraceae): effects on host plant's size, genotype, and resistant conspecific neighbors. Ph. D. dissertation, Cornell University, Ithaca, NY.

Krieger, R. I., P. P. Feeny, and C. F. Wilkinson. 1971. Detoxication enzymes in the guts of caterpillars: an evolutionary answer to plant defenses? *Science* 172:579–581.

Krischik, V. A. and R. F. Denno. 1983. Individual, population, and geographic patterns in plant defense. In *Variable Plants and Herbivores in Natural and Managed Systems*, eds. R. F. Denno and M. S. McClure, pp. 463–512. Academic Press, New York.

Lande, R. 1975. The maintenance of genetic variability by mutation in a polygenic character with linked loci. *Genet. Res.* 26:221–234.

Lande, R. 1977. The influence of the mating system on the maintenance of genetic variability in polygenic characters. *Genetics* 86:485–498.

Lande, R. 1979. Quantitative genetic analyses of multivariate evolution, applied to brain:body size allometry. *Evolution* 33:402–416.

Lande, R. 1980. The genetic covariance between characters maintained by pleiotropic mutations. *Genetics* 94:203–215.

Lande, R. and S. J. Arnold. 1983. The measurement of selection on correlated characters. *Evolution* 37:1210–1226.

Lawton, J. H. and D. R. Strong. 1981. Community patterns and competition in folivorous insects. *Am. Nat.* 118:317–338.

Lees, A. D. 1966. The control of polymorphism in aphids. *Adv. Insect Physiol.* 3:207–277.

Lerner, I. M. 1958. *The Genetic Basis of Selection*. John Wiley, New York.

Levin, D. A. 1976. The chemical defenses of plants to pathogens and herbivores. *Annu. Rev. Ecol. System.* 7:121–159.

Levins, R. and R. MacArthur. 1969. An hypothesis to explain the incidence of monophagy. *Ecology* 50:910–911.

Lewontin, R. C. 1964. The interaction of selection and linkage. II. Optimum models. *Genetics* 50:757–782.

Lewontin, R. C. 1974. *The Genetic Basis of Evolutionary Change.* Columbia University Press, New York.

Louda, S. M. 1983. Seed predation and seedling mortality in the recruitment of a shrub, *Hymenopappus venetus* (Asteraceae), along a climatic gradient. *Ecology* 64:511–521.

Luckinbill, L. S., and M. J. Clare. 1985. Selection for life span in *Drosophila melanogaster. Heredity* 55:9–18.

Maddox, G. David, and R. B. Root. 1990. Ecological genetics of resistance to insects in *Solidago altissima*: Genetic correlations among resistances. In *Ecology and Evolution of Plant Resistance*, eds. R. S. Fritz and E. L. Simms, Univ. of Chicago Press, Chicago (in press).

Marquis, R. J. 1984. Leaf herbivores decrease fitness of a tropical plant. *Science* 226:537–539.

Maschinsky, J. and T. G. Whitham. The continuum of plant responses to herbivory: the influence of plant association, nutrient availability and timing. *Am. Nat.* 134:1–19.

McKey, D. 1974. Adaptive patterns in alkaloid physiology. *Am. Nat.* 108:305–320.

Miller, James S. 1987. Host-plant relationships in the Papilionidae (Lepidoptera): parallel cladogenesis or colonization? *Cladistics* 3:105–120.

Mitter, C. and D. R. Brooks. 1983. Phylogenetic aspects of coevolution. In *Coevolution*, eds. D. J. Futuyma and M. Slatkin, pp. 65–98. Sinauer, Sunderland, Massachusetts.

Mitter, C., D. Futuyma, J. Schneider, and J. Hare. 1979. Genetic variation and host plant relations in a parthenogenetic moth. *Evolution* 33:777–790.

Mooney, H. A. and E. L. Dunn. 1970. Photosynthetic systems of mediterranean-climate shrubs and trees of California and Chile. *Am. Nat.* 104:447–453.

Moran, N. 1981. Intraspecific variability in herbivore performance and host quality: a field study of *Uroleucon caligatum* (Homoptera: Aphididae) and its *Solidago* hosts (Asteraceae). *Ecol. Entomol.* 6:301–306.

Moran, N. 1986. Morphological adaptation to host plants in *Uroleucon* (Homoptera: Aphididae). *Evolution* 40:1044–1050.

Morrow, P. P. and V. C. LaMarche. 1978. Tree ring evidence for chronic insect suppression of productivity in subalpine *Eucalyptus. Science* 201:1244–1246.

Murphy. P. A., J. T. Giesel, and M. N. Manlove. 1983. Temperature effects on life history variation in *Drosophila simulants. Evolution* 37:1181–1192.

Nass, H. G. 1972. Cyanogenesis: its inheritance in *Sorghum bicolor*, *Sorghum sudanense*, *Lotus*, and *Trifolium repens*: a review. *Crop Science* 12:503–506.

Nei, M., T. Maruyama, and R. Chakraborty. 1975. The bottleneck effect and genetic variability in populations. *Evolution* 29:1–10.

Ohsaki, N. and Y. Sato. 1990. Avoidance mechanisms of three *Pieris* butterfly species against the parasitoid wasp *Apanteles glomeratus*. *Ecological Entomology* 15:169–176.

Ohsaki, N. and Y. Sato. 1991. Host plant choice of three *Pieris* butterflies influenced by specialist parasitoids and host quality. *Ecology* (in review).

Oster, G. and P. Alberch. 1982. Evolution and bifurcation of developmental programs. *Evolution* 36:444–459.

Owen, D. F. 1980. How plants may benefit from the animals that eat them. *Oikos* 35:230–235.

Owen, D. F. and R. G. Wiegert. 1976. Do consumers maximize plant fitness? *Oikos* 27:488–492.

Owen, D. F. and R. G. Wiegert. 1982. Grasses and grazers: is there a mutualism? *Oikos* 38:258–259.

Paige, K. N. and T. G. Whitham. 1987. Overcompensation in response to mammalian herbivory: the advantage of being eaten. *Am. Nat.* 129:407–416.

Painter, R. H. 1958. Resistance of plants to insects. *Annu. Rev. Entomol.* 3:267–290.

Parker, M. A. and R. B. Root. 1981. Insect herbivores limit distribution of a native composite, *Machaeranthera canescens*. *Ecology* 62:1390–1392.

Parsons, P. A. 1977. Genotype-temperature interaction for longevity in natural populations of *Drosophila simulans*. *Exper. Geront.* 12:241–244.

Price, G. R. 1970. Selection and covariance. *Nature* 227:520–521.

Price, P. W. 1983. Hypotheses on organization and evolution in herbivorous insect communities. In *Variable Plants and Herbivores in Natural and Managed Systems*, eds. R. F. Denno and M. S. McClure, pp. 559–598. Academic Press, New York.

Price, P. W., C. E. Bouton, P. Gross, B. A. McPherson, J. N. Thompson, and A. E. Weis. 1980. Interactions among three trophic levels: influence of plants on interactions between insect herbivores and natural enemies. *Annu. Rev. Ecol. Syst.* 11:41–65.

Price, T., M. Kirkpatrick, and S. J. Arnold. 1988. Directional selection and the evolution of breeding date in birds. *Science* 240:798–799.

Price, P. W., M. Westoby, B. Rice, P. R. Atsatt, R. S. Fritz, J. N. Thompson, and K. Mobley. 1986. Parasite mediation in ecological interactions. *Annu. Rev. Ecol. Syst.* 17:487–505.

Rausher, M. D. 1978. Search image for leaf shape in a butterfly, *Science* 200:1071–1073.

Rausher, M. D. 1979. Coevolution in a sample plant-herbivore system. Ph. D. dissertation, Cornell University, Ithaca, NY.

Rausher, M. D. 1981. Host plant selection by *Battus philenor* butterflies: The roles of predation, nutrition, and plant chemistry. *Ecol. Monogr.* 51:1–20.

Rausher, M. D. 1983. Ecology of host-selection behavior in phytophagous insects. In *Variable Plants and Herbivores in Natural and Managed Systems*, eds. R. F. Denno and M. S. McClure, pp. 223–257. Academic Press, New York.

Rausher, M. D. 1984a. Tradeoffs in performance on different hosts: evidence from within- and between-site variation in the beetle *Deloyala guttata*. *Evolution* 38:582–595.

Rausher, M. D. 1984b. The evolution of habitat preference in subdivided populations. *Evolution* 38:596–608.

Rausher, M. D. 1985. Variability for host preference in insect populations: mechanistic and evolutionary models. *J. Insect Physiol.* 31:873–889.

Rausher, M. D. 1992. The evolution of habitat selection. III. The evolution of avoidance and counterresistance. *Evolution of Insect Pests: The Pattern of Variations*, ed. K. C. Kim. John Wiley, New York.

Rausher, M. D. and R. Englander. 1987. The evolution of habitat preference. II. Evolutionary genetic stability under soft selection. *Theor. Pop. Biol.* 31:116–139.

Rausher, M. D. and P. Feeny. 1980. Herbivory, plant density, and plant reproductive success: the effect of *Battus philenor* on *Aristolochia reticulata*. *Ecology* 61:905–917.

Rausher, M. D. and E. L. Simms. 1989. The evolution of resistance to herbivory in *Ipomoea purpurea*. I. Attempts to detect selection. *Evolution* 43:563–572.

Reardon, P. O., C. L. Leinwever, and L. B. Merrill. 1972. The effect of bovine saliva on grasses. *J. Anim. Sci.* 34:897–898.

Reese, C. J. C. 1969. Chemoreceptor specificity associated with choice of feeding site by the beetle *Chrysolina brunsvicensis* on its foodplant, *Hypericum hirsutum*. *Entomol. Exp. Appl.* 12:565–583.

Rehr, S. S., P. P. Feeny, and D. H. Janzen. 1973. Chemical defense in Central American non-ant acacias. *J. Anim. Ecol.* 42:405–416.

Renwick, J. A. A. 1988. Comparative mechanisms of host selection by insects attacking pine trees and crucifers. In *Chemical Mediation of Coevolution*, ed. K. C. Spencer, pp. 303–316. Academic Press, New York.

Rhoades, D. F. 1979. Evolution of plant chemical defense against herbivores. In *Herbivores. Their Interaction With Secondary Plant Metabolites*, eds. G. A. Rosenthal and D. H. Janzen, pp. 4–54. Academic Press, London.

Rhoades, D. F. 1983. Herbivore population dynamics and plant chemistry. *Variable Plants and Herbivores in Natural and Managed Systems*, eds. R. F. Denno and M. C. McClure, pp. 155–220. Academic Press, London.

Rhoades, D. F. and R. G. Cates. 1976. Toward a general theory of plant antiherbivore chemistry. *Rec. Adv. Phytochem.* 10:168–213.

Robinson, T. 1974. Metabolism and function of alkaloids in plants. *Science* 184:430–435.

Rose, H. A. 1985. The relationship between feeding specialization and host plants to aldrin epoxidase activities of midgut homogenates in larval lepidoptera. *Ecol. Entomol.* 10:455–467.

Roughgarden, J. 1979. *Theory of Population Genetics and Evolutionary Ecology: An Introduction.* MacMillan, New York.

Russell, W. A. and S. A. Eberhart. 1970. Effects of three gene loci in the inheritance of quantitative characters in maize. *Crop Sci.* 10:165–169.

Sato, Y. and N. Ohsaki. 1987. Host-habitat location by *Apanteles glomeratus* and effect of food-plant exposure on host-parasitism. *Ecol. Entomol.* 12:291–297.

Schneider, J. C. 1980. The role of parthenogenesis and female aptery in microgeographic, ecological adaptation in the fall cankerworm, *Alsophila pometaria* Harris (Lepidoptera: Geometriae). *Ecology* 61:1082–1090.

Schopf, T. J. M. 1982. A critical assessment of punctuated equilibria. I. Duration of taxa. *Evolution* 36:1144–1157.

Schroeder, L. A. 1976. Energy, matter and nitrogen utilization by the larvae of the monarch butterfly *Danaus plexippus*. *Oikos* 27:259–264.

Schroeder, L. A. 1977. Energy, matter and nitrogen utilization by the larvae of the milkweed tiger moth *Euchaetias egle*. *Oikos* 28:27–31.

Scriber, J. M. 1973. Latitudinal gradients in larval feeding specialization of the world Papilionidae (Lepidoptera). *Psyche* 80:355–373.

Scriber, J. M. 1988. Tale of the tiger: beringial biogeography, binomial classification, and breakfast choices in the *Papilio glaucus* complex of butterflies. In *Chemical Mediation of Coevolution*, ed. K. C. Spencer, pp. 241–301. Academic Press, New York.

Scriber, J. M. and P. Feeny. 1979. Growth of herbivorous caterpillars in relation to feeding specialization and to growth form of their food plants. *Ecology* 60:829–850.

Scriber, J. M. and F. Slansky, Jr. 1981. The nutritional ecology of immature insects. *Annu. Rev. Entomol.* 26:183–212.

Service, P. M. and M. R. Rose. 1985. Genetic covariation among life-history components: the effect of novel environments. *Evolution* 39:943–945.

Siegler, D. S. 1977. Primary roles for secondary compounds. *Bioch. Syst. Ecol.* 5:195–199.

Siegler, D. and P. W. Price. 1976. Secondary compounds in plants: primary functions. *Am. Nat.* 110:101–105.

Simberloff, D., B. J. Brown, and S. Lowrie. 1978. Isopod and insect root borers may benefit Florida mangroves. *Science* 201:630–632.

Simms, E. L. 1992. Costs of resistance to herbivory. In *Ecology and Evolution of Plant Resistance*, eds. R. S. Fritz and E. L. Simms. Univ. of Chicago Press, Chicago, (in press).

Simms, E. L. and M. D. Rausher. 1987. Costs and benefits of plant resistance to herbivory. *Am. Nat.* 130:570–581.

Simms, E. L. and M. D. Rausher. 1989. The evolution of resistance to herbivory in *Ipomoea purpurea*. II. Natural selection by insects and costs of resistance. *Evolution* 43:573–585.

Smiley, J. 1978. Plant chemistry and the evolution of host specificity: new evidence from *Heliconius* and *Passiflora*. *Science* 201:745–747.

Smith, R. H. 1964. Variation in the monoterpenes of *Pinus ponderosa* laws. *Science* 143:1337–1338.

Spencer, K. C. 1988. Chemical mediation of coevolution in the *Passiflora-Heliconius* interaction. In *Chemical Mediation of Coevolution*, ed. K. C. Spencer, pp. 167–240. Academic Press, New York.

Spickett, S. G. 1963. Genetic and developmental studies of a quantitative character. *Nature* 199:870–873.

Spickett, S. G. and J. M. Thoday. 1966. Regular responses to selection 3. Interaction between located polygenes. *Gen. Res.* 7:96–121.

Stahl, E. 1988. Pflanzen und Schnecken. Biologische Studie über die Schutzmittel der Pflanzen gegen Schneckenrass. *Jena. Zeit. Med. u. Naturw.* 22:557–684.

Stanley, S. M. 1979. *Macroevolution: Pattern and Process*. W. H. Freeman, San Francisco.

Stanley, S. M. 1982. Macroevolution and the fossil record. *Evolution* 36:460–473.

Strong, D. R., J. H. Lawton, and T. R. E. Southwood. 1984. *Insects on Plants*. Harvard Univ. Press, Cambridge, MA.

Sturgeon, K. B. 1979. Monoterpene variation in ponderosa pine xylem resin related to western pine beetle predation. *Evolution* 33:803–814.

Suolmalainen, E. 1950. Parthenogenesis in animals. *Adv. Genet.* 3:193–253.

Suolmalainen, E., A. Saura, J. Lokki, and T. Teeri. 1980. Genetic polymorphism in parthenogenetic animals, part 9: absence of variation within parthenogenetic aphid clones. *Theor. Appl. Genet.* 57:129–132.

Sutherland, O. 1969a. The role of crowding in the production of winged forms by two strains of a pea aphid, *Acyrthosiphon pisum*. *J. Insect Physiol.* 15:1385–1410.

Sutherland, O. 1969b. The role of the host plant in the production of winged forms by two strains of the pea aphid, *Acyrthosiphon pisum*. *J. Insect. Physiol.* 15:2179–2201.

Swain, T. 1977. Secondary compounds as protective agents. *Annu. Rev. Plant Physiol.* 28:479–501.

Tauber, M. J., C. A. Tauber, and S. Masaki. 1986. *Seasonal Adaptations of Insects*. Oxford Univ. Press, New York.

Thompson, J. N. 1986. Constraints on arms races in coevolution. *Trends in Ecol. Evol.* 1:105–107.

Turelli, M. 1984. Heritable genetic variation via mutation-selection balance: Lerch's zeta meets the abdominal bristle. *Theoret. Pop. Biol.* 25:138–193.

Turelli, M. 1985. Effects of pleiotropy on predictions concerning mutation-selection balance for polygenic traits. *Genetics* 111:165–195.

Turelli, M. 1988. Phenotypic evolution, constant covariances, and the maintenance of additive variance. *Evolution* 42:1342–1347.

Uyenoyama, M. K. 1986. Inbreeding and the cost of meiosis: the evolution of selfing in populations practicing biparental inbreeding. *Evolution* 40:388–404.

Varley, G. C., G. R. Gradwell, and M. P. Hassell. 1974. *Insect Population Ecology: An Analytical Approach.* Univ. of California Press, Berkeley.

Verschaffelt, E. 1910. The cause determining the selection of food in some herbivorous insects. *Proc. Acad. Sci. Amsterdam* 13:536–542.

Via, S. 1984. The quantitative genetics of polyphagy in an insect herbivore. II. Genetic correlations in larval performance within and across host plants. *Evolution* 38:896–905.

Waldbauer, G. P. 1968. The consumption and utilization of food by insects. *Adv. Insect Physiol.* 5:229–289.

Wallace, B. 1968. Polymorphism, population size, and genetic load. In *Population Biology and Evolution*, ed. R. C. Lewontin, pp. 87–108. Syracuse Univ. Press, Syracuse.

Weber, G. 1985. Genetic variability in host plant adaptation of the green peach aphid, *Myzus persicae. Entomol. Exp. Appl.* 38:49–56.

Wehrhan, C., and R. W. Allard. 1965. The detection and measurement of the effects of individual genes involved in the inheritance of a quantitative character in wheat. *Genetics* 51:109–119.

Whitham, T. G. 1978. Habitat selection by *Pemphigus* aphids in response to resource limitation and competition. *Ecology* 59:1164–1176.

Whitham, T. G. 1980. The theory of habitat selection: examined and extended using *Pemphigus* aphids. *Am. Nat.* 115:449–466.

Whittaker, R. H. and P. P. Feeny. 1971. Allelochemics: chemical interactions between species. *Science* 171:757–770.

Wiklund, C. 1975. The evolutionary relationship between adult oviposition preferences and larval host plant range in *Papilio machaon* L. *Oecologia* 18:185–197.

Williams, K. S. and L. E. Gilbert. 1981. Insects as selective agents on plant vegetative morphology: egg mimicry reduces egg laying by butterflies. *Science* 212:467–469.

Windle, P. N. and E. H. Franz. 1979. The effects of insect parasitism on plant competition: greenbugs and barley. *Ecology* 60:521–529.

Wright, S. 1969. *Evolution and the Genetics of Populations. Vol. 2. The Theory of Gene Frequencies.* Univ. of Chicago Press, Chicago.

Wright, S. 1982. Character change, speciation, and the higher taxa. *Evolution* 36:427–433.

Yoo, B. H. 1980. Long-term selection for a quantitative character in large replicate populations of *Drosophila melanogaster*. I. Response to selection. *Genet. Research* 35:1–17.

3

Biochemicals: Engineering Problems for Natural Selection

M. Berenbaum and *D. Seigler*

University of Illinois

While living organisms are remarkably conservative with respect to the chemicals required for carrying out basic physiological functions, they are exuberantly idiosyncratic with respect to secondary metabolism, that is, the production of chemical substances that are not involved in the fundamental life processes. One of the most striking aspects of plant secondary metabolism, for example, is its astounding diversity; at least 20,000 individual compounds have been isolated and characterized from angiosperms alone. Yet, although it is remarkable that such great diversity exists, it is at the same time remarkable that there are commonalities underlying the production of secondary metabolites irrespective of taxon. In all likelihood, these commonalities reflect constraints on biosynthesis imposed by opposing selection pressures or by phylogenetic inertia.

Just as variation in primary metabolic processes has the capacity to effect differential survival, variation in secondary metabolism can also affect fitness. The biosynthetic profiles of extant organisms represent adaptive compromises imposed by both the biotic and abiotic environment. Availability and suitability of raw materials with which to construct secondary metabolites, phylogenetic constraints on biosynthetic capabilities, and energetic and toxicological costs associated with different patterns of construction act to limit the type and number of secondary chemicals that can be produced; natural selection exerted by other organisms, including predators, mutualists, pathogens, or competitors profoundly influences the type and number of phytochemicals that can be maintained by individual species.

While certainly ubiquitous, production of secondary metabolites is not universal. Whereas bacteria and fungi are known to produce thousands of secondary metabolites, relatively few are known from blue-green algae; secondary metabolism in many marine invertebrate phyla appears to be minimal, but coral reef coelenterates produce a remarkable array of compounds (Zaehner et al., 1982). Depending upon ecological conditions, natural selection can act either as a brake or as an accelerator with respect to the evolution of secondary biosynthetic pathways in organisms. In this chapter, discussions focus on two groups of organisms that are both speciose and chemically diverse–plants and herbivorous insects. Each taxon comprises over a quarter of a million species, and, collectively, plant-insect interactions are likely the predominant ecological interaction on the planet. Due to the extensive interactions between these taxa in both ecological and evolutionary time, commonalities of chemical diversity and speciosity in these two taxa are more than coincidental—they are instead the result of "intimate contact and reciprocal selection between arthropods and plants" (Rodriguez and Levin, 1975).

3.1. Availability and Suitability of Elements

According to Richardson (1977): "When an engineer decides what metals will be used in a new aircraft, his choice is governed by suitability and economy. Only some metals possess the desired properties, and of the possible alternatives some are more expensive than others. If we translate cost into relative abundance, it is evident from the composition of protoplasm that life has evolved on earth under the same pair of constraints faced by the engineer with regard to raw materials." In terms of availability, most secondary metabolites are constructed primarily of carbon, hydrogen, oxygen, and nitrogen; these four elements are among the most abundant in the universe (Table 3.1). Therefore, it is not altogether surprising that these elements are so abundant in living things.

Availability, however, is only one criterion for incorporation into biochemicals; chemical reactivity is a crucial arbiter of the relative value of elements for construction purposes. For example, helium is second only to hydrogen in cosmic abundance, yet it is of no importance to the structure and function of living things because, as a "noble" gas, it is exceedingly unreactive. With its ability to bond to four other atoms by sharing a pair of electrons, the carbon atom is quite reactive and is capable of reacting with many other elements. Moreover, it is capable of reacting with itself to form relatively stable rings and chains. Due to the tetrahedral configuration of these electron pairs, three-dimensional structure is possible (Lehninger, 1972). Together with its solubility in water (in the form of carbon dioxide or bicarbonate), it is well-suited to the biosynthetic needs of living organisms. Silicon, for example, is also tetravalent, but its oxides are less water-soluble and its polymers less stable than those of carbon, so it is therefore less suitable for water-rich living organisms. Although carbon constitutes only 0.16% to 0.19% of the outer lithosphere of the earth, and only 0.03% of the atmosphere, it constitutes 25% of plant

Table 3.1. Relative abundance of elements in the universe, earth layers, and living matter. Values are percentages of the total number of atoms present.*

Universe		Crust Only (Outer Lithosphere)		Hydro-sphere		Atmo-sphere		Plant Material (Wood)	
H	91	O	47–60	H	66	N	78	H	50
He	9	Si	21–28	O	33	O	21	O	25
O	0.06	Al	6–8	Cl	0.3	A	0.93	C	25
N	0.04	Fe	2–5	Na	0.3	H	0.5	N	0.27
C	0.02	Ca	2–4	Mg	0.03	C	0.03	Ca	0.07
Si	0.003	Na	c.2.5	S	0.02			K	0.05
Ne	0.003	K	1.4–2.5	Ca	0.006			Si	0.03
Mg	0.002	Mg	1.8–2.2	K	0.006			Mg	0.03
Fe	0.002	Ti	0.3–0.5	C	0.001			P	0.03
S	0.001	H	0.2–3.0	Br	0.0005			S	0.02
		C	0.16–0.19	B	0.0002			Al	0.02

*From Richardson (1977).

material in toto (Richardson, 1977) and is a universal component of secondary metabolites (Table 3.1).

Oxygen, hydrogen, and nitrogen are also common components of secondary metabolites. Aside from their abundance, they have chemical properties that render them particularly suitable for the construction of chemical substances that mediate interactions with other organisms. Oxygen, hydrogen, and nitrogen, by sharing a pair of electrons, can form covalent bonds; C, N, and O can share either one or two pairs of electrons to form single or double bonds, respectively. Due to the low atomic weight of these elements, the bonds formed are very strong as well; bond strength is an inverse function of atomic weight (Morrison and Boyd, 1970). Oxygen and nitrogen are both electronegative; thus, they are capable of forming polar bonds. Carbon- and nitrogen-containing compounds are therefore likely to interact with water, another polar molecule and the medium in which most, if not all, life processes take place.

Although certain elements are well-suited for primary physiological processes, they may be present in insufficient supply to allow their use by plants in secondary metabolic processes. Phosphorus is a necessary component of nucleosides and nucleotides that carry chemical energy in cells, that transport monomers involved in biosynthesis of numerous cell constituents, and that form the backbone of the nucleic acids DNA and RNA. Available phosphorus is present in soil in very low amounts and is thought to be the element in most limited supply to plant life globally. Because phosphorus is so essential to primary physiological processes, it may be less available for incorporation into plant secondary metabolites (Coley et al., 1985), although, in the form of high-energy phosphate bonds, it may be involved in their biosynthesis.

The availability of elements to a plant determines not only the type of secondary metabolites that can be synthesized, but also the amount. Nitrogen availability has been shown in several taxa to affect secondary metabolism. Lupines associated with nitrogen-fixing bacteria, for example, produce substantially larger amounts of alkaloids than do lupines without well-established endosymbionts or lupines growing in nitrogen-poor soil (Johnson et al., 1988; Bentley and Johnson, 1991). Similarly, *Sarracenia* growing in low-nitrogen bogs do not produce N-containing secondary metabolites typical of the genus (Romeo et al., 1977). While elemental deficiencies can reduce production of certain secondary metabolites, fertilization or an otherwise elevated supply of elements can increase production. Nitrogen fertilization increases production of cyanogenic glycosides in clover and sorghum (DeWaal, 1972; Drolsom, 1972), and sulfate supplementation of crucifers can increase production of S-containing mustard oil glycosides (references reviewed in Chew and Rodman, 1979). Even carbon availability can conceivably limit secondary metabolite production. While enrichment of atmospheric carbon dioxide levels failed to increase production of carbon-based secondary metabolites in *Mentha piperita* (terpenes; Lincoln and Couvet, 1989) and in *Plantago lanceolata* (iridoid glycosides; Fajer et al., 1989), carbon availability to plants for secondary metabolism, as regulated by rates of photosynthetic carbon fixation, may indeed affect rates of production of secondary metabolites.

Bryant et al. (1983) proposed that the ratio of carbon to nutrients available to plants affects patterns of allocation of these elements to growth and to secondary metabolism. Basically, the theory predicts that, in situations where nutrients (generally nitrogen supply) limit photosynthesis and growth, excess carbon is produced and plants can synthesize larger quantities of carbon-based secondary metabolites (compounds lacking nitrogen). Consistent with this theory is the observation that shaded plants or plant parts, presumably with lower photosynthetic rates, often contain lower levels of carbon-based metabolites than do plants or plant parts with greater light exposure (Waterman and Mole, 1989; Bryant, 1987; Zangerl & Berenbaum, 1987). Conversely, when plants are not nutrient-limited, there is little excess carbon for production of secondary metabolites; rather, nitrogen-containing secondary compounds can be produced. The production of volatile terpenes in *Heterotheca subaxillaris* (Asteraceae), for example, is inversely related to nitrate availability (Mihaliak and Lincoln, 1985). Briggs (1990) demonstrated that *Lotus corniculatus* plants, when fertilized with nitrogen, produced higher tannin levels than did plants that were not fertilized and were therefore dependent entirely on symbiotic bacterial N-fixation. Maintaining N-fixing symbionts in these plants consumes almost 20% of an individual's carbon budget. Nitrogen fertilization, however, did not affect production of cyanide, a nitrogen-containing secondary metabolite.

The existence of opportunistic incorporation of available elements is illustrated by a comparison of secondary metabolites from terrestrial and marine plants. One of the more pronounced differences is the comparative ubiquity of halogen-containing compounds in marine plants (Bakus et al., 1986), in comparison with terrestrial plants. Sea water is rich in bromine, chlorine, and other halogens in comparison with fresh water or the lithosphere (Table 2.1); the frequency with which halogenated compounds occur in marine plants may reflect opportunistic incorporation (Richardson, 1977).

Phytochemical opportunism may also be responsible for "relict biochemistry," or the presence of unusual elements in primitive taxa, such as vanadium in tunicates (Pirie, 1959). At the time of its evolutionary origin, an organism may have developed a dependency upon an element that was relatively abundant in the atmosphere or lithosphere. Such relict species may be rare in contemporary floras and faunas due to selection pressure against organisms whose demand for certain rare elements exceeded their availability as environmental composition changed.

Herbivorous insects are almost entirely dependent upon their food-plants for their nutritional intake. While some species can obtain certain elements from microbial symbionts, such as termites with nitrogen-fixing endosymbionts (Prestwich et al., 1988), and others can actually obtain limiting elements directly from the environment, such as "puddling" butterflies that take up sodium ions from mud puddles (Adler and Pearson, 1982), these are the exceptions rather than the rule. The elemental composition of arthropod defensive secretions (exclusive of proteinaceous venoms, which can arguably be called primary metabolites if they are used for food procurement in addition to defense) is by and large restricted to C, H, O, and N. Sulfides, for example, are restricted in distribution to only two genera of ponerine ants (Blum, 1981).

Phytochemical opportunism is responsible for a considerable portion of the diversity in secondary metabolism in herbivorous insects in that many species sequester secondary compounds (or their metabolites) from host plants. Sequestration of plant secondary compounds—deposition into specialized tissues for the purpose of detoxification—has been documented in no fewer than six orders of arthropods (Rothschild, 1972). Sequestration is generally selective, with retention of ingested plant secondary metabolites dependent upon such physicochemical properties as polarity, size, structure, solubility, toxicity, and transportability of the molecules. Such sequestration is dependent upon the evolution of insensitivity in the insect to the toxic effects of these plant secondary metabolites (Bowers, this volume).

3.2. Primary vs Secondary Metabolism

3.2.1. In Plants

Primary and secondary metabolism are not always easily distinguishable in plants. Angiospermous plants have based biosynthetic pathways for both primary and secondary metabolism on relatively few "building blocks" (Waterman and Mole, 1989). The number of possible pathways in plants is limited, to a certain extent, by phylogenetic history; ancestry determines, at least in part, the metabolic machinery available for selection.

Despite the biochemical diversity exhibited by plants, it has been suggested that the earliest living cells required only about 30 biomolecules, including amino acids, purine and pyrimidine bases, sugars, fatty acids, glycerol, and choline (Lehninger, 1972). All secondary metabolism derives ultimately from pathways involved in primary physiological processes, such as respiration or photosynthesis (Figure 3.1). Ohno (1970) proposed that secondary metabolic pathways arise via duplication events for genes involved in primary pathways and subsequent mutation of the doubled genes.

In general, secondary metabolism begins with the production of glucose 6-phosphate, resulting from the pentose phosphate cycle, and its subsequent glycolytic breakdown. Acetyl CoA is produced by oxidative decarboxylation of the glycolysis product pyruvate; it participates directly in the biosynthesis of fatty acids and polyketides. In plants and bacteria, malonyl units are attached to acetate in sequence until a chain length of 16 carbons (palmitate) is reached. The C-16 acid is then elongated to C-18. In eukaryotic organisms, the C-18 acid (as stearyl ACP) is oxidized to oleic acid, an unsaturated fatty acid. Acetyl CoA also participates in the biosynthesis of amino acids via the tricarboxylic acid cycle.

Condensation of three acetyl CoA molecules yields mevalonic acid, which, via decarboxylation, gives rise to two interconvertible 5-carbon isoprenoid units, isopentenyl pyrophosphate (IPP) and dimethylallylpyrophosphate (DMAPP), that form the basis of terpene synthesis. Terpenes are the largest known group of secondary metabolites and are found in all organisms. Monoterpenes, C-10 compounds derived from condensation of one unit of DMAPP and one of IPP, are widespread and probably occur in all plants and fungi. Included among the terpenoid constituents of plants are primary metabolites such as the C-20 diterpene phytohormone gibberellic acid, triterpene-derived sterols such as sitosterol and stigmasterol, and C-40 tetraterpene accessory pigments such as the carotenoids and xanthophylls. In addition, there are many well-known and widely distributed secondary metabolites from all terpene classes in many plant taxa.

Glycolysis of glucose-6-phosphate yields the 3-carbon compound phosphoenol pyruvate, which can combine with erythrose-4-phosphate (another product of the pentose phosphate shunt) to form shikimic acid, a key intermediate in both primary and secondary metabolism. Shikimic acid is a precursor in the formation of aromatic amino acids, which serve as primary metabolites. In addition, these amino acids are themselves precursors in the biosynthesis of cyanogenic glycosides, glucosinolates, and certain alkaloids, as well as intermediates via deamination in the synthesis of secondary metabolites such as phenylpropanoids, lignans, and coumarins.

3.2.2. In Insects

Primary metabolism in herbivorous insects is, as is the case for plants, inextricably intertwined with secondary metabolism. For example, much insect secondary metabolism derives from amino acid metabolism, such as quinone synthesis from tyrosine, formic acid from serine, other organic acids from valine and isoleucine, and alkyl sulfides from methionine (Blum, 1981).

The ubiquity of particular classes of arthropod secondary metabolites may be directly related to their similarity to primary metabolism; the less derived an end product, the more likely it is to be synthesized by a wide variety of taxa. Benzoquinones are widespread in arthropod defensive secretions; they are known from over 100 species in seven orders (Rodriguez and Levin, 1985). That they are ubiquitous may be attributable to the fact that all arthropods synthesize quinones in the process of producing a sclerotized exoskeleton; cuticular tanning involves the formation of cross-links between proteins and ortho-quinones derived from tyrosine (Chapman, 1971). Hydrocarbon compounds are found in no fewer than eight arthropod orders (Table 3.2); this widespread distribution may reflect the fact that many arthropods have developed hydrocarbon biosynthetic pathways for cuticular waterproofing (Blum, 1981).

The absence of a highly developed biosynthetic pathway may restrict the distribution of particular classes of secondary metabolites in arthropods. Insects are dependent upon dietary sterols for the production of primary metabolites containing a cholestane nucleus (e.g., ecdysone). Although many species can produce mono-, sesqui-, or di-terpenes from mevalonic acid (as do plants), they appear to lack enzymes that cyclize squalene oxide en route to cholesterol (Blum, 1981). Production of steroidal defensive compounds in arthropods is restricted to several families in the order Coleoptera (Blum, 1981). Among these families, two (Dytiscidae and

Table 3.2. Distribution of secondary metabolites (chemical defenses) in arthropods.*

	Hydro-carbon	Alcohol	Aldehyde	Ketones	Carboxylic Acid	Quinones	Esters	Lactones	Phenols	Steroids	Alkaloids
Arachnida											
Phalangida	+	+	+	+	−	+	−	−	+	−	−
Uropygi	−	−	−	−	+	−	−	−	−	−	−
Crustacea											
Isopoda	−	+	−	−	−	−	−	−	−	−	−
Diplopoda	−	−	+	−	+	+	+	−	+	−	+
Chilopoda	−	−	−	−	+	−	−	−	−	−	−
Insecta											
Dictyoptera	−	+	+	+	+	+	−	+	+	−	−
Isoptera	+	+	+	+	−	+	+	−	−	−	−
Phasmida	−	−	+	−	−	−	−	−	−	−	−
Orthoptera	−	−	−	+	−	+	−	+	+	−	+
Dermaptera	−	−	−	−	−	+	−	−	−	−	−
Hemiptera	+	+	+	+	+	−	+	−	−	−	−
Neuroptera	+	−	−	−	−	−	−	−	−	−	−
Coleoptera	+	+	+	+	+	+	+	+	+	−	+
Lepidoptera	+	+	−	+	+	−	+	−	−	−	+
Hymenoptera	+	+	+	+	+	−	+	+	−	−	+
Trichoptera	−	−	−	−	−	−	−	−	+	+	+

*From Blum (1981).

Lampyridae) are entirely carnivorous and may derive the cholestane skeleton from ingested prey. In the remaining family, the Chrysomelidae, production of autogenous steroidal defensive compounds is restricted to the genus *Chrysolina*.

3.3. Phylogenetic Constraints on Secondary Metabolism

3.3.1. In Plants

Although it is impossible to know with any certainty the chemical characteristics of early land plants ancestral to contemporary angiosperms, it is not unreasonable to assume that these early land colonists were of algal origin. Prokaryotic organisms (bacteria and blue-green algae) are unlikely ancestors. Contemporary representatives produce a relatively small number of compounds other than primary metabolites. These secondary metabolites include polyketides, peptides, amino acid derivatives, fatty acids (including monoenes), and some terpenoids; many of the more conspicuous secondary metabolic pathways of plants (e.g., flavonoid biosynthesis) are completely absent in prokaryotic organisms (Table 3.3). The relative rarity of cyclized oxygen-containing terpene derivatives (oxidized sterols and xanthophylls) in prokaryotes suggests that terpene synthesis in these organisms evolved under conditions in which oxygen was less available than at present (Swain and Cooper-Driver, 1981).

The chemistry of extant algal groups may provide some insight into the chemical composition of ancestral algal groups. Contemporary brown (Phaeophyceae), red (Rhodophyceae), and green (Charophyceae) algae can produce a variety of secondary metabolites. These include terpenoids, fatty acids other than common C-16 or C-18 types, simple nitrogenous compounds derived from amino acid pathways, polyketides, and some simple phenolic compounds.

Tri- and tetra-terpenoids are widespread in algae, as are steroids; while sesqui- and di-terpenes are common, monoterpenes are by comparison rare (Bakus et al., 1986; Paul and Fenical, 1987; Paul et al., 1987; Hay, 1987). Both oxygenated and non-oxygenated carotenoids (tetraterpenoids) are also found in many algal taxa, although carotenoids from marine algae tend to be more complex and variable than those from terrestrial algae (Bakus et al., 1986). Based on the widespread occurrence of terpenes in extant algal groups, early land plants almost certainly had the ability to synthesize a similar spectrum of compounds. At least some of these terpenoids were involved in primary physiological processes; many extant primitive plants, for example, possess the hormones abscisic acid (a sesquiterpene) and gibberellic acid (a diterpene) (Swain and Cooper-Driver, 1981), and carotenoids in general serve as accessory pigments and antioxidants in photosynthetic organisms.

Table 3.3. Taxonomic distribution of secondary metabolites in non-animals*

Taxon	Mono-terpene	Sesqui-terpene	Diterpene	Triterpene	Tetra-terpene	Lunularic Acid	C-glycosides	Flavones	Flavonols
Bacteria				+	+				
Algae	(+)	(+)	(+)?	+	+	+			
Fungi	(+)	(+)	+	+	+				
Bryophytes		+		+	+	+	+		
Tracheophytes									
Psilopsida				+	+				
Lycopsida				+	+		+	+	
Sphenopsida				+	+		(+)	+	+
Pteropsida									
Filicinae		(+)	+	+	+		+	+	+
Gymnospermae	+	+	+	+	+		+	+	+
Angiospermae	+	+	+	+	+	(+)	+	+	+

*From Hegnauer (1969–1973); Swain (1978).
+ = present throughout the taxon.
(+) = reported in only a few species.

While alkaloids, condensed tannins, and lignins, all secondary metabolites associated with terrestrial plants, are unknown in all algal groups, some green algae, in particular *Chara* and *Nitella*, possess the ability to synthesize and accumulate flavonoids, which are otherwise known only from higher (terrestrial) plants.

Land plants did not appear until the Silurian period; these early land plants were probably descended from intertidal or amphibious algae (Swain and Cooper-Driver, 1981). Plants colonizing terrestrial environments faced several physical challenges. A major abiotic stress in terrestrial environments is ultraviolet radiation (UV), which is damaging in many ways to DNA, proteins, and many cofactors (Berenbaum, 1988). These wavelengths are largely attenuated by water and thus probably did not effect significant mortality in aquatic organisms. Flavonoids and carotenoids characteristic of vascular plants and charophyte algae (*Chara* and *Nitella*) may have served as sunscreens for early land plants. The flavonoids of bryophytes are relatively complex and resemble those of many vascular plants (Swain and Cooper-Driver, 1981). The bryophytes share *C*-glycosyl flavonoids with most other primitive plant groups as well as a number of angiosperms (Table 3.2). The presence of phenolic pigments in the cuticle of an extant psilophyte (*Tmesipteris*) (Caldicott et al., 1975) may indicate that these compounds were important as UV screens in early land plants as well.

Desiccation was another distinct abiotic challenge for hitherto exclusively aquatic taxa. One possible early step in colonization of terrestrial habitats was the evolution of desiccation-resistant spores surrounded by sporopollenin, a waterproof polymer. While sporopollenin comprises less than half the content of spore walls of chlorophyte algae and fungi, it

Table 3.3.   (continued)

Tri-OH	O-CH$_3$ C-CH$_3$	Antho-cyanins	Iso-flavonids	Pro-antho-cyanins	Coumarins	Iridoid-glycosides	Gluco-sinolates	Cyanogenic-glycosides	Hydrolyzable tannins
					(+)				
					(+)				
	(+)			+					
+		(+)		+					
+	+	+		+	(+)			+	
+	+	+		+	(+)			(+)	
+	+	+	+	+	+	+	+	+	+

comprises over 90% of the spore walls of lycopods and undoubtedly prolongs the viability of such spores in xeric conditions (Swain and Cooper-Driver, 1981). Vegetative parts gained protection with a waxy cuticle, poorly developed in bryophytes but extensive in even primitive tracheophytes (Knoll, 1986). Cutin, the primary constituent of the waxy cuticle, characteristically is found overlaying the epidermis of aerial parts of all land plants. Cutin synthesis derives directly from primary fatty acid metabolism via ω-hydroxylation of C16 and C18 fatty acids, over and above normal hydroxylation at C9 or C10 (Swain and Cooper-Driver, 1981). The structure of cuticular components in primitive plants, however, differs from that of more advanced taxa. The cutin of mosses, lycopods, and ferns characteristically consists of hydroxyhexadecanoic or octadecanoic acids, rather than mono- or di-hydric alcohols characteristic of higher plants (Swain, 1978). In addition, these early land plants likely possessed foliage that contained a series of low molecular weight volatile compounds based on acetate metabolism, including ketones, aldehydes, alcohols, acids, and hydrocarbons derived from metabolized fatty acids.

Rigidity and support were also problematical on land, where gravitational forces are exerted to a greater extent than in water. Rigidity was particularly important for successful water transport in terrestrial systems. A system of rigid conducting cells, or tracheids, facilitated active water transport throughout the plant. These cells owed their rigidity to lignins, products of phenylpropanoid biosynthesis. Lignins arise by polymerization of substituted cinnamic acid derivatives. P-coumaric acid derivatives form the basis for psilopsid lignins, while lycopod lignins, like lignins of higher plants, consist of methoxylated cinnamic acid monomers such as ferulic and sinapic acids (Swain, 1978). By the Mid-Devonian (387 mya), the dominant plant groups—protogymnosperms, lycophytes, and ferns—in all probability produced lignin.

Thus, abiotic stresses associated with terrestrial environments, such as desiccation, gravity, and ultraviolet radiation, may well have conferred a selective advantage on primitive plants with the ability to synthesize a variety of secondary metabolites—terpene and hydrocarbon derivatives for waterproofing, phenylpropanoid derivatives for rigidity and support, and flavonoids and phenolics as ultraviolet radiation shields. Concomitant with abiotic mortality factors in terrestrial environments were biotic mortality factors such as microbes, fungi, and, beginning in the Devonian, arthropods. There may have been strong selection to retain metabolic pathways whose products conferred resistance not only to physical mortality factors but to biological mortality factors as well. Biocidal properties of secondary metabolites may have been exaptive or adaptive; in either case, their survival value no doubt contributed to their retention through evolutionary time. Plant fossils dating to the Devonian show evidence of fungal hyphae and mycelia (Kevan et al., 1975); some Devonian *Rhynia* fossils have sporangia which include arthropod fragments.

Swain (1978) has speculated that the evolution of aromatic acid acylation of cell-wall polysaccharides may have provided protection to early land plants against both biotic and abiotic mortality factors. Cell-wall polysaccharides in some forms of algae and bacteria are acylated with sulfur or acetic acid; presumably, celluloses and hemicelluloses acylated in a novel fashion by phenolic acids in plants were more resistant to coeval fungal acyl sulfatases and acetylases. At the same time, aromatic acid acylation of polysaccharides may have facilitated the evolution of lignification; cross-linkage of adjacent aromatic acid substituents (e.g., ferulyl dimers) in the polysaccharide matrix may have initiated aromatic (substituted cinnamic) acid polymerization to form lignin.

The ability to produce 3-hydroxy substituted flavonoids, characteristic of sphenopsids as well as higher plants, greatly increased the diversity as well as the utility of flavonoid constituents. For example, 3-hydroxylation made possible the biosynthesis of condensed tannins, long-chain polymers of pro-(leuco-)anthocyanidins. These highly hydroxylated compounds are capable of binding to proteins (via H-bonds) and thus are toxic to a wide variety of organisms, including bacteria, fungi, viruses, and insects (Swain, 1977). Among the dominant plant groups of the Carboniferous, condensed tannins were probably produced by sphenopsids (Swain, 1979) but not by Psilopsida and Lycopsida; no modern members of these groups produce tannins. Protogymnosperms most likely contained condensed tannins (Swain and Cooper-Driver, 1981), which were probably influential in the successful establishment of woody plants, particularly with respect to resisting microbial attack. Swain and Cooper-Driver (1981) speculate that tree lycopods may have succumbed to attacks of pathogenic fungi in the early Carboniferous, whereas tannin-containing protogymnosperms of similar size survived.

Critical to the successful establishment of early plants in terrestrial environments, then, were high-molecular-weight complex polymers, such as lignin and condensed tannins. As plants evolved, the macromolecules of relatively primitive plants were replaced by or supplemented with biosynthetically-related micromolecular compounds. Contemporary gymnosperms manufacture large quantities of tannins, stilbenes, lignans, lignins, and resinous diterpenes as products of secondary metabolism, whereas representatives of many families of angiosperms produce relatively small quantities of such derivatives of phenylpropanoid metabolism as coumarins, cyanogenic glycosides, and alkaloids.

Even within the angiosperms, there are phylogenetic shifts in secondary metabolism, from macromolecular to micromolecular metabolites. Whereas condensed tannins are largely restricted to primitive woody taxa, they are relatively rare among advanced primarily herbaceous taxa (Bate-Smith, 1962); in contrast, low-molecular-weight flavonoid derivatives, such as aurones and flavanones, are restricted to more advanced angiosperms (Hegnauer, 1969–1973). The same is true for low-molecular-weight derivatives of lignin biosynthesis intermediates; primarily herbaceous (non-lignified) angiosperm families, such as the Apiaceae and Asteraceae, produce a profusion of coumarin derivatives, including furanocoumarins and pyranocoumarins, absent or rare in more primitive angiosperms (Berenbaum, 1991). While alkaloids—secondary metabolites that are effective biocides individually rather than as polymers—are known from a few primitive taxa (e.g., modern *Lycopodium* and *Equisetum*), their diversity and ubiquity are greatest in the angiosperms.

Whereas large polymers may have been well-suited as protective agents against abiotic stresses, these smaller molecules may have provided superior protection against biotic mortality factors. In general, more advanced groups of defensive compounds can undergo rapid turnover and presumably could be produced at lower cost than compounds such as lignins, tannins, and resinous terpenes (Swain, 1979). The fact that the biosynthetic precursors of many of these smaller molecules are intermediates in the biosynthesis of polymeric compounds (Figure 3.1) suggests that secondary metabolism in early land plants may have been constrained by energetic or toxicological considerations.

3.3.2. In Insects

Phylogenetic trends in secondary metabolism in arthropods are not as readily discernible as they are in plants. Secondary metabolism is not a universal characteristic in all arthropod taxa (Table 3.3). Among chelicerate arthropods, secondary metabolism is relatively rare. Phalangids are

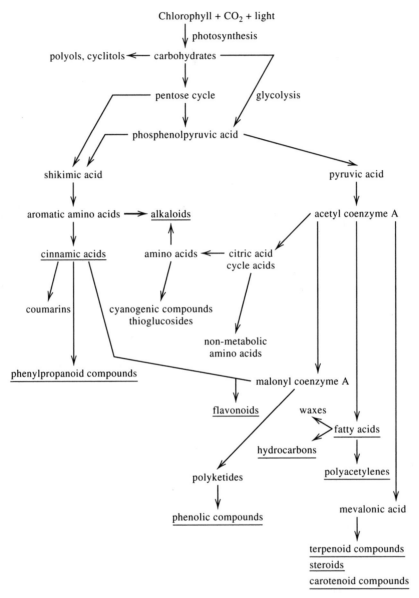

Figure 3.1. Products of secondary plant metabolism. Modified from Geissman and Crout (1969).

unique among arachnids in the diversity of secondary metabolites produced within the order. Venom production within the Arachnida cannot strictly be considered secondary metabolism because venoms are used not only for defense but also for prey capture. Among the mandibulate arthropods, secondary metabolism is well-developed in diplopods, chilopods, and insects; the aquatic crustaceans are not particularly known for secondary metabolite production, and the terrestrial pauropods and symphylans are simply not well-known at all; the absence of records of secondary metabolism likely reflects a lack of inquiry. Certain biosynthetically-related secondary metabolites are known from a broad cross-section of the phylum; hydrocarbons, for example, are reported from Opiliones in the Arachnida as well as from eight orders of insects (Blum, 1981) (Table 3.3). In contrast, monoterpene production appears limited to insects. While quinonoids are characteristic of a diversity of arthropods, there are phylogenetic differences in the biosynthesis of these secondary metabolites; whereas millipedes are dependent upon amino acids for preformed aromatic rings in the synthesis of quinonoids, many insects can produce alkylated quinones via the acetate-malonate pathway (Blum, 1981).

Within the class Insecta there are some conspicuous absences of chemical defense that may be of phylogenetic significance. Chemical defenses are apparently absent in the apterygote orders Protura, Collembola, Diplura, and Thysanura. The lack of reports from these orders may reflect a relative paucity of investigations, since some of these hexapods are tiny, soil-dwelling, and relatively inconspicuous; the absence of records from the more prominent collembolans and thysanurans is not as easily dismissed as the result of a lack of interest. In addition, chemical defenses are conspicuously absent from the paleopterous insect orders Odonata and Ephemeroptera. The absence of obvious chemical defenses from the more primitive orders, and, correspondingly, the ubiquity of chemical defenses in several of the indisputably more recent orders (Hymenoptera, Lepidoptera) suggest that ancestral hexapods may have lacked chemical defenses and their descendants developed them independently from other arthropod groups.

The distribution of secondary metabolism among insects (and among herbivorous insects in particular) may be more influenced by ecology than by phylogeny. Generally, chemically-defended species are long-lived (and thus more likely to encounter predators), lacking alternate defense mechanisms (e.g., saltatory locomotion), and conspicuous (e.g., not concealed or internal feeders of plant tissues) (Pasteels et al., 1988). In that sociality increases conspicuousness by virtue of numbers, it is not surprising that secondary metabolism is highly developed in social species. By the same token, ectoparasites, which are unlikely to encounter predators in large numbers or in high frequency in their natural habitats, as a group appear to lack chemical defenses. The parasitic orders Mallophaga, Anoplura, and Siphonaptera are conspicuously absent, for example, from Table 3.3. One possible selection pressure leading to the proliferation of chemical defenses in insects may have been colonization of terrestrial habitats and the evolution of wings (and flight). While flight improved food-finding capabilities and dispersal ability, allowing insects to exploit a tremendous variety of new habitats, it may also have increased conspicuousness to predators. Whereas early terrestrial insects (e.g., Protodonata) may have relied on rapid flight to elude predators, chemical defenses may have evolved in slower, smaller organisms. Low-molecular-weight chemical signals may have been too inefficient in aquatic systems to be functional as predator deterrents; diffusion coefficients of small molecules, with a molecular weight of less than 300, tend to be 1,000 to 10,000 times greater in air than in water (Wilson, 1970).

3.4. Energetic Constraints on Secondary Metabolism

3.4. In Plants

Although a plant may possess the biosynthetic capacity to manufacture secondary metabolites, expression of a particular biosynthetic pathway may be contingent upon ecological conditions. A topic of extensive (and often acrimonious) debate in recent years is whether or not production of secondary metabolites entails an energetic cost to the plant (see Rausher, Sabelis and Dicke, this volume). Ostensibly, chemical energy is required to synthesize, transport, and store secondary metabolites, as well as to synthesize and transport enzymes involved in their biosynthesis (Chew and Rodman, 1979).

There is certainly suggestive evidence in support of such costs. Negative genetic correlations can be cited, for example, between yield (growth or seed production) and secondary metabolite content (Zangerl and Bazzaz, 1991). Such negative genetic correlations have been documented between alkaloid production and yield in tobacco (Matsinger et al., 1989; Baldwin et al., 1990) and furanocoumarin production and potential seed production (number of secondary rays) in wild parsnip (Berenbaum et al., 1986). Studies documenting a negative correlation between growth or dry weight and secondary metabolite production include terpenes in *Pinus monticola* (Hanover, 1966), cyanogenesis in *Trifolium repens* (Foulds and Grime, 1972), pyrrolizidine alkaloid production in *Senecio jacobaea* (Vrieling, 1990), alkaloids in *Phalaris arundinacea* (Ostrem, 1987), and tannins in *Cecropia peltata* (Coley, 1986). In *Lotus corniculatus*, seed production is associated with a decrease in tannin and cyanide production (Briggs and Schultz, 1990).

Additional evidence of an energy requirement for secondary metabolite production can be found in the form of linkages between photosynthetic activity and secondary metabolism. Reduction of photosynthesis by shading in foliage of wild parsnip also reduces furanocoumarin production (Zangerl and Berenbaum, 1987); reduced light intensity similarly lowers production of glandular trichome exudate in alfalfa, and increasing day length increases production of 2-tridecanone in tomato foliage (Kennedy et al., 1981).

Within-plant distributions of secondary metabolites are generally consistent with allocation patterns imposed by tissue value. Mooney and Gulman (1982) suggested that the value of a leaf to a plant is equivalent to its relative contribution to carbon fixation as well as to its replacement cost. Many plants produce higher quantities of secondary metabolites in young foliage, which has higher value than does older foliage (Mihaliak and Lincoln, 1985). In reproductive tissues as well, secondary metabolite production is commensurate with replacement costs. McKey (1974) noted that tomatine content of tomato ovaries increases after pollination, a pattern consistent with allocation of secondary metabolites according to tissue value. Nitao and Zangerl (1987) demonstrated that removal of developmentally more advanced floral units in *Pastinaca sativa* had a progressively greater impact on fitness; concomitant with this increase in tissue value was an increase in furanocoumarin content of the floral units. Even within a single life stage, more advanced floral units contained greater amounts of furanocoumarin.

3.4.2. In Insects

Very little information is available on the magnitude (or existence) of energetic costs associated with either production of autogenous defensive compounds or sequestration of plant-derived compounds in herbivorous insects. It has been suggested (Rowell-Rahier and Pasteels, 1986) that the reason sequestration of plant secondary metabolites is not more widespread among arthropods than it appears is that sequestration effects a metabolic toll on the sequesterer. Little evidence for such a constraint, however, is available. In species that sequester cardenolides from asclepiadaceous hosts, several investigators failed to demonstrate physiological costs of sequestration (Smith, 1977; Isman, 1977; Dixon et al., 1978). The fact that *Oncopeltus fasciatus* (large milkweed bug) sequesters cardenolides largely by a physical, rather than metabolic, processes suggests that costs may be minimized in such a fashion by sequestering species (Blum, 1981). Sequestration of plant secondary metabolites may even represent an energetic saving over biosynthesis of autogenous defense compounds, but definitive data are few and far between (Pasteels et al., 1988).

Another potential cost savings for sequestering species may be in the form of reduced need for detoxicative metabolism. Inducibility of detoxification enzymes has been suggested to have evolved as an energy-saving process (Harshman et al., 1991) in that oxidative and other forms of metabolism are energy-consumptive. Attempts to quantify actual costs of detoxicative metabolism, however, are few (Berenbaum, 1986; Neal, 1987).

3.5. Toxicological Constraints

Aside from energetic constraints, there may also be toxicological constraints on production of secondary metabolites. Many secondary metabolites act on general cellular targets; furanocoumarins, for example, cross-link DNA, thiophenes disrupt membranes, and nonprotein amino acids interfere with protein synthesis and function (Berenbaum, 1988). Quinones, produced by both plants and insects, bind nonspecifically to proteins and amino acids and thus can interfere with enzyme synthesis and function (Rodriguez and Levin, 1975). The production of such biocidal compounds in living tissue may therefore present storage problems (Fowden and Lea, 1979).

Many plant secondary metabolites are extruded onto plant surfaces (e.g., Zobel and Brown, 1990) and thereby present no toxic threat to plant cells or tissues. Toxic substances are also present in some plant tissues in insoluble form, as is the case for plants producing raphides of calcium oxalate. Polymerization can also reduce the activity of biocidal metabolites, as is the case with tannins in ripening fruits (Goldstein and Swain, 1963).

In all likelihood, the most common solution to problems of autotoxicity in plants is localization of toxic materials in special storage organs or organelles. The size and construction costs of such storage depots may impose an upper limit on the amount of secondary metabolism possible. In *Pastinaca sativa* (parsnip), for example, furanocoumarins in seeds are restricted to vittae (oil tubes) on the seed coat; positive genetic correlations exist between the size of vittae and the amount of furancoumarin produced (Zangerl et al., 1989). Selection for increased furanocoumarin production, then, is genetically linked to increased vitta size, a trait that exhibits additive genetic variation (heritability = 0.50; Zangerl et al., 1989). Similarly, the number of glandular hairs, in which terpenes are sequestered, is a heritable trait in several other species (Mehlenbacher et al., 1984).

The toxicity of sequestered or compartmentalized plant metabolites may be reduced further by physical separation of precursors of toxic products. Many cyanogenic plants cannot produce hydrogen cyanide unless cells or tissues are damaged; glucosides liberate HCN only after contact with cellular β-glucosidases, which are physically segregated from their substrates in intact cells (Seigler, 1991).

Insects face many of the same storage problems as plants in terms of secondary metabolite production. One of the evolutionary innovations that undoubtedly contributed to the proliferation of chemical defenses in arthropods, in comparison with annelid ancestors, was the evolution of the sclerotized cuticle. Not only did a rigid waterproof cuticle facilitate the colonization of terrestrial environments by facilitating locomotion and reducing the risks of desiccation (Hinton, 1976), it also facilitated the retention of toxic substances within the body. An "especially impervious integument" (Eisner, 1970) may also have permitted arthropods to withstand the effects of their own secretions after discharge. It is unlikely, for example, that the highly exothermic explosive reaction by which *Brachinus* species (bombadier beetles) produce their quinonoid defensive secretions could have evolved in any taxon without a "reactor gland" lined with relatively nonreactive material.

3.6. Biotic Selection Pressures: Interactions That Decrease Plant Fitness

The structure and bioactivity of secondary metabolites in any given species are likely to reflect, to some degree, differing (and possibly conflicting) selection pressures. Whereas in insects, secondary metabolites are universally regarded as defensive in function (generally against predators, although some, particularly in social species, may be antimicrobial in function; Duffey, 1976), secondary metabolites are thought to serve a variety of functions in plants; some of these functions are of an internal regulatory nature (regulation of plant growth, energy or material storage, detoxification, or transport facilitation; Chew and Rodman, 1979). There is a general consensus, however, that at least some of the selection pressures affecting secondary metabolism may be the result of interactions with biotic agents (see Rausher, this volume). These interactions involve not only deterrence of detrimental organisms (e.g., pathogens or herbivores) but also attraction of desirable organisms (e.g., those involved in pollination or seed dispersal). These diametrically opposed ecological functions can at times operate at cross-purposes.

Plant secondary metabolites have been demonstrated repeatedly to deter or destroy herbivores and pathogens (Rosenthal and Janzen, 1979), and insect defensive secretions are well-known to deter both vertebrate and invertebrate predators (Eisner, 1970; Blum, 1981). In order to be effective against an enemy, a toxin or deterrent must have certain properties that permit it to interact with target sites. These include properties that promote uptake or adsorption of a molecule, interaction with a receptor site (e.g., by hydrogen bonding, or by hydrophobic or electrostatic interactions), and reaction (transduction, action potential, or biochemical lesion).

While some toxins interact with specific receptors (e.g., nicotine and acetylcholine receptors, phytoestrogens and estrogen receptors), others interact with macromolecules that are present in a wide variety of organisms and are thus broadly biocidal. Cyanide, produced by both plants and insects, exerts its acute toxic effect by combining with metalloxyporphyrin-containing enzymes such as cytochrome oxidase (Poulton, 1983). Because these enzymes are almost universal eukaryotic cell constituents, hydrogen cyanide has a general toxic effect on most living organisms. Quinones, also produced by both plants and insects, complex nonspecifically with proteins and disrupt enzyme function. The similarities between plant and insect secondary metabolites (to the extent that both taxa in many instances produce the same compounds) may reflect similar selection pressures in that both groups of organisms are susceptible to predation by invertebrate and vertebrate predators (Rodriguez and Levin, 1975).

If there is a difference between plant and insect defensive compounds, it is likely due to the fact that plants, as modular organisms, can withstand a substantial amount of tissue loss to a predator without dying; with the possible exception of butterfly wings, few, if any, body parts of arthropods are expendable. Accordingly, arthropod defensive secretions may be more acutely toxic or irritating than plant defensive secretions, so as to deter predation before major damage is inflicted.

Herbivores and pathogens may act as selective agents on plant chemistry by acting as differential mortality agents. There is considerable evidence that plant pathogens differentially infect genotypes within a host species and that differential susceptibility is associated with genetically controlled differences in chemistry (Day, 1974). Evidence that herbivorous insects can distinguish among genotypes within a species based on chemistry is less compelling; there are, however, studies that have demonstrated differential resistance that is both chemically based and genetically controlled (and therefore available for selection) (Berenbaum et al., 1986).

Irrespective of the importance of insects or other herbivores as selective agents on plants and, by extension, plant chemistry, plant chemistry can certainly, by causing differential mortality, act as a selective agent on herbivore behavior and physiology. Resistance to plant toxins takes many forms in insects; these include biochemical metabolism, target site insensitivity, and behavioral resistance (Brattsten and Ahmad, 1986; Isman, this volume). Indeed, one of the principal systems for detoxifying plant chemicals in most herbivores are the cytochrome P450 monooxygenases. These broadly substrate-specific, membrane-bound enzymes effect oxidative reactions that in general render lipophilic compounds more water-soluble; this conversion allows an organism to excrete a potentially toxic substance, as well as reduces the probability that an ingested lipophilic compound can bind to or interfere with lipid-rich organ systems. Approximately 400 million years ago, based on sequence data from extant forms, a tremendous explosion occurred in the diversity and abundance of forms of cytochrome P450. This explosion may have been due to novel encounters between terrestrial plants and animals (Gonzalez and Nebert, 1990). This date also corresponds to the appearance in the fossil record of neopterous insects, the contemporary descendants of which produce a wide variety of secondary metabolites. Development of detoxicative metabolism may have been accompanied by a proliferation of biosynthetic metabolic pathways.

Cytochrome P450 metabolism represents only one form of resistance manifested by herbivores to plant chemicals. Other detoxicative enzyme systems exist (Brattsten, 1979), as do genetically-based avoidance behaviors (Tallamy, 1985). Differences in resistance mechanisms among herbivores suggest that secondary metabolites are differentially effective against a spectrum of herbivores. Compounds that are toxins to generalist feeders may well serve as feeding stimulants for specialists with an enhanced capacity to metabolize those compounds (Fraenkel, 1959). For example, while xanthotoxin, a furanocoumarin, is toxic to the southern armyworm *Spodoptera eridania* at dietary concentrations of 100 ppm or less, it stimulates growth of the adapted specialist *Papilio polyxenes* at concentrations in excess of 1,000 ppm (Berenbaum, 1990). By the same token, while concentrations of 0.01% (100 ppm) angelicin, another furanocoumarin, in foliage reduce fecundity of *P. polyxenes*, they have no demonstrable effect on another specialist, the parsnip webworm *Depressaria pastinacella* (Berenbaum and Feeny, 1981; Nitao, 1989). Such differential efficacy of closely related compounds against herbivores may be one reason that individual plant species tend to produce series of biosynthetically related compounds.

3.7. Biotic Selection Pressures: Interactions That Enhance Plant Fitness

Since the vast majority of angiosperms are effectively sedentary throughout most of their lives, many rely upon biotic agents to transport both pollen and seed. In order to recruit such dispersal agents, attractants are often utilized. In order to be maximally effective as attractants, plant secondary metabolites must be detectable by the sensory systems of biotic agents. A plant is more likely to benefit from an association with a pollen dispersal agent if it produces a long distance attractant to which the pollen dispersal agent can respond. In general, signal compounds are more likely to be effective if they are relatively volatile—in the molecular weight range of 100–300. Many, if not most, essential oil components fall into this range. Once at a sensillum, molecules must have properties conducive to diffusion through pores to the dendrite. Finally, in order to effect a behavioral response, a signal chemical must interact with a receptor protein. Dipole-dipole interactions are important in allowing a compound to bind to a receptor. Thus, many plant volatiles that are used by insects for chemoorientation share certain structural characteristics, such as the presence of oxygen functionalities like aldehyde or ketone groups. That many plant volatiles share more than a passing resemblance to many insect sex pheromones may result from the common selection pressure to increase long-range attraction (Rodriguez and Levin, 1975).

Historically, the most important biotic selective agents on angiosperms may well be insects, not only by virtue of their ability to cause mortality, but also by virtue of their ability to enhance plant fitness by acting as pollen dispersal agents. Although there is evidence to suggest that insect pollinators can act as selective agents on floral morphology (Galen, 1989), there is less evidence to demonstrate that insect pollinators can act as selective agents on floral chemistry. Intra- and inter-specific differences in floral pigment chemistry have been attributed to selection pressure for maintaining appropriate pollinator associations. Sufficient selection pressure can effect changes in color phenotype distributions in as little as two generations (Baker and Hurd, 1968). Whereas *Penstemon centranthifolius* has open-tubed red flowers and *P. grinellii* has two-lipped blue flowers, *P. spectabilis*, a presumptive hybrid of the two species, has purplish-blue flowers; the maintenance of the hybrid populations is attributable to the fact that the three species are associated with different pollinators— *P. centranthifolius* with hummingbirds, *P. grinellii* with carpenter bees, and *P. spectabilis* with *Pseudomaris* wasps (Grant, 1971). In another example, scarlet gilia (*Ipomopsis aggregata*) shifts seasonally from dark to light flower colors coincident with hymmingbird emigration from and sphingid moth immigration to the area; sphingids preferentially visit light-colored flowers (Paige and Whitham, 1986).

Studies of interactions between orchids and euglossine bees lend credence to the idea that insect pollinators are important selective agents on floral volatile chemistry. Orchids pollinated by euglossine bees produce odoriferous compounds in glandular structures called osmophores. Male euglossine bees orient to these compounds and, once at the flower, actively collect fragrance compounds with the use of tarsal brushes. These compounds are then transferred to the hind tibia and are used in courtship and pheromone production (Dodson, 1975; Williams, 1982; Dressler, 1982).

In the process of collecting odoriferous compounds from male flowers, male euglossine bees trigger the release of a sticky pollinarium, which adheres to their bodies and can be carried to a female flower and dislodged into the stigmatic pocket, effecting pollination (Dodson, 1962). Although euglossine-pollinated orchids produce a wide variety of odoriferous compounds, each species produces a unique combination of fragrances (Dressler, 1982). Some constituents act as general attractants for many euglossine bee species, whereas others attract only a subset of potential pollinators, and still others act as repellents (Williams, 1982). Accordingly, each species of orchid attracts only a fraction of the euglossine species in an area. For example, populations of the orchid *Catasetum dilectum* have different floral chemistry in Panama and Ecuador. In Panama, cineole is the principal component of floral scent and is attractive to *Euglossa asarophora*, the major pollinator; in Ecuador, flowers produce primarily benzyl acetate, which is attractive to the local pollinator *Eulaema cingulata* (Hills et al., 1972).

Although the chemical constituents have not been identified, a scent polymorphism in sky pilot, *Polemonium viscosum*, an alpine plant of the southwestern U.S., has been attributed to pollinator selection pressures (Galen et al., 1987; Galen, 1989). Individuals of *P. viscosum* possess either a "sweet" or "skunky" odor. In lower elevations, "skunky" individuals predominate, whereas at higher elevations "sweet" individuals predominate. At lower elevations, most floral visitors are flies, which prefer "skunky" flowers; at higher elevations, over 75% of pollinator visits are from the bumble bee *Bombus kirbyellus*, which strongly prefers "sweet" flowers. "Sweet" flowers may be selected against at lower elevations, not only because of the reduced availability of bumble bee pollinators, but also because of the abundance of nectar-thieving ants, which are attracted to sweet flowers (Galen et al., 1987).

3.8. Coevolution and Chemistry

In a classic paper, Ehrlich and Raven (1964) suggested that the diversification of angiospermous plants, as well as the concomitant diversification of angiospermous plant chemistry, was largely the result of interactions with herbivorous insects. According to Ehrlich and Raven (1964), random genetic events leading to novel biosynthetic pathways and products could confer a selective advantage to a plant producing those products by reducing its losses to herbivory; released from constraints imposed by herbivores, taxa possessing the defensive novelty could undergo an adaptive radiation. Eventual evolution of resistance, by virtue of random genetic events that conferred a selective advantage on the herbivore, led in turn to adaptive radiation in herbivore taxa. Such adaptive radiations may have encompassed co-opting plant secondary metabolites (via sequestration) for self-defense. These stepwise radiations, generated by biochemical novelties, were proposed as the means by which angiosperms and insects have become the dominant life forms on the planet today.

It is entirely possible, however, that insect mutualists, rather than antagonists, of plants had as much (or more) to do with the diversification of secondary chemicals in plants, particularly angiosperms. Insect herbivory has been a continuous source of mortality for plants, since Devonian times, yet contemporary representatives of acient plant taxa (presumably under selection by insect herbivores for a considerably longer time than contemporary representatives of more recent taxa) lack the phytochemical diversity of mole modern taxa. It can be argued that primitive herbivorous insects perhaps did not exert sufficient selection pressure on primitive plants to lead to biochemical diversification; however, the appearance of such advanced taxa as the Coleoptera antedates the appearance of angiosperms in the fossil record.

The flower is "the single characteristic that sets the angiosperms off from all other groups" (Raven et al., 1976). It is not unreasonable to examine features of the angiosperm flower (and the mature ovary, or fruit) as the key to understanding the biochemical diversification of angiosperms. One of the most conspicuous features of the angiosperm flower is that in many cases it is, unlike most plant parts, designed to attract rather than repel insects. Although Kevan et al. (1975) suggested that insects or other arthropods may have served as dispersers of spores in Devonian times and may have been instrumental as selective agents promoting the evolution of heterospory, insects serve as biotic transport agents for angiosperm pollen and fruits on an evolutionarily unprecedented scale.

Attraction of appropriate pollinators and maintenance of pollinator fidelity (in order to ensure successive visits of conspecifics) may require greater chemical complexity than simply repelling or even killing herbivores that consume plant tissue. Opportunistic responses of herbivorous insects to plant volatiles may have then initiated cycles of stepwise evolution. Indeed, long distance olfactory orientation to host plants by herbivore, is known mostly from holometabolous orders (e.g., Coleoptera and Lepidoptera; Mitchell, 1988); the more ancient hemipteroid and orthopteroid orders, which underwent radiations prior to the evolution of the angiosperms, do not for the most part display directed movement in response to olfactory signals. These same holometabolous orders are involved in orientation to floral (or foliar) volatiles and in pollination. By the same token, many of the volatile constituents of angiosperms to which these holometabolous insects orient (Table 3.3) (mono- and sesquiterpenes, phenylpropanoids) are unknown or rare in more primitive plants. It is difficult to determine *a posteriori* if olfactory host orientation promotes dietary specificity or vice versa. The ability to respond to specific compounds detected at great distances actually makes host specificity a tenable enterprise, particularly if hosts are widely dispersed.

While Darwin considered the origin of the angiosperms an "abominable mystery," the biochemical diversity of the angiosperms is no less mysterious. At least three scenarios can be constructed to account for the tremendous explosion of phytochemical diversity in angiosperms.

Scenario 1. By virtue of their reproductive system, angiosperms experienced unprecedented success in colonizing habitats (or adopting lifestyles) hitherto unavailable to other taxa. A system for fertilization that was independent of water for sperm transport as well as the construction of water-resistant seeds to protect embryos from desiccation allowed angiosperms to establish in many xeric habitats, including arctic and alpine as well as desert regions. Occupying such habitats may have necessitated the elaboration of secondary metabolites to protect against abiotic mortality sources (e.g., high UV intensities in alpine or arctic habitats). While angiosperms are not unique in occupying such habitats (Raven et al., 1976), abiotic selection pressures may be sufficiently variable qualitatively as well as quantitatively (e.g., UV variation in intensity as well as wavelength) to result in diverse biochemical responses in plants.

Scenario 2. Biochemical diversity in angiosperms may well be the result of exposure to herbivory by arthropods with hitherto unprecedented selective impact. The selective impact of herbivores may have been greater when angiosperms were evolving than at any other time in the evolutionary history of plants for at least two reasons. On one hand, angiosperms may be more vulnerable by virtue of design. In monocarpic short-lived plants, destruction of reproductive parts (such as flowers) has immediate evolutionary consequences for a particular genotype. Enclosure of many seeds in a fruit (designed to be attractive to animals in the case of frugivore-dispersed species) leaves a genotype vulnerable to a significant loss of fitness if that fruit should be attacked by a herbivore. On the other hand, since their origin angiosperms have interacted extensively with more recent herbivores that, by virtue of physiology or behavior, may be able to exert greater negative selection than other herbivorous insect taxa. The Lepidoptera, for example, is a recent order that is almost entirely phytophagous and is associated to a large degree with angiosperms. Herbivory in this order is facilitated by sophisticated behavioral and physiological mechanisms for detecting and detoxifying chemicals (e.g., cytochrome P450 monooxygenases; Berenbaum and Isman, 1988), requiring the proliferation by angiosperms of novel and more potent toxins and deterrents.

Scenario 3. Angiospermous plants utilize biotic agents for sexual reproduction and dispersal to a hitherto unprecedented extent. The elaboration of attractants and rewards that are detectable and distinguishable by pollinators in particular resulted from differential reproduction of genotypes, depending upon their relative success at promoting visitation and pollen transport to a conspecific. In nonreproductive plant tissues, chemical attractants may have secondarily (or concomitantly) assumed defensive functions, possibly by increases in levels of production (e.g., while many monoterpenes are attractants, the same compounds are antifeedants for many insect herbivores; Table 3.4).

Distinguishing among these three scenarios is challenging to say the least, as is the case for most evolutionary questions, particularly since the scenarios may not be mutually exclusive. It is worth emphasizing, however, that, that there are several plausible scenarios to account for the biochemical diversification of angiosperms. It is certainly true that antagonistic interactions can result in differential fitness; plant genotypes that are not chemically defended are not represented in future generations (Ehrlich and Raven, 1964; Swain and Cooper-Driver, 1981). It is, however, no less true that mutualistic interactions can also result in differential fitness; plant genotypes that can attract a pollinator and induce it to transport pollen to a conspecific will be represented in greater numbers than less attractive genotypes.

Very little attention has been paid to the importance of angiosperms in the chemical diversification of insects. At least part of the diversity may be derivative—insects specialized on chemically distinctive host plants have the opportunity to sequester a greater variety of defensive compounds. Diversification of endogenous, in addition to exogenous, toxins seems to have taken place in herbivorous insects as well. In the Lepidoptera, for example, papilionid and notodontid larvae appear to synthesize carboxylic acids and their derivatives independently of host chemistry (Blum, 1981). The greater nutritional suitability of angiospermous plant tissues as food for insects (Scriber and Slansky, 1981) may have contributed to the evolution of secondary metabolism in holometabolous herbivorous insects. Utilization of angiospermous hosts, with succulent and protein-rich foliage and reproductive parts (relative to nonangiosperm plants; Slansky and Rodriguez, 1987), may have allowed herbivorous insects on these hosts to achieve unprecedented population densities. Greater conspicuousness to predators, by virtue of numbers, may have in turn led to the diversification of secondary metabolism in these insects for defensive purposes.

The tremendous structural and biosynthetic diversity of both angiospermous plants and the insects that feed on them is at least partly due to certain life history resemblances in these taxa. Both are notable for the

Table 3.4. Attractant/deterrent properties of volatile plant constituents for insects.*

Compound	Attractant for:	Deterrent for:
Anethole	Pales weevil, Japanese beetle, black swallowtail	*Drosophila*
Camphene	Scolytids	House fly
Carvone	House fly, *Cavariella aegopodii*, Black swallowtail,	House fly, fall armyworm
Caryophyllene oxide	Boll weevil	Beet armyworm, leafcutter ants
Cineole	*Euplassia, Euglossa*	House fly
Citral	Silkworm	House fly, *Tribolium castaneum, Drosophila*
Citronellal	Pales weevil	House fly, *Drosophila*
Coumarin	Vegetable weevil, sweet clover weevil	*Epicauta, Hyperallpostica*
Geraniol	Scolytids, honey bee, Japanese beetle	House fly, silkworm
Limonene	Scolytids, boll weevil	Cat flea, Western pine beetle, house fly
Linalool	Silkworm	House fly, *Drosophila*
α-pinene	Douglas fir beetle, spruce budworm, scolytids, boll weevil, pales weevil	Euglossine bees
Terpineol	Scolytids	Silkworm, *Drosophila, Dermatophagoides, Tyrophagus*

*From Koul, 1982; Brattsten, 1983; Metcalf, 1987; Ryan, 1988; Inagaki and Ishida, 1987; Yozo et al., 1985.

ability of some members to colonize disturbed habitats and to exploit ecologically "difficult" habitats. It is almost a certainty, however, that a significant portion of the diversification of secondary metabolism in these two taxa simply reflects the fact that these compounds mediate a tremendous variety of interactions between plants and herbivorous insects and, due to reciprocal adaptive evolution, each interaction has the potential for generating even greater diversity.

Acknowledgements

We thank C. Labandeira, E. MacLeod, T. Phillips, and the students who enrolled in Paleobiology of Plant-Insect Interactions for valuable discussion; A. Zangerl, E. McCloud, J. Conner, S. Strauss, and M. Cohen for insightful comments on the manuscript; and N.S.F. 8818205 to M.R.B for financial support during the preparation of this manuscript.

References

Adler, P. and D. Pearson. 1982. Why do male butterflies visit mud puddles? *Can. J. Zool.* 60:322–324.

Baker, H. G. and P. D. Hurd. 1968. Intrafloral ecology. *Annu. Rev. Entomol.* 13:385–414.

Bakus, G. J., N. M. Targett, and Bruce Schulte. 1986. Chemical ecology of marine organisms: an overview. *J. Chem. Ecol.* 12:951–987.

Baldwin, I., C. L. Sims, and S. E. Kean. 1990. The reproductive consequences associated with inducible alkaloidal responses in wold tobacco. *Ecology* 71:252–262.

Bate-Smith, E. C. 1962. The phenolic constituents of plants and their taxonomic significance. I. Dicotyledons. *J. Linn. Soc. Bot.* 58:95–173.

Bentley, B. and N. D. Johnson. 1991. Plants as food for herbivores: the roles of nitrogen fixation and carbon dioxide enrichment. In *Plant-Animal Interactions: Evolutionary Ecology in Tropical and Temperate Regions*, eds. P. W. Price, T. M. Lewinsohn, G. W. Fernandes, and W. W. Benson, pp. 257–272. John Wiley, New York.

Berenbaum, M. R. 1986. Target site insensitivity in plant-insect interactions. In *Molecular Aspects in Insect-Plant Associations*, eds. L. Brattsten and S. Ahmad, pp. 257–272. Plenum Press, New York.

Berenbaum, M. R. 1988. Effects of electromagnetic radiation on insect-plant interactions. In *Plant Stress-Insect Interactions*, ed. E. A. Heinrichs, pp. 167–186. John Wiley, New York.

Berenbaum, M. R. 1990. Evolution of specialization in insect-umbellifer associations. *Annu. Rev. Entomol.* 35:319–343.

Berenbaum, M. R. 1991. Plant consumers and plant secondary metabolites: past present, and future. In *Oxford Surveys In Evolutionary Biology*, eds. J. Antonovics and D. Futuyma. pp. 289–307. Oxford Univ. Press, Oxford.

Berenbaum, M. R. and P. P. Feeny. 1981. Toxicity of angular furanocoumarins to swallowtails: escalation in the coevolutionary arms race. *Science* 212:927–929.

Berenbaum, M. R. and M. B. Isman. 1989. Herbivory in holometabolous and hemimetabolous insects: contrasts between Orthoptera and Lepidoptera. *Experientia* 45:229–236.

Berenbaum, M. R., A. R. Zangerl, and J. K. Nitao. 1986. Constraints on chemical coevolution: wild parsnip and the parsnip webworm. *Evolution* 40:1215–1228.

Blum, M. S. 1981. *Chemical Defenses of Arthropods*. Academic Press, New York.

Brattsten, L. B., 1979. Biochemical defense mechanisms in herbivores against plant allelochemicals. In *Herbivores: Their Interaction with Secondary Plant Metabolites*, eds. G. Rosenthal and D. Janzen. pp. 199–270. Academic Press, New York.

Brattsten, L. B. 1983. Cytochrome P450 involvement in the interactions between plant terpenes and insect herbivores. In *Plant Resistance to Herbivores*, ed. P. A. Hedin, pp. 173–195. ACS Symp. Ser. 208, Washington, DC.

Brattsen, L. B. and S. Ahmad. 1986. *Molecular Aspects of Insect-Plant Associations*. Plenum Press, New York.

Briggs, M. A. 1990. Chemical defense production in *Lotus corniculatus* L. I. The effects of nitrogen source on growth, reproduction and defense. *Oecologia*. 83:27–31.

Briggs, M. A. and J. C. Schultz. 1990. Chemical defense production in *Lotus corniculatus* L. II. Tradeoffs among growth, reproduction and defense. *Oecologia* 83:32–37.

Bryant, J. 1987. Feltleaf willow-snowshoe hare interactions: plant carbon/nutrient balance and floodplain succession. *Ecology* 68:1319–1327.

Bryant, J., F. S. Chapin, and D. R. Klein. 1983. Carbon/nutrient balance of boreal plants in relation to vertebrate herbivory. *Oikos* 40:357–368.

Caldicott, A. B., B. R. T. Simoneit, and G. Eglinton. 1975. Alkanetriols in psilophyte cutins. *Phytochem.* 14:2223–2228.

Chapman, R. F., 1971. *The Insects Structure and Function*. Elsevier, New York.

Chew, F. and J. Rodman. 1979. Plant resources for chemical defense. In *Herbivores: Their Interaction with Secondary Plant Metabolites*, eds. G. Rosenthal and D. Janzen, pp. 271–307. Academic Press, New York.

Coley, P. D. 1986. Costs and benefits of defense by tannins in a neotropical tree. *Oecologia* 70:238–241.

Coley, P. D., J. P. Bryant, and S. Chapin. 1985. Resource availability and plant antiherbivore defense. *Science* 230:895–899.

Day, P. R. 1974. *Genetics of Host-Parasite Interaction*. W. H. Freeman, San Francisco.

DeWaal, D. 1942. Het Cyanophore Karakter van WitteKlaver, *Trifolium repens* L. Ph.D. dissertation, Agricultural University, Wageningen, The Netherlands.

Dodson, C. H. 1962. Pollination and variation in the subtribe Catasetinae (Orchidaceae). *Ann. Mo. Bot. Gard.* 49:35–56.

Dodson, C. H. 1975. Coevolution of orchids and bees. In *Coevolution of Animals and Plants*, eds. L. E. Gilbert and P. H. Raven, pp. 91–99. Univ. of Texas Press, Austin.

Dressler, R. L. 1982. Biology of the orchid bees (Euglossini). *Annu. Rev. Ecol. Syst.* 13:373–394.

Drolsom, P. N. 1972. Evaluation of hydrocyanic acid potential of sorghum. *Rep. 27th Annl. Corn and Sorghum Res. Conf. Publ. No. 27*, pp. 29–34. American Seed Trade Association, Washington, D.C.

Duffey, S. S. 1977. Arthropod allomones: chemical effronteries and antagonists. *Proc. XV Int. Cong. Ent. Wash. DC, Aug. 1976* 15:323–394.

Ehrlich, P. and P. Raven. 1964. Butterflies and plants: a study in coevolution. *Evolution* 18:586–608.

Eisner, T. 1970. Chemical defense against predation in arthropods. In *Chemical Ecology*, eds. E. Sondheimer and J. B. Simeone, pp. 157–218. Academic Press, New York.

Fajer, E., M. D. Bowers, and F. Bazzazz. 1989. The effects of enriched carbon dioxide atmospheres on plant-insect herbivore interactions. *Science* 243:1198–1200.

Foulds, W. and J. P. Grime. 1972. The response of cyanogenic and acyanogenic phenotypes of *Trifolium repens* to soil moisture supply. *Heredity* 21:181–187.

Fowden, L. and P. J. Lea. 1979. Mechanisms of plant avoidance of autotoxicity by secondary metabolites, especially by nonprotein amino acids. In *Herbivores: Their Interactions with Secondary Plant Metabolites* eds. G. Rosenthal and D. Janzen. pp. 135–160. Academic Press, New York.

Fraenkel, G. 1959. The raison d'etre of secondary plant substances. *Science* 129:1466–1470.

Galen, C. 1989. Measuring pollinator-mediated selection on morphometric floral traits: bumblebees and the alpine sky pilot *Polemonium viscosum. Evolution.* 43:882–887.

Galen, C., K. Zimmer, and M. E. Newport. 1987. Pollination in floral scent morphs of *Polemonium viscosum*: a mechanism for disruptive selection on flower size. *Evolution* 4:599–606.

Geissman, A. and D. H. G. Crout. 1969. *Organic Chemistry of Secondary Plant Metabolism*. Freeman Cooper, San Francisco.

Goldstein, J. and T. Swain. 1963. Changes in tannins in ripening fruits. *Phytochemistry* 2:371–383.

Gonzalez, F. J. and D. W. Nebert. 1990. Evolution of the P450 gene superfamily. *Trends in Genetics* 6:182–186.

Grant, V. 1971. *Plant Speciation*. Columbia University Press, New York.

Gulmon, S. L. and H. Mooney. 1983. Costs of defense and their effects on plant productivity. In *On the Economy of Plant Form and Function*, ed., T. Givnish, pp. 681–698. Cambridge Univ. Press, Cambridge.

Hanover, J. W. 1966. Genetics of terpenes I. Gene control of monoterpene levels in *Pinus monticola. Heredity* 21:73–84.

Harshman, L. G., J. A. Ottea, and B. D. Hammock. 1991. Evolved environment-dependent expression of detoxication enzyme activity in *Drosophila melanogaster. Evolution* 45:791–795.

Hay, M. E., W. Fenical, and K. Gustafson. 1987. Chemical defense against diverse coral-reef herbivores. *Ecology* 68:1581–1591.

Hegnauer, R. 1969–1973. Chemotaxonomie der Pflanzen. Birkhäuser Verlag, Basel.

Hills, H. G., N. H. Williams, and C. H. Dodson. 1972. Floral fragrances and isolating mechanisms in the genus *Catasetum* (Orchidaceae). *Biotropica* 4:61–76.

Hinton, H. 1976. Enabling mechanisms. *Proc. XVI Int. Cong. Ent.* 1976:71–83.

Isman, M. B. 1977. Dietary influence of cardenolides on larval growth and development of the milkweed bug, *Oncopeltus fasciatus*. *J. Insect Physiol.* 23:1183–1187.

Johnson, N., L. Rigney, and B. L. Bentley. 1988. Short-term changes in alkaloid levels following leaf damages in lupines with and without symbiotic nitrogen fixation. *J. Chem. Ecol.* 15:2425–2434.

Kennedy, G. G., R. T. Yamamato, M. B. Dimock, W. G. Williams, and J. Bordner. 1981. Effect of day length and light intensity on 2-tridecanone levels and resistance in *Lycopersicon hirsutum* f. *glabratum* to *Manduca sexta*. *J. Chem. Ecol.* 7:707–716.

Kevan, P. G., W. G. Chaloner, and D. B. O. Savile. 1975. Interrelationships of early terrestrial arthropods and plants. *Paleobiology* 18:391–418.

Knoll, A. 1986. The early evolution of land plants. In *Land Plants: Notes for a Short Course*. ed. T. W. Broadhead, pp. 45–63. Univ. Tennessee Dept. Geol. Sci.: Knoxville.

Koul, O. 1982. Insect feeding deterrents in plants. *Ind. Rev. Life Sci.* 2:97–125.

Lehninger, A. L. 1972. *Biochemistry*. Worth Pub., New York.

Lincoln, D.E. and D. Couvet. 1989. The effect of carbon supply on allocation to allelochemicals and caterpillar consumption of peppermint. *Oecologia* 78:112–114.

Matsinger, D. F., E. A. Wernsman, and W. W. Weeks. 1989. Restricted index selection for total alkaloids and yield in tobacco. *Crop Sci.* 29:74–77.

Mehlenbacher, S. A., R. L. Plaisted, and W. M. Tingey. 1984. Heritability of trichome density and droplet size in interspecific potato hybrids and relationship to aphid resistance. *Crop Sci.* 24:320–322.

McKey, D. 1974. Adaptive patterns in alkaloid physiology. *Am. Nat.* 108:305–320.

Metcalf, R. L. 1987. Plant volatiles as insect attractants. *CRC Critical Rev. in Plant Sciences* 5:251–301.

Mihaliak, C. A. and D. E. Lincoln. 1985. Growth pattern and carbon allocation to volatile leaf terpenes under nitrogen-limiting conditions in *Heterotheca subaxillaris* (Asteraceae). *Oecologia*. 66:423–426.

Mitchell, B. K. 1988. Adult leaf beetles as models for exploring the chemical basis of host plant recognition. *J. Insect Physiol.* 34:213–226.

Morrison, R. T. and R. N. Boyd. 1970. *Organic Chemistry*. Allyn & Bacon, Boston.

Neal, J. J. 1987. Metabolic costs of mixed-function oxidase induction in *Heliothis zea*. *Entomol. Exp. Appl.* 43:175–179.

Nitao, J. K. 1989. Enzymatic adaptation in a specialist herbivore for feeding on furanocoumarin-containing plants. *Ecology* 70:629–635.

Nitao, J. K. and A. R. Zangerl. 1987. Floral development and chemical defense allocation in wild parsnip (*Pastinaca sativa*). *Ecology* 68:521–529.

Ohno, S. 1970. *Evolution by Gene Duplication*. Springer-Verlag, New York.

Ostrem, L. 1987. Studies on genetic variation in reed canarygrass, *Phalaris arundinacea* L. I. Alkaloid type and concentration. *Hereditas* 107:235–248.

Paige, K. N. and T. Whitham. 1986. Individual and population shifts in flower color by scarlet gilia: a mechanism for pollinator tracking. *Science* 227:315–317.

Pasteels, J. M., M. Rowell-Rahier, and M. J. Raup. 1988. Plant-derived defense in chrysomelid beetles. In *Novel Aspects of Insect-Plant Interactions* eds. P. Barbosa and D. Letourneau, pp. 235–272. John Wiley, New York.

Paul, V. J. and W. Fenical. 1987. Natural products chemistry and chemical defense in tropical marine algae of the phylum Chlorophyta. *Bioorg. Marine Chem.* 1:1–29.

Paul, V. J., M. E. Hay, J. E. Duffy, W. Fenical, and K. Gustafson. 1987. Chemical defense in the seaweed *Ochtodes secundiramea* (Montagne) Howe (Rhodophyta): effects of its monoterpenoid components upon diverse coral roof herbivores. *J. Exp. Mar. Biol. Ecol.* 114:249–260.

Paul, V. J., M. M. Littler, D. S. Littler, and W. Fenical. 1987. Evidence for chemical defense in tropical green alga *Caulerpa ashmeadii* (Caulerpaceae: Chlorophyta): isolation of new bioactive sesquiterpenoids. *J. Chem. Ecol.* 13:1171–1184.

Pirie, N. 1959. Chemical diversity and the origins of life. In *The Origin of Life on the Earth*, eds. F. Clark and R. C. M. Synge, pp. 76–83. Pergamon Press, Oxford.

Poulton, J. E. 1983. Cyanogenic compounds in plants and their toxic effects. In *Handbook of Natural Toxins* Vol. 1, eds. R. F. Keeler and A. T. Tu, pp. 117–157. Marcel Dekker, New York.

Prestwich, G., B. Bentley, and E. Carpenter. 1980. Nitrogen sources for neotropical nasute termites: fixation and selective foraging. *Oecologia* 46:397–401.

Raven, P. H., R. F. Evert, and H. Curtis, 1976. *Biology of Plants*. Worth Pub., New York.

Richardson, J. L. 1977. *Dimensions of Ecology*. Williams and Wilkins, Baltimore.

Rodriguez, E. and D. Levin. 1975. Biochemical parallelisms of repellents and attractants in higher plants and arthropods. *Rec. Adv. Phytochem.* 10:214–267.

Romeo, J. T. J. D. Bacon, and T. Mabry. 1977. Ecological considerations of amino acids and flavonoids in *Sarrecenia* species. *Biochem. Syst. Ecol.* 5:117–120.

Rosenthal, G. A. and D. H. Janzen. 1979. *Herbivores their Interactions with Secondary Plant Metabolites*. Academic Press, New York.

Rothschild, M. 1972. Some observations on the relationship between plants, toxic insects, and birds. In *Phytochemical Ecology*, ed. J. B. Harborne, pp. 2–12. Academic Press, New York.

Rowell-Rahier, M. and J. M. Pasteels, 1986. Economics of chemical defense in Chrysomelinae. *J. Chem. Ecol.* 12:1189–1203.

Ryan, M. and O. Byrne. 1988. Plant-insect coevolution and inhibition of acetyl-cholinesterase. *J. Chem. Ecol.* 14:1965–1975.

Scriber, J. M. and F. Slansky, Jr. 1981. The nutritional ecology of immature insects. *Annu. Rev. Entomol.* 26:183–211.

Swain, T. 1977. Secondary products as protective agents. *Annu. Rev. Plant Physiol.* 28:479–501.

Swain T. 1978. Plant-animal coevolution: a synoptic view of the Paleozoic and Mesozoic. In *Biochemical Aspects of Plant and Animal Coevolution*, ed. J. Harborne. Academic Press, London.

Swain, T. and G. Cooper-Driver. 1981. Biochemical evolution in early land plants. In *Paleobotany, Paleoecology and Evolution*, Vol. 1, ed. K. J. Niklas, pp. 103–134. Praeger, New York.

Tallamy, D. 1985. Squash bettle feeding behavior: an adaptation against induced cucurbit defenses. *Ecology* 66:1574–1579.

Vrieling, K. 1990. Costs and benefits of alkaloids of *Senecio jacobaea* L. Ph.D. dissertation, Leiden, Leiden University.

Waterman, P. G. and S. Mole, 1989. Extrinsic factors influencing production of secondary metabolites in plants. In *Insect-Plant Interactions*, Vol. 1, ed. E. Bernays, pp. 107–134. CRC Press, Boca Raton, FL.

Williams, N. H. 1982. The biology of orchids and euglossine bees. In *Orchid Biology: Reviews and Perspectives*, Vol. 3, ed. J. Arditti, pp. 119–171. Cornell Univ. Press, Ithaca, NY.

Wilson, E. O. 1970. Chemical communication within animal species. In *Chemical Ecology*, eds. E. Sondheimer and J.B. Simeone, pp. 133–155. Academic Press, New York.

Yozo, I., A. Cho, S. Togashi, and M. Asashima. 1985. Effects of exposure to volatile substances from plants on the behavior of *Drosophila*. *Yokohamashiritsu Daigaku Ronso Shizen Kagaku Keiretsu* 36:15–33 (CA 106(7):45697f).

Zaehner, H., H. Anke, and T. Anke. 1982. Evolution and secondary pathways. In *Secondary Metabolism and Differentiation in Fungi*, eds. J. Benet and A. Ciegler, pp. 153–171. Marcel Dekker, New York.

Zangerl, A. R. and F. Bazzaz. 1991. Theory and pattern in plant defense allocation. In *Ecology and Evolution of Plant Resistance*, eds. R. Fritz and E. Simms, University of Chicago Press, Chicago.

Zangerl, A. R. and M. R. Berenbaum. 1987. Furanocoumarins in wild parsnip: effect of photosynthetically active radiation, ultraviolet light, and soil nutrients. *Ecology* 68:516–520.

Zangerl, A. R., M. R. Berenbaum, and E. Levine. 1989. Genetics of seed chemistry and morphology in *Pastinaca sativa* (Umbelliferae). *J. Heredity* 80:404–407.

Zobel, A. M. and S. A. Brown. 1990. Dermatitis-inducing furanocoumarins on the leaf surfaces of rutaceous and umbelliferous plants. *J. Chem. Ecol.* 16:693–700.

4

Costs and Benefits of Chemical Information Conveyance: Proximate and Ultimate Factors

Marcel Dicke   Agricultural University Wageningen
and
Maurice W. Sabelis   University of Amsterdam

4.1. Introduction

It is one thing to determine the chemical nature of information transfer, but it is another to assess its function. There are unambiguous procedures to identify a chemical structure and to determine the producer and potential receivers. However, are there such methods to elucidate the role of chemical information transfer? In principle, it should be possible to quantify costs and benefits to both producer and receiver, but there are two major hurdles to overcome in order to perform a comprehensive cost-benefit analysis.

1.   To determine costs and benefits, *all* possible interactions need to be considered. Organisms are part of complex food webs. Each organism interacts with others, either conspecific or not, either at the same trophic level or not. If one considers just two participants in the transfer of information, then it is entirely possible for the cost-benefit balance to be negative to the producer. For example, if one considers signal production by a herbivore and its effect on searching behavior of a predator, then there are only associated costs. However, to understand why it is produced, other interactions should be taken into account, such as information transfer to sexual partners.

2.   Costs and benefits of a trait are not constant in magnitude, but may well change in the course of evolution. Consider the following example. The first host plant to produce volatile chemicals in response to herbivore attack may have lured other herbivores as well as their predators. If there were to be a net benefit, e.g. in the case that the host plant were to die anyway from attack by the herbivores, then genes coding for the trait of releasing volatiles would spread, thereby reinforcing selection on (a) the herbivore's enemies to respond to the plant's signal and (b) enemies of other herbivores not to respond and even avoid these plants. Thus, to one herbivore, the host plants having the trait become less profitable as they represent enemy-dense space, whereas other herbivores may profit from including these plants in their diet as they represent enemy-free space.

This may then lead to host plant shifts in the herbivore and changes in responses of predators to lures provided by the host plant. The result of all this may well be cyclic changes in the host plant signal, host plant preference of the herbivore and response to the host plant's signal by the predator. In other words, costs and benefits may be frequency- or density-dependent.

We contend that the function of chemical information conveyance is not easily assessed. Yet, virtually all terminology on information transfer is wholly or partly based on assumptions of costs and benefits to producer as well as receiver (Nordlund and Lewis, 1976; Dicke and Sabelis, 1988a). To enable the use of such terminology, important simplifications have to be made. Costs and benefits result from all interactions of an organism in which the chemical is involved. Yet, for pragmatic reasons, it is generally agreed to consider costs and benefits of no more than two interactants, a producer and a receiver. Polymorphic production and responses should be disregarded for the same reason. Costs and benefits are thus to be assessed between two interactants with each having a well defined strategy of food searching, reproduction, and survival.

Surprisingly little attention has been paid to the nature of costs and benefits. Generally, there seems to be a trend in the literature to consider proximate factors only (i.e., the mechanistic aspects). Why ultimate factors are often ignored is an unanswered question, however. There is no general reason to assume that one is easier to assess than the other. Let us first briefly review the history of terminology on information-conveying chemicals (section 4.2), then consider the various ways in which information can be used and misused in complex food webs, (sections 4.3 and 4.4) and finally illustrate how costs and benefits might be quantified by discussing a specific example.

In this chapter, we frequently use anthropomorphic terms such as "spies," "stowaways," and "conspiracy." We do not ascribe cognition or purpose to animals and plants, nor do we consider natural selection to have a purpose. We use anthropomorphic terms with the sole purpose of defining the set of possible strategies in the context of game theory. What strategies will evolve is another question, the solution of which may be found, for example, by the use of game theory. Instead, we question how to analyze costs and benefits of each possible strategy in the information game.

4.2. Terminology of Information-Conveying Chemicals

Many terms have been designated for chemicals that convey information between organisms (Brown et al., 1970; Whittaker and Feeny, 1971; Nordlund, 1981). Nordlund and Lewis (1976) were the first to integrate terms into a system describing phenomena in interspecific and intraspecific

Table 4.1. Infochemical terminology (Dicke and Sabelis, 1988a).

Infochemical: A chemical that, in the natural context, conveys information in an interaction between two individuals, evoking in the receiver a behavioral or physiological response.

   Pheromone: An infochemical that mediates an interaction between organisms of the same species whereby the benefit is to the origin-related organism ([+, −] pheromone), to the receiver ([−, +] pheromone), or to both ([+, +] pheromone).

   Allelochemical: An infochemical that mediates an interaction between two individuals that belong to different species.

      Allomone: An allelochemical that is pertinent to the biology of an organism (organism 1) and that, when it contacts an individual of another species (organism 2) evokes in the receiver a behavioral or physiological response that is adaptively favorable to organism 1, but not to organism 2.

      Kairomone: An allelochemical that is pertinent to the biology of an organism (organism 1) and that, when it contacts an individual of another species (organism 2) evokes in the receiver a behavioral or physiological response that is adaptively favorable to organism 2, but not to organism 1.

      Synomone: An allelochemical that is pertinent to the biology of an organism (organism 1) and that, when it contacts an individual of another species (organism 2) evokes in the receiver a behavioral or physiological response that is adaptively favorable to both organism 1 and 2.

---

interactions. They used "semiochemical" as the umbrella term, which comprises both information-conveying chemicals and toxins. We (Dicke and Sabelis 1988a) emphasized the special status of information-conveying chemicals: this category differs from toxins and nutrients in that the former are not *themselves* detrimental or beneficial, but may be through the responses they elicit. We termed information-conveying chemicals "infochemicals" (Table 4.1); these may be regarded as a subcategory of semiochemicals.

Various infochemical categories have been distinguished. They are based on (1) whether the interaction is intra- or interspecific (pheromones vs. allelochemicals), (2) which costs and benefits fall to each of the two interacting organisms, and (3) the identity of the producer and the receiver. We (Dicke and Sabelis 1988a) argued that the latter criterion should be dropped. Several examples were presented, showing that organisms once thought to be producers later on were found to be associated with other organisms (e.g., microorganisms) that were responsible for production or that they induced the production in another organism. By taking the producer criterion too strictly, terminology would exclude an important class of ecologically significant interactions—those between

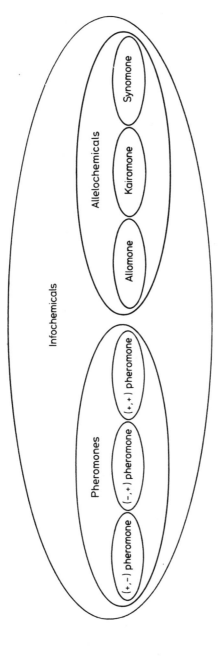

Figure 4.1. Structure of infochemical terminology (Dicke and Sabelis, 1988a).

receiver and organisms closely associated with the producer (Dicke and Sabelis, 1988a).

We also argue that it does not make sense to employ a terminology for allelochemicals with strong roots in cost-benefit thinking, whereas the older terminology on pheromones has nothing to do with cost and benefit at all. Clearly, just as different species may affect each other either positively or negatively, so may different individuals of the same species (cf. Rutowski, 1981; Alcock, 1982). To neglect these aspects in pheromone terminology is to close one's eyes to many of the interactions that are pertinent to biology.

Therefore, we proposed three pheromone subcategories ($+ -$, $- +$, $+ +$), in exact analogy to the allelochemical subcategories allomone, kairomone, and synomone (Fig. 4.1; Table 4.1) (Dicke and Sabelis, 1988a).

Semiochemical and infochemical terminologies have only regarded interactions in which at least one of the two interactants benefits (Nordlund and Lewis, 1976; Dicke & Sabelis, 1988a). Recently, Whitman (1988) added a fourth allelochemical category in which neither interactant benefits: antimones. The idea is valuable; in analogy to the rationale for kairomones and ($-$, $+$) pheromones, infochemicals that benefit an organism in one interaction may also have side effects in other interactions where they are to the detriment of both interactants. However, the examples presented by Whitman (1988) are shaky; it is not clear that the effects described can be attributed to information-conveyance, rather than mechanical or physical characteristics. The phenomenon will need more investigation to provide good examples (Takabayashi and Takahashi, 1990). No examples of such interactions are known for pheromones yet.

It is important to note that all terms are context-specific rather than chemical-specific (Nordlund and Lewis, 1976; Dicke and Sabelis, 1988a); for example, a plant-produced chemical that attracts herbivores is a kairomone, but the same compound may serve as a synomone in the interaction between plant and predator of the herbivore or as a pheromone in plant-plant interactions.

4.3. Legitimate Use of Information

If infochemicals benefit both interactants, the information transfer will be maintained by natural selection acting upon both the receiver and the emitter. We regard that as a "legitimate use of information".

Several forms of legitimate use of information may be distinguished.

4.3.1. Cooperation

Cooperation occurs when the emitter asks for help, which the receiver gives; this interaction is to the benefit of both interactants. An example involving a pheromone is attraction of a mate by a female moth. The female increases chances for reproduction by producing the pheromone and the male does so by responding to it. Thus, the pheromone benefits emitter and receiver. It is a (+ , +) pheromone.

Examples for allelochemicals can be found in the class of synomones (see Nordlund et al., 1988, and Whitman, 1988, for reviews). A well-known example is the production of volatiles by flowers that attract pollinators (Harborne, 1988; Bertin, 1989).

4.3.2. Conspiracy

Conspiracy is a form of cooperation between two organisms, which is obviously detrimental to a third organism. When attacked by predators, aphids release a volatile alarm pheromone that causes nearby aphids to disperse (Nault and Phelan, 1984). Some aphid species are tended by ants that obtain sugars in the form of honeydew from the herbivore. Nault et al. (1976) observed that ant-associated (myrmecophilous) aphids disperse less readily in response to the alarm pheromone than do non-myrmecophilous species. These authors describe ant responses to aphid alarm pheromone as raising their antennae, opening their mandibles, and attacking the aphid predator. Also, the dispersive response of aphids upon perception of the alarm pheromone is reduced when ants are actually tending aphids. These data show that aphids and ants conspire against aphid predators. In aphid-ant interactions, the alarm pheromone thus acts as a synomone. An example of conspiracy between plants and predatory mites, to which herbivorous mites fall victim, will be given below (sections 4.6–4.9).

4.3.3. Notification

Notification occurs when the emitter leaves a note (a mark) to inform another individual about local conditions that may lead the receiver to refrain from some type of behavior. The receiver will evaluate the information and "decide" what the best response will be (Mangel and Roitberg, 1989; Roitberg and Mangel, 1989). If that response also benefits the emitter, the mark is legitimately used.

Many insect species mark hosts or foraging areas (van Lenteren, 1981; Prokopy, 1981; Prokopy et al., 1984). An illustrative example, showing the characteristics of *notification* is available for host-marking by the parasitoid *Venturia canescens* (Hubbard et al., 1987; Marris et al., 1989). These wasps parasitize larvae of the flour moth *Plodia interpunctella*. Each individual host can support complete development of only a single parasitoid larva. Upon oviposition, the parasitoids mark a host with a mixture of chemicals originating from the Dufour's gland. This chemical mark may deter oviposition during subsequent encounters. A female wasp prefers to oviposit into an unparasitized host. However, when such hosts are rare wasps may "decide" to neglect the chemical mark and superparasitize. If the ovipositions are sufficiently close together, then the probability of the second egg winning the competition may be great enough to warrant this risk. If a female decides to superparasitize, then doing this in a selective manner by avoiding hosts containing her own or closely-related offspring, will certainly be the best strategy. This was well demonstrated in the case of *V. canescens* (Hubbard et al., 1987; Marris et al., 1989). The tendency to superparasitize is negatively correlated with the relatedness between first and second parasitoid. Chemical analysis of Dufour's secretion demonstrated clone-specificity (Marris et al., 1989). Behavioral data indicate that the chemical mark may also be individual-specific (Hubbard et al., 1987). A model study indicated that this strategy of *V. canescens* to avoid superparasitism in hosts that contain an egg of closely-related conspecifics is evolutionarily stable over a considerable range of conditions. For a more elaborate treatise on evolution of marking pheromones, see Roitberg and Mangel (1989), and Roitberg (this volume).

4.4. Illegitimate Use of Information

The communicative interactions described above may be exploited by non-intended receivers upon interception of the information or by illegitimate emitters who manipulate the response of legitimate receivers to their own benefit. In both cases, the exploiters rely upon a functional communicative system. Examples of such illegitimate use are summarized below.

4.4.1. Illegitimate Information Reception

Illegitimate information reception comprises systems in which the emitter is exploited by the receiver. Examples will thus be found among kairomones and $(-, +)$ pheromones. Two types of non-intended receivers are distinguished.

4.4.1.1. Spies

Spies are illegitimate receivers that respond to an infochemical, to their own benefit and usually to the detriment of the producer. Many examples are known, such as the use of a noctuid host's sex pheromone as a kairomone by the egg parasitoid *Trichogramma evanescens*. Presence of the host's sex pheromone, an indication that reproducing hosts are nearby, suppresses parasitoid dispersal and initiates searching activities (Noldus, 1989). Clerid predators of bark beetles use their prey's pheromones in prey location (Wood, 1982). Espionage is also employed by many species of herbivorous insects during location of suitable host plants for oviposition or feeding (Visser, 1986).

Apart from these situations where individuals of a higher trophic level spy on a lower trophic level, the reverse situation may also occur. This phenomenon is well-known for mammals, such as hares that avoid areas contaminated with fox urine (Sullivan and Crump, 1986). Examples for arthropods are scarce as yet (Whitman, 1988). Some examples are avoidance of stonefly predators by mayflies after perception of a kairomone (Malmqvist, 1987) and induction of defensive behavior in Crustacea upon perception of a kairomone of an approaching sea star predator (Glynn, 1980). Since knowledge of pheromones of parasitic and predatory arthropods is increasing (Tietjen and Rovner, 1982; Aldrich et al., 1984a, 1984b, 1986; Eller et al., 1984), it may be a fruitful approach to search for responses amongst their prey species.

Spies may also occur within a species. Marking pheromones may receive legitimate receivers (see section 4.3.3), but exploitation may occur as well. Several *Encarsia* species are adelphoparasitoids: diploid larvae are primary parasitoids and haploid larvae are parasitoids of diploid conspecifics (Gerling, 1966; Nguyen and Sailer, 1987). Unfertilized females of the adelphoparasitoid *Encarsia smithi* discriminate between unparasitized and parasitized hosts (citrus blackfly larvae). They lay haploid eggs in parasitized hosts, but do not oviposit in unparasitized hosts (Nguyen and Sailer, 1987). Such host discrimination behavior has also been described for other *Encarsia* species and is accomplished during antennal contacts with the host (van Lenteren et al., 1976, 1980). The marking pheromone involved is clearly a $(-, +)$ pheromone if used for espionage by the adelophoparasitoid resulting in deposition of a haploid egg that parasitizes the primary parasitic larva.

4.4.1.2. Stowaways

Stowaways are illegitimate receivers that refrain from producing and/or releasing a certain infochemical as long as other, nearby individuals produce it. This strategy saves energy and may also be seen as a way to reduce the risk of attracting spying enemies. Males of many insect species are attracted by sex pheromones produced by conspecific males (Fletcher 1968; Hardee et al., 1969; Alcock, 1975; Harris and Todd, 1980). It is highly probable that the responders exploit the emitters by lying silently in wait to steal females that are attracted to the caller (cf. Alcock, 1975; Alcock et al., 1978). This situation has been speculatively described for tropical euglossine bees, where males are olfactorily attracted by conspecific males that are actively attracting females. There are good reasons to assume that production of sex pheromone is expensive for these bees (Alcock et al., 1978), which would favor the stowaway strategy.

4.4.2. Illegitimate Information Emission

Illegitimate information emission concerns interactions in which the receiver is exploited by the emitter. Examples are thus to be found among allomones and (+, −) pheromones. Two types of illegitimate emitters are distinguished:

4.4.2.1. Cheaters

Cheaters are illegitimate emitters that mimic infochemicals of other organisms to exploit a communicative system. For example, bolas spiders produce mimics of moth sex pheromones and eat the attracted male moths (Eberhard, 1977; Stowe et al., 1987; Yeargan, 1988). Instead of being a producer, the illegitimate emitter may also acquire the infochemical from the organisms that are exploited. Ants recognize colony members by touching each other's cuticle with the antennae. The myrmecophilous beetle *Myrmecaphodius excavaticollis* can move freely in colonies of the ant *Solenopsis richteri* and obtains food from the workers. This is possible because the beetles rapidly acquire hydrocarbons of their hosts during ant-beetle contacts. The beetle's armored exterior may help the beetle to survive initial aggressive contacts that no longer occur once the beetle has acquired the cuticle hydrocarbons (Vander Meer and Wojcik, 1982). Similar phenomena are known for other myrmecophiles (Hölldobler, 1971) and for a termitophile (Howard et al., 1980).

Cheating is also employed by plants. Orchids of the genus *Ophrys* do not produce nectar and thus do not provide pollinators with a reward. Instead, they produce mimics of sex pheromones of *Andrena* bees that pollinate the orchids after landing (Kullenberg and Bergström, 1976; Borg-Karlson, 1986; Borg-Karlson and Tengö, 1986). Some other plant species have a similar strategy for attracting pollinators (Harborne, 1988). Wild potato plants emit aphid alarm pheromone which makes aphids reluctant to land on these plants (Gibson and Pickett, 1983).

4.4.2.2. Boasters

Boasters are illegitimate emitters that produce an "inflated" signal. Situations may occur where an individual increases the emission rate to emphasize the message. Upon infestation of a dead (defenseless) tree, bark beetles produce a pheromone that attracts conspecific males and females (Birch, 1984). The arrival of competitor females reduces the food available for the brood of the emitter (thus it is a [−, +] pheromone in female-female interactions; Table 4.1) (Raffa and Berryman, 1983). In addition, bark beetles may also attack healthy trees with intact resin defenses. Under these circumstances, risk of death from resin entrapment is reduced for an initial settler or her progeny if she is joined by additional females. Thus, it is beneficial for an early colonist to attract females under these circumstances, but for dispersing females, it will be most rewarding to find dead trees. The initial colonist might exploit the dispersing females to make the healthy tree safer for themselves at the expense of the attracted females. Therefore, it is interesting to see that females of the mountain pine beetle, *Dendroctonus ponderosae*, produce more aggregation pheromone (mimicking higher levels of infestation and thus higher probabilities of tree death and lower probability of beetle death) when burrowing in hosts that have relatively large amounts of protective resin (Vité and Pitman, 1968; Raffa and Berryman, 1983). In this particular interaction, the pheromone is a $(+, -)$ pheromone.

4.4.3. Communication Disrupted From Both Sides

Communication disrupted from both sides refers to interactions in which neither emitter nor receiver benefit from the interaction; they relate to antimones. Sound examples of this phenomenon are still scarce (Whitman, 1988); an intriguing one is the following. *Apanteles kariyai* is a specialist parasitoid of the common armyworm, *Pseudaletia separata*. The wasps are arrested by a contact kairomone which is present in feces, feeding traces and exuviae of the host (Takabayashi et al., 1985). Upon subsequent antennal contact with an armyworm larva, a wasp immediately inserts her ovipositor and deposits eggs. The kairomone consists of 2,5-dialkyltetrahydrofuran homologs (Takabayashi and Takahashi, 1986); these components are also present in a non-host noctuid, *Acantholeucania loreyi*, that feeds on the same plant species as *P. separata* (Takabayashi and Takahashi, 1990). Feces of *A. loreyi* elicit a behavioral response similar to that upon contact with armyworm feces. The parasitoid also inserts her ovipositor into an *A. loreyi* larva if contacted after examination of feces. However, this does not result in parasitoid offspring. Whether this is due to the parasitoid not ovipositing or to oviposited eggs not developing is not known (Takabayashi and Takahashi, 1990).

In the *A. loreyi-A. kariyai* interaction, the contact allelochemical serves as an antimone; it does not benefit the emitter, nor the receiver. Both noctuid species are polyphagous herbivores, that occur together. This would favor the ability of the parasitoid to distinguish between the two. Whether the antimonally-mediated interaction between *A. loreyi* and *A. kariyai* is the result of evolution or whether it is a case of artificial sympatry is not clear as yet (Takabayashi and Takahashi, 1990).

4.5. How to Minimize Energetic Costs of Infochemicals

Apart from costs in terms of illegitimate use of infochemicals, information-transmitting organisms are also faced with energetic costs in terms of production, transport, storage, and release. In addition, there are maintenance costs to be spent on synthesis of enzymes needed in the above-mentioned processes. Calculation of energetic costs is difficult, if not impossible, for several of the component costs. Biosynthetic costs is the component that can usually be quantified most reliably, and this has been done for several plant-produced compounds (Chew and Rodman 1979; Gulmon and Mooney, 1986; Lambers and Rychter, 1989). However, these are still rough estimates rather than accurate quantifications of cost. To enable accurate quantification, one needs to know, for example, the biosynthetic pathway actually used, the precursors from which biosynthesis starts, the actual number of required and produced molecules of ATP and/or $NADH_2$. To our knowledge, no quantification of energetic costs has been made for animal-produced infochemicals.

Whatever the energetic costs of infochemical production are, it is conceivable that organisms are selected for minimizing these costs (cf. section 4.10 below). Several characteristics of infochemical emission show how organisms may reduce energetic costs. Production and/or emission of an infochemical is usually restricted. They may depend on the following conditions.

1. *Time of the day.* Some flower fragrances are produced exclusively during the day or the night, which is correlated with the activity of their pollinators (Matile and Altenburger, 1988; Altenburger and Matile, 1988; Harborne, 1988). Similarly, many insects restrict release of sex pheromone ("calling") to a certain time of the day (Tamaki, 1985).

2. *Circumstances.* Production and/or emission of infochemicals may be restricted to external conditions that warrant effectiveness, such as being in an environment in which the legitimate receiver may be expected. Sex pheromone production of *Heliothis phloxiphaga* is dependent on presence of host plants. Corn silk volatiles affect the release of a hormone that initiates pheromone production (Raina, 1988). Perception of plant volatiles is also a prerequisite for pheromone release by the polyphemus moth, *Antheraea polyphemus* (Riddiford, 1967; Riddiford and Williams, 1967) and the sunflower moth *Homoeosoma electellum* (McNeil and Delisle, 1989). The release of a herbivore-deterring synomone by *Sorghum* plants occurs upon herbivore damage; cyanogenic glycosides that are stored in the cell vacuole are degraded enzymatically upon destruction of compartmentation by herbivory. The resulting cyanide deters the herbivores (Woodhead and Bernayes, 1977).

3. *Acquisition of the infochemical or a precursor*. Instead of producing the infochemical *de novo*, an organism may collect a precursor. Males of several species of three groups of Lepidoptera (Danainae, Ithomiinae, and Arctiidae) collect pyrrolizidine alkaloids from which they synthesize pheromones. Males without these pheromones are not accepted for mating by females (Boppré, 1986; Schneider, 1987). Male boll weevils, *Anthonomus grandis* can produce sex pheromone after feeding on a variety of fruits, but cotton squares produce the most attractive males (Hardee, 1970, in Tamaki, 1985). Another example is that of the myrmecophilous beetle *Myrmecaphodius excavaticollis* that acquires hydrocarbons from host ants and thus obtains protection in the ant nest (Vander Meer and Wojcik, 1982).

These mechanisms that minimize costs of infochemical release are beneficial not only from an economic standpoint, but also because they minimize risks of exploitation by illegitimate receivers. Because illegitimate use of information may involve members of other trophic levels, it is important to investigate information transfer in multitrophic systems. In the remainder of this chapter, we will present energetic and evolutionary aspects of information transfer in one tritrophic system, consisting of plants, spider mites, and their predators.

4.6. Induced Indirect Defense in a System Consisting of Predatory Mites, Herbivorous Mites, and Their Host Plants

Spider mites (Fig. 4.2) are polyphagous herbivores that reach pest status in many agricultural crops (see Helle and Sabelis 1985a for a review). They insert their stylets in the leaves, inject saliva, and ingest parenchymous cell contents (Tomczyk and Kropczynska, 1985). An adult female gives rise to a colony from which daughters disperse upon reaching adulthood to settle nearby and initiate new colonies. Spider mites overexploit their food source in the absence of predators, but local populations are decimated if discovered by predators such as phytoseiid mites (Fig. 4.3) (see Helle and Sabelis 1985b for a review). Predatory mites disperse on wind currents and probably cannot control where they land. Thus, the probability of landing in a spider mite colony or on a spider mite-infested plant may be very small. However, after landing, volatile infochemicals are used in making foraging decisions, such as whether to stay or to leave, and where or how long to search (for a review, see Sabelis & Dicke 1985). Because predatory mites overexploit their prey locally, it may be envisaged that any plant genotype that increases chances of predatory-mite invasion into spider mite colonies will have a higher fitness than conspecifics without this

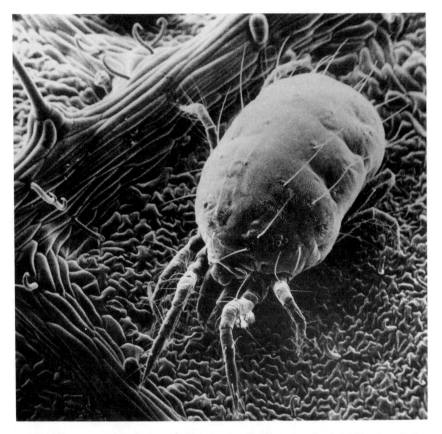

Figure 4.2. Scanning electron microscope photograph of adult female of the herbivorous spider mite *Tetranychus urticae*. (Sabelis, 1981).

ability. This phenomenon of defense through natural enemies of herbivores is referred to as indirect defense, in contrast to direct defense aimed at the herbivore itself (Price et al., 1980; Dicke and Sabelis, 1988b).

Much research has been done on a system consisting of lima bean plants (*Phaseolus lunatus*), the two-spotted spider mite *Tetranychus urticae* (Fig. 4.2), and the predatory mite *Phytoseiulus persimilis* (Fig. 4.3). These predatory mites distinguish between *T. urticae*-infested and uninfested lima bean plants by means of olfaction. The volatile kairomone involved in this prey-predator interaction is mainly emitted from the leaves after infestation. Upon removal of spider mites and their visible products, previously infested leaves remained attractive to predatory mites for several hours, whereas the spider mites removed from the leaves were

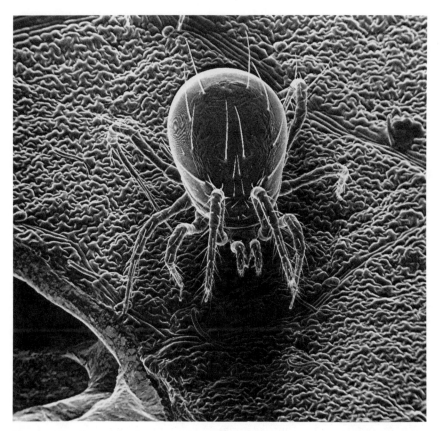

Figure 4.3. Scanning electron microscope photograph of adult female of the predatory mite *Phytoseiulus persimilis.* (Sabelis, 1981).

themselves not attractive (Sabelis and Van de Baan, 1983; Sabelis et al. 1984a). Several attempts have been made to obtain a more detailed knowledge of the origin of the kairomone by fragmentation of the spider mite colony or extracting the leaf surface after spider-mite feeding had occurred (Sabelis et al. 1984a). This showed that all feeding spider mite stages contributed to kairomone production and that the infested leaf was the main kairomone source. Very little kairomonal activity was present in feces or in the spider mites themselves. No kairomonal activity could be demonstrated in plants treated in several ways, unless spider mite-feeding had recently occurred. Although this investigation did not result in a final elucidation of the origin of the kairomone, it indicated that both spider mite and host plant are involved in its production. Recent evidence indicates that the kairomone is emitted by the plant; uninfested leaves of spider mite-infested plants are more attractive to predatory mites than uninfested leaves of uninfested plants (Dicke et al., 1990b).

Predatory mites distinguish between different plant-spider mite combinations by olfaction (Dicke and Groeneveld, 1986; Dicke et al., 1986; Dicke, 1988a). The kairomone is spider mite species-specific; predatory mites distinguish between plants infested by different spider mite species (Sabelis and Van de Baan, 1983). Moreover, the kairomone is also plant species-specific (Takabayashi et al., 1990; Dicke et al., 1990b).

Chemical investigations (Dicke, 1988b; Dicke et al. 1990a) revealed that large amounts of volatiles are emitted from artificially-damaged plants or spider mite-infested plants (Fig. 4.4, Table 4.2). Identification of these compounds revealed only chemicals that are well-known from the plant kingdom (Table 4.2). Four of the compounds attracted *P. persimilis* females. These were identified in the headspace of *T. urticae*-infested lima bean plants, but were not found in the headspace of intact or artificially-damaged plants. These kairomone components are the terpenes linalool and (E)-β-ocimene, the methylene terpene 4,8-dimethyl-1,3(E),7-nonatriene and the phenolic compound methyl silicylate (Fig. 4.5). These classes of compounds are known to be produced by plants but not by animals. Linalool, (E)-β-ocimene and methyl salicylate are known from many plant species. However, the methylene terpene deserves special attention, because it has only recently been identified in plants (Maurer et al. 1986; Kaiser 1987). It was not detected in steam distillates of uninfested lima bean plants (H. van Bokhoven and T. A. van Beek, unpublished data).

While none of the four kairomone components were found in the volatile blend emitted by undamaged or artificially-damaged lima bean plants (many analyses during several years; Dicke et al., 1990a) and undamaged plants were not attractive to *P. persimilis* (Sabelis and Van de Baan, 1983), most most recent analyses consistently revealed small amounts of the four predator-attracting compounds in the headspace of undamaged or artificially-damaged lima bean plants. The amounts are of the same order of magnitude for clean and artificially-damaged plants, but much lower (up to 200 times) than in spider mite-infested plants. Also, a slight attractiveness of undamaged lima beans to *P. persimilis* was recorded in recent experiments (Takabayashi et al., 1991; Takabayashi and Dicke, 1992). The plants used were never damaged by spider mites and have never been in the same room as spider mite-infested plants. Thus, these recent data show that spider mite damage is not necessary for production of the allelochemicals, but that spider mite damage, in contrast to artificial damage, increases the emission rate enormously. The discrepancy between recent data and those by Dicke et al. (1990a) for undamaged and artificially-damaged plants is currently under investigation.

Although this chemical evidence corroborates the biological evidence that the plant is involved in kairomone production, it remains unknown *how* this is accomplished. For example, the plant may produce kairomone precursors, from which production of the kairomone components is stimulated by spider mite salivary enzymes (see Gäbler et al. 1991 for recent experimental support); compare the production of herbivore repellents in *Sorghum* plants (Woodhead and Bernays, 1977; section 4.5), but note that the infochemical in the plant-mite system is much more specific. Alternatively, *de novo* synthesis by the plant may be stimulated when infestation by spider mites occurs.

Simulation models of local population dynamics gave exceedingly poor predictions when the predators were assumed to search at random as they do in prey-infested leaf areas. However, if the predators, upon leaving the prey-infested leaf area, are assumed to return immediately, the simulation models gave reasonably good predictions (Sabelis and Van der Meer, 1986). Behavioral studies have demonstrated that the volatile-kairomone gradient present at the edge of the patch affects such returning behavior (Sabelis et al., 1984b) and that, even when starved, the predatory mite *P. persimilis* did not emigrate on wind currents as long as kairomone was present (Sabelis and Afman, 1984; Sabelis and Janssen, 1991). Because the volatile kairomone appears to be essential in the limitation of spider mite populations, it is important to study the role of the plant in its production in more detail, thus improving knowledge of this indirect, inducible defense mechanism.

4.7. Effect of Induced Infochemicals on Spider Mites

The spider mite *T. urticae* is attracted by a volatile kairomone of undamaged lima bean plants (Dicke, 1986). However, spider mites disperse from an odor plume of lima bean plants infested by conspecifics. In the context of intraspecific spider mite interactions, the volatile infochemical involved is a pheromone, in this case a $(+, +)$ dispersing pheromone (Table 4.1). The behavioral reaction depends upon the ratio of pheromone to kairomone. At a low ratio (low degree of infestation), the mites are attracted; at a high ratio (high degree of infestation), they are repelled. From biological evidence alone, Dicke (1986) argued that the kairomone which attracts predatory mites and the dispersing pheromone might have components in common. Indeed, from the combined data of Dabrowski and Rodriguez (1971), Dicke (1986), and Dicke et al. (1990a), it can be

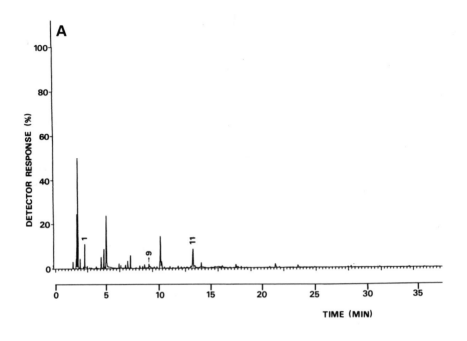

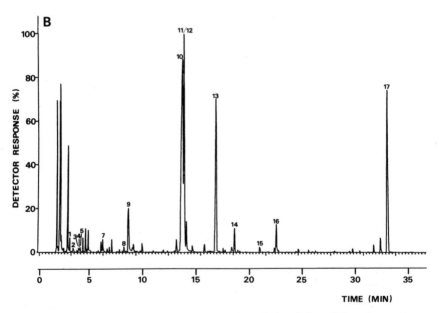

Figure 4.4. Gas chromatogram of volatiles from (A) uninfested lima bean plants and (B) spider-mite infested lima bean plants, collected on Tenax-TA. Peak numbers correspond to numbers given in Table 4.2. Unlabeled peaks are atmospheric or instrumental background peaks. (Dicke et al., 1990a).

Table 4.2. Volatiles collected from intact, artificially-damaged, and *T. urticae*-infested lima bean plants.*

		Identified in		
Peak Number[†]	Tentative Identification	*T. urticae.* Infested Plants	Artificially-Damaged Plants	Undamaged Plants
1	2-butanone	+	+	+
2	2-methyl-propan-1-ol	+	−	−
3	1-butanol	+	−	−
4	3-pentanone	+	−	−
5	1-penten-3-ol	+	−	−
6	hexanal	−	+	+
7	unidentified	+	+	−
8	2-hexenal	+	+	+
9	(Z)-3-hexen-1-ol	+	+	−
10	(Z)-3-hexen-1-yl acetate	+	+	−
11	1-octen-3-ol	+	+	+
12	(E)-β-ocimene	+	−	−
13	4,8-dimethyl-1,3(E),7-nonatriene	+	−	−
14	linalool	+	−	−
15	(Z)-3-hexen-1-yl butyrate	+	+	−
16	methyl salicylate	+	−	−
17	4,8,12-trimethyl-1,3(E), 7(E),11-tridecatetraene	+	−	−

*From Dicke et al., 1990a.

[†]Refers to peaks in Figure 4.4.

deduced that at least one component of the spider mite pheromone is also a component of the predator's kairomone: linalool (Fig. 4.6). Methyl salicylate is not, while no data are available for the response of *T. urticae* to (E)-β-ocimene and 4,8-dimethyl-1,3(E), 7-nonatriene.

4.8. Who Controls The Production and Release of the Infochemical?

Does the volatile infochemical function primarily in plant-predator interaction or in intraspecific spider mite communication?

The preceding sections indicate how several interactions in this system are mediated by volatile infochemicals whose release is induced by spider mite feeding. An important question is: do predatory mites "spy" on spider mite communication or do spider mites leave as soon as the plant's

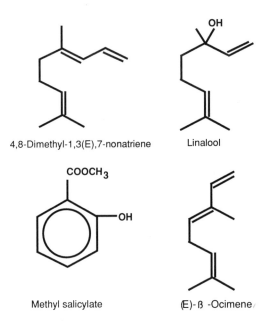

4,8-Dimethyl-1,3(E),7-nonatriene   Linalool

Methyl salicylate   (E)-β-Ocimene

Figure 4.5. Structure of volatile-synomone components that singly attract *P. persimilis* females.

"cry for help" gets too loud? This is essentially a question about who controls production and/or release of the chemicals.

1. The spider mites might control production of the infochemical to inform conspecifics about local density, and thus about food quantity and prospects for competition. But the spider mites would not need a volatile compound for this because information can also be conveyed by non-volatile chemicals or by contacts between individuals, without the costs of volatiles or the associated risks of attracting predatory mites.

2. The plant might control production of volatiles to recruit predatory mites as bodyguards. The volatile nature of the chemicals is then indispensable for fast spread of the advertisement: success of *induced* indirect defense depends heavily on rapidly recruited defenders. In this case, the spider mites would do best by using the volatiles as dispersing pheromone. It seems unlikely that the mites could avoid the feeding-dependent release of the infochemical, unless they can make their feeding unrecognizable for the plant.

These considerations suggest that the volatile infochemical primarily functions in plant-predatory mite interactions and that the spider mite response is secondary.

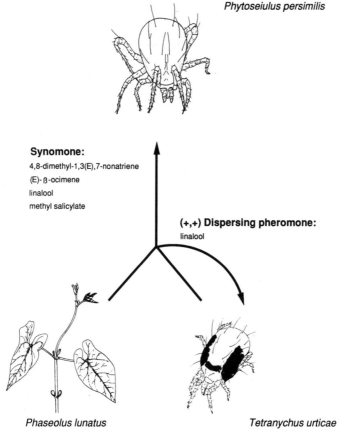

Figure 4.6. Infochemicals in a tritrophic system of predatory mites, spider mites and their host plants.

In the remainder of this chapter, we will discuss the volatile infochemical in plant-predatory mite interactions, and thus the term synomone is appropriate in this context (cf. Dicke and Sabelis, 1988a).

4.9. Costs of Bodyguard Recruitment

Attraction of predatory mites as bodyguards has clear benefits to plants that are attacked by such ravaging herbivores as spider mites. To state that the production of volatiles is an evolved response to herbivores may seem farfetched and not the most parsimonous explanation. Clearly, it may well be a response evolved for entirely other reasons which, in the context of

defense against herbivores, had the additional benefit of attracting predators. This benefit, though originating as a side effect, increases the selective advantage of this trait and is unlikely to be without any evolutionary
consequences (e.g., with respect to fine-tuning the alarm system). Hence,
whether originated as a side effect or evolved in itself, the most relevant
question to be answered is what costs are involved in the alarm system and
how these compare with the benefits gained.

4.9.1.   Instantaneous Costs Related to Infochemical Production and Release

As mentioned above (section 4.5), costs are very difficult to quantify
comprehensively. The figures used in the calculations below are all rough
estimates, so that the biosynthesis costs should be considered as an order
of magnitude, rather than the exact values.

The most abundant synomone components, (E)-β-ocimene and 4,8-di-
methyl-1,3(E),7-nonatriene, are each released in amounts of ca. 25 μg/16
h/500 g leaves (fresh weight) (Dicke et al., 1990a). The least abundant
synomone components, methyl salicylate and linalool, are released in ca.
10 times lower quantities. Production costs of terpenes are 4.7 g CO_2/g
terpene, which is about twice as much as the production costs of leaves
(dry weight basis) (Gulmon and Mooney, 1986). Thus, the energy spent on
production of each of the most abundant synomone components per 24 h
could alternatively have been used for production of 75 μg of leaf tissue
(dry weight). Assuming that fresh matter consists of 15% dry matter, this
would be equivalent to a relative growth rate of 75 μg/75 g/day, which is
10^{-6} day^{-1}. Maximum relative growth rates can attain values of 0.15 to
0.30 day^{-1} (Lambers and Dijkstra, 1987) and thus production of the
infochemical leads to a reduction in relative growth rate of 3 to 6* 10^{-4}%
per major synomone component per day. Thus, assuming equal costs on a
weight basis for all four synomone components, total biosynthetic costs
would amount to 0.65 to 1.3* 10^{-3}% of leaf production per day.

Some comments have to be made on the basic figure used in this
calculation, the release rate. This has been determined for detached leaves
in the dark. If kairomone release through stomata is substantial, the
assessed release rate may be an underestimate. Furthermore, nothing is
known about release rates from leaves attached to the plant, but if the
plant is actively involved, the release rate may be higher. Also, the given
release rate is an average: a large variation was observed, 10–100 μg/16
h/500 g leaves for the major components, most likely related to leaf
condition (recently infested or consumed to a large extent).

Biosynthetic costs have been related to maximum relative growth rate.
Thus, the reduction of growth rate resulting from infochemical production,
as calculated above, is a minimum value, and it could be proportionately

higher at slower growth rates (e.g., when photosynthate is removed by tetranychids).

Although biosynthetic costs of the synomone seem to be low, they are but one of the defense costs paid by the plant. Other costs relate to maintaining the biosynthetic pathway, avoiding autotoxicity, storage (of precursors), and metabolic turnover (Givnish, 1986). Costs of turnover may be especially high, since high metabolic turnover rates of terpenes have been recorded for other plant species (Burbott and Loomis, 1969; Croteau and Loomis, 1972; Croteau and Johnson, 1984 but see Mihaliak et al., 1991). If turnover of synomone precursors is indeed high, the functions of terpenes should be further elucidated. In addition to metabolic turnover in the plant, high terpene losses occur upon volatilization, which contributes considerably to the energetic costs paid by the plant.

Synomone production is not the only defense method, and spider mites are not the only herbivores; they are but one of an array of defense systems and herbivores, respectively. Thus, it is quite likely that the total costs for all possible defense systems and herbivores would exceed the actual "defense budget."

In conclusion, there are good reasons to expect that plants economize on energy spent in defense, even though the energy for each defense system may be small.

4.9.2. Future Costs of Infochemical Production

Assuming exponential growth of leaf biomass, Gulmon and Mooney (1986) showed that small costs paid in an early phase of growth result in a marked decrease in the final biomass at the end of the growing season. The later the investment in defense is made, the less the effect on accumulated leaf weight. These authors found, for a realistic set of parameters, that an early defended plant accumulated 21% less leaf weight than a late defended plant. Hence, inducible defenses can be more economical than constitutive ones, provided the extra costs associated with inducibility (if any) are sufficiently low.

4.9.3. Costs in Relation to Investment Strategies of Other Plants

Significant energetic costs related to infochemical production may occur. On one hand, plants may take part in an ever-continuing "arms race" to increase the production of infochemicals as long as this leads to an increased probability of detection by predators of the herbivores. On the other hand, there will be an upper release rate at which predator attraction is maximized. There may be neighboring plants which respond to the infochemical by suppressing their own infochemical production and investing in alternative defensive systems, thereby profiting from the predators attracted by neighboring plants. Sabelis and De Jong (1988) constructed a model of predator-herbivore-plant interaction with two plant types, one that invests in synomone production and one that does not. The model applied to a patchy environment and was based on the assumption that a limited number of producers were required for the stowaways, sharing the patch, to profit. Analysis of the model showed that plant populations are expected to be polymorphic with respect to synomone production, under a wide range of conditions. Attempts to assess genetic variability of plants in the field have not yet been made, but for plants of economic importance some first indications have been obtained. For example, bean plant (*Phaseoulus vulgaris*) varieties differ in attraction of predatory mites after plant infestation with spider mites (Fig. 4.7) and different apple cultivars infested by the same spider mite species release different chemical blends (Takabayashi et al., 1991). Examples also exist for constitutively-produced synomones (Elzen et al. 1985, 1986).

4.9.4. Ultimate Costs and Benefits by Evolutionary Boomeranging

Induction of synomone production may well be a more stable character on an evolutionary time scale than is constitutive production. Suppose that the costs of "SOS" signals are low, would it then pay for the plant to produce volatile synomones continually? Presumably not, because this would lead to a high frequency of predator responses without rewards in terms of prey. Any mutant plant with a slightly different signal, produced *only* after induction by the herbivore would be a more reliable interactant to the predator and could spend the saved energy for other purposes. It may therefore spread through the response evolved in the predator population. Exactly the same argument applies to the coevolution of herbivore species-specific chemicals produced by the host plant and the responses evoked in the predators. It is interesting to note that inducibility is associated with some degree of specificity of the synomone; predatory mites differentiate between spider mite species on the same host plant species (Sabelis and Van de Baan, 1983). Differences in the chemical composition of the blend of volatiles released in such cases have also been observed (Takabayashi et al., 1991). Such a specificity of the synomone will ensure recruitment of the appropriate natural enemy species.

These predictions on the evolution of inducible responses seem plausible, but caution should be exercized regarding conclusions before we know more about the costs involved in developing and maintaining an inducible system. So little is known about the elicitors of the response and how the plant "recognizes" the elicitors and then responds, that it is premature to make any quantitative statement as to when inducible defenses might be expected.

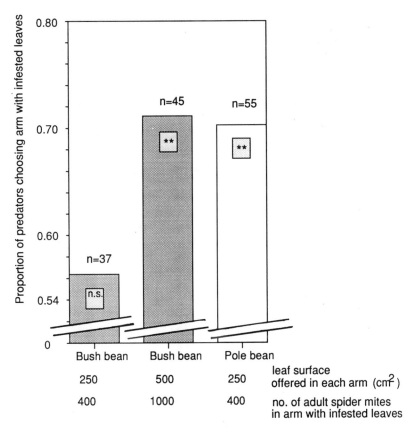

Figure 4.7. Response of *P. persimilis* in Y-tube olfactometer experiments with different *Phaseolus vulgaris* cultivars infested by *T. urticae*. Infested leaves were offered versus uninfested leaves of the same cultivar. *Phaseolus vulgaris* cultivars: (1) Bush bean "dubbele zonder draad," nr. 697 and (2) pole-bean "Westlandse dubbele" nr. 671, Turkenburg Ltd., The Netherlands.** = $P \leq 0.01$; n.s. = $p > 0.05$, sign test for differences from $50:50$ distribution of predators over the two arms. *n* represents the number of predatory mites tested. From Dicke et al., 1990b.

4.10. Effects of Small Fitness Differences on Evolutionary Rates

The true cost of defense, in evolutionary terms, is reduced reproductive success. To the extent that reproductive success is affected by reproductive output, total leaf growth will be its primary determinant. Total biomass at the start of the reproductive phase is expected to be strongly related to reproductive output. Small costs due to infochemical release may have a large impact on this biomass (cf. Gulmon and Mooney, 1986) and thus also on reproductive output.

Even if costs of synomone production were to be small, this does not imply that they are irrelevant for natural selection. A simple model discussed by Haldane and others (see Roughgarden, 1979) shows that for the one-locus-two-allele (a, A) case, with intermediate heterozygosity, non-overlapping generations, random mating, and no mutations, it takes ca. 1,000 generations (t) to change the frequency of A from 0.01 to 0.99:

$$t \approx -\ln(0.01)\left(1/\ln(v_{Aa}/v_{aa}) - 1/\ln(v_{Aa}/v_{AA})\right)$$

where v = viability and $\ln(v_{Aa}/v_{aa}) = -\ln(v_{Aa}/v_{AA}) = 0.01$. This shows that replacement of a by A is dominated by an exponential phase, thus causing the number of generations required for replacement to be quite low on an evolutionary time scale.

4.11. Future Prospects

In the tritrophic system of plants, spider mites and predatory mites, infochemicals that are emitted by plants upon spider mite damage mediate interactions between (1) plant and spider mite, (2) spider mite and predatory mite (the infochemical is spider mite species-specific) and (3) plant and predatory mite. In addition, first indications have recently been obtained that these infochemicals also affect plant-plant interactions. Uninfested plants that stand downwind from spider mite-infested plants are attractive to predatory mites (compared to control plants that stood either upwind from infested plants or downwind from uninfested plants), which is possibly caused by infochemical production by the receiving *uninfested* plants (Bruin et al., 1992; Dicke et al., 1990b; Takabayashi et al., in prep.). Chemical information mediates interactions within and between species as well as between trophic levels of this system. Thus, this tritrophic system provides a good model for continuing investigations on evolutionary aspects of information transfer, especially if genetic aspects related to the chemical differences can be incorporated. In addition, initial calculations of costs related to production of the infochemicals have raised many question on economic aspects.

Some of the questions to be answered are:

1. How do plants respond to bodyguard-attracting neighbors? Do stowaways occur, and when is their strategy applied? Do plants distinguish between infochemicals of infested conspecifics and other infested plants? Evidence for information transfer between conspecific plants has been presented for several plant species (Rhoades, 1983, 1985; Baldwin and Schultz, 1983; Zeringue, 1987; Bruin et al., 1992; Dicke et al., 1990b), but

no evidence for information transfer between plants of different species is known to us.

2. How do plants contribute to synomone production upon attack by spider mites: by *de novo* production after spider mite attack, or by having precursors available that are degraded by spider mites? Do plants produce a different synomone after receiving an infochemical from nearby spider mite-infested plants?

3. So far, four chemicals have been identified as synomone components. This was done in olfactometer tests, where one chemical was offered against a solvent control (Dicke, 1988b; Dicke et al., 1990a). However, many more chemicals are emitted after spider mite damage. These do not elicit a behavioral response in the predatory mites when offered singly. Nevertheless, they may be synomone components that are synergistic to the other four chemicals. If this is so, production costs will be much higher than calculated in this chapter. The methylene terpene 4,8,12-trimethyl-1,3(E),7(E),11-tridecatetraene deserves special attention. It is an uncommon compound, which has only recently been identified in other plants (Maurer et al., 1986; Kaiser, 1987). It is released in large quantities after spider mite infestation, but not after artificial damage (Dicke et al., 1990a). This compound does not elicit a behavioral response of *P. persimilis* females when offered singly (Dicke et al., 1990a).

4. Current calculations of production costs reveal low values: ca. $10^{-3}\%$ reduction of relative growth rate (on a daily basis). However, many costs have not been included because of a lack of knowledge. These comprise maintenance, storage and turnover costs, but also future costs in terms of photosynthesis reduction. Investigation of biosynthesis will be essential for quantification of turnover and maintenance costs. However, even when costs related to synomone production and release are low indeed, this does not mean that they are insignificant for the plant's biology on an evolutionary time scale.

5. A model study based on the assumption that production costs of volatile synomones are important revealed that evolutionary stable polymorphisms may occur under a wide range of conditions (Sabelis and De Jong, 1988). Finding "stowaways" that rely on neighbors rather than on production by themselves will offer new prospects for research. Their defense success could be compared in the presence and absence of synomone producers and the reproductive success of stowaways and synomone producers could be compared in the absence of spider mites.

6. Several plant species, for example tomato plants, defend themselves through direct defense against spider mites, even when this severely hampers predatory mite performance (Van Haren et al. 1987). This offers possibilities for gaining insight in economic aspects of indirect defense; for example: do plants with an efficient direct defense expend any energy on indirect defense?

4.12. Epilogue

What makes information-conveying chemicals so deeply interesting is that they themselves do not represent the raw material for body building nor for building bodies of progeny; they merely represent information which may or may not be utilized. Their very existence challenges biologists to state hypotheses on advantages and disadvantages to producer or inducer and receiver. This gives rise to teleological thinking, but for good reasons, namely to state and test hypotheses and in doing so, gain a better understanding of risks, costs, and benefits to those who are taking part in the transfer of information. For the same reasons, it is important to develop a terminology based on cost-benefit criteria, but this terminology should not be too complex and operational in that the criteria can be applied in an unambiguous way. We suggested in an earlier paper (Dicke and Sabelis, 1988a) to skip the origin criterion and include both producers and inducers of information flow in the terminology. Moreover, we suggest that the terminology be based on a cost-benefit analysis of no more than two interactants at a time. In doing so, classifications should include the possibility of cost-benefit balances to be negative to one or even both participants in information transfer. Problems may arise in identifying all possible costs and benefits, whether related to proximate or ultimate factors. This, we believe, is a serious problem indeed, but an interesting one; one that is central to any approach in behavioral ecology.

References

Alcock, J. 1975. Male mating strategies of some philanthine wasps (Hymenoptera: Sphecidae). *J. Kansas Entomol. Soc.* 48:532–545.

Alcock, J. 1982. Natural selection and communication among bark beetles. *Fla. Entomol.* 65:17–32.

Alcock, J., E. M. Barrows, G. Gordh, L. J. Hubbard, L. Kirkendall, D. W. Pyle, T. L. Ponder, and F. G. Zalom. 1978. The ecology and evolution of male reproductive behaviour in the bees and wasps. *Zool. J. Linn. Soc.* 64:293–326.

Aldrich, J. R., J. P. Kochansky, C. B. Abrams. 1984a. Attractant for a beneficial insect and its parasitoids: pheromone of the predatory soldier bug, *Posidus maculiventris* (Hemiptera: Pentatomidae). *Environ. Entomol* 13:1031–1036.

Aldrich, J. R., W. R. Lusby, J. P. Kochansky, and C. B. Abrams. 1984b. Volatile compounds from the predatory insect *Posidus maculiventris* (Hemiptera: Pentatomidae): male and female metathoracic scent gland and female dorsal abdominal gland secretions. *J. Chem. Ecol.* 10:561–568.

Aldrich, J. R., W. R. Lusby, and J. P. Koschansky. 1986. Identification of a new predaceous stink bug pheromone and its attractiveness to the eastern yellowjacket. *Experientia* 42:583–585.

Altenburger, R. and P. Matile. 1988. Circadian rhythmicity of fragrance emission in flowers of *Hoya carnosa* R. Br. *Planta.* 174:248–252.

Baldwin, I. T. and J. C. Schultz. 1983. Rapid changes in tree leaf chemistry induced by damage: evidence for communication between plants. *Science* 221:277–279.

Bertin, R. I. 1989. Pollination biology. In *Plant-Animal Interactions*, ed. W. G. Abrahamson, pp. 23–86. McGraw-Hill, New York.

Birch, M. C., 1984. Aggregation in bark beetles. In *Chemical Ecology of Insects*, eds. W. J. Bell and R. T. Cardé, pp. 331–353. Chapman and Hall, London.

Boppré, M. 1986. Insects pharmacophagously utilizing defensive plant chemicals (Pyrrolizidine Alkaloids). *Naturwiss.* 73:17–26.

Borg-Karlson, A. K. 1986. Chemical basis for the relationship between *Ophrys* orchids and their pollinators. III. Volatile compounds of species in the *Ophrys* sections Fuciflorae and Bombiliflorae as insect mimetic attractants/excitants. *Chemica Scripta* 27:313–325.

Borg-Karlson, A. K. and J. Tengö. 1986. Odor mimetism? Key substances in *Ophrys lutea-Andrena* pollination relationship (Orchidaceae: Andrenidea). . Chem. Ecol. 12:1927–1941.

Brown, W. L., T. Eisner, and R. H. Whittaker. 1970. Allomones and kairomones: transspecific chemical messengers. *BioScience* 20:21–22.

Bruin, J., M. Dicke, and M. W. Sabelis. 1992. Plants are better protected against spider mites after exposure to volatiles from infested conspecifics. *Experientia* (in press).

Burbott A. J. and W. D. Loomis. 1969. Evidence for metabolic turnover of monoterpenes in peppermint. *Plant Phys.* 44:173–179.

Chew, F. S. and J. E. Rodman. 1979. Plant resources for chemical defense. In *Herbivores. Their Interaction with Secondary Plant Metabolites*, eds. G. A. Rosenthal and D. H. Janzen, pp. 271–307. Academic Press, New York.

Croteau, R. and M. A. Johnson. 1984. Biosynthesis of terpenoids in glandular trichomes. In *Biology and Chemistry of Plant Trichomes*, eds. E. Rodriguez, P. L. Healy, and I. Mehta, pp. 133–185. Plenum Press, New York.

Croteau, R. and W. D. Loomis. 1972. Biosynthesis of mono- and sesquiterpenes in peppermint from mevalonate-2-^{14}C. *Phytochem.* 11:1055–1066.

Dabrowski, Z. T. and J. G. Rodriguez. 1971. Studies on resistance of strawberries to mites. 3. Preference and nonpreference responses of *Tetranychus urticae* and *T. turkestani* to essential oils of foliage. *J. Econ. Entomol.* 64:387–391.

Dicke, M. 1986. Volatile spider-mite pheromone and host-plant kairomone, involved in spaced-out gregariousness in the spider mite *Tetranychus urticae*. *Physiol. Entomol.* 11:251–262.

Dicke, M. 1988a. Prey preference of the phytoseiid mite *Typhlodromus pyri*: 1. Response to volatile kairomones. *Exp. Appl. Acarol.* 4:1–13.

Dicke, M. 1988b. Infochemicals in Tritrophic Interactions. Origin and Function in a System Consisting of Predatory Mites, Phytophagous Mites and Their Host

Plants. Ph. D. dissertation, Agricultural University, Wageningen, The Netherlands.

Dicke, M. and A. Groeneveld. 1986. Hierarchical structure in kairomone preference of the predatory mite *Amblyseius potentillae*: dietary component indispensable for diapause induction affects prey location behaviour. *Ecol. Entomol.* 11:131–138.

Dicke, M. and M. W. Sabelis. 1988a. Infochemical terminology: should it be based on cost-benefit analysis rather than origin of compounds? *Funct. Ecol.* 2:131–139.

Dicke, M. and M. W. Sabelis, M. W. 1988b. How plants obtain predatory mites as bodyguards. *Neth. J. Zool.* 38:148–165.

Dicke, M., M. W. Sabelis, and A. Groeneveld. 1986. Vitamin A deficiency modifies response of predatory mite *Amblyseius potentillae* to volatile kairomone of two-spotted spider mite, *Tetranychus urticae*. *J. Chem. Ecol.* 12:1389–1396.

Dicke, M., T. van Beek, M. A. van Posthumus, N. Ben Dom, H. van Bokhoven, and Æ. de Groot. 1990a. Isolation and identification of volatile kairomone that affects acarine predator prey interactions: involvement of host plant in its production. *J. Chem. Ecol.* 16:381–396.

Dicke, M., M. W. Sabelis, J. Takabayashi, J. Bruin, and M. A. Posthumus. 1990b. Plant strategies of manipulating predator-prey interactions through allelochemicals: prospects for application in pest control. *J. Chem. Ecol.* 16:3091–3118.

Eberhard, W. G. 1977. Aggressive chemical mimicry by a bolas spider. *Science* 198:1173–1175.

Eller, F. J., R. J. Bartelt, R. L. Jones, H. M. Kulman. 1984. Ethyl (Z)-9-hexadecenoate, a sex pheromone of *Syndipnus rubiginosus*, a sawfly parasitoid. *J. Chem. Ecol.* 10:291–300.

Elzen, G. W., H. J. Williams, A. A. Bell, R. D. Stipanovic, and S. B. Vinson. 1985. Quantification of volatile terpenes of glanded and glandless *Gossypium hirsutum* L. cultivars and lines by gas chromatography. *J. Agric. Food Chem.* 33:1079–1082.

Elzen, G. W., H. J. Williams, and S. B. Vinson. 1986. Wind tunnel flight responses by the hymenopterous parasitoid *Campoletis sonorensis* to cotton cultivars and lines. *Entomol. Exp. Appl.* 42:285–289.

Fletcher, B. S. 1968. Storage and release of a sex pheromone by the Queenland fruit fly, *Dacus tryoni* (Diptera: Trypetidae). *Nature* 219:631–632.

Gäbler, A., W. Boland, U. Preiss, and H. Simon. 1991. Stereochemical studies on homoterpene biosynthesis in higher plants; Mechanistic, phylogenetic and ecological aspects. *Helv. Chim. Acta.* 74:1–17.

Gerling, D. 1966. Biological studies on *Encarsia formosa* (Hymenoptera: Aphelinidae). *Ann. Entomol. Soc. Am.* 59:142–143.

Gibson, R. W. and J. A. Pickett. 1983. Wild potato repels aphids by release of aphid alarm pheromone. *Nature* 302; 608–609.

Givnish, T. J. 1986. Economics of biotic interactions. In *On the Economy of Plant Form and Function*, (ed. T. J. Givnish,) pp. 667–680. Cambridge Univ. Press, Cambridge, England.

Glynn, P. W. 1980. Defense by symbiotic crustacea of host corals elicited by chemical cues from predator. *Oecologia* 47:287–290.

Gulmon, S. L. and H. A. Mooney. 1986. Costs of defense and their effects on plant productivity. In *On the Economy of Plant Form and Function*, ed. T. J. Givnish, pp. 681–698. Cambridge Univ. Press, Cambridge, England.

Harborne, J. B. 1988. *Introduction to Ecological Biochemistry*, 3rd ed. Academic Press, London.

Hardee, D. D., W. H. Cross, and E. B. Mitchell. 1969. Male boll weevils are more attractive than cotton plants to boll weevils. *J. Econ. Entomol.* 62:165–169.

Haren, R. J. F. van, M. M. Steenhuis, M. W. Sabelis, and O. M. B. De Ponti. 1987. Tomato stem trichomes and dispersal success of *Phytoseiulus persimilis* relative to its prey *Tetranychus urticae. Exp. Appl. Acarol.* 3:115–121.

Harris, V. E. and J. W. Todd. 1980. Male-mediated aggregation of male, female and 5th-instar southern green stink bugs and concomitant attraction of a tachinid parasite, *Trichopoda pennipes. Entomol. Exp. Appl.* 27:117–126.

Helle, W. and M. W. Sabelis. 1985a. *Spider Mites. Their Biology, Natural Enemies and Control*. World Crop Pests Vol. 1A. Elsevier, Amsterdam.

Helle, W. and M. W. Sabelis. 1985b. *Spider Mites. Their Biology, Natural Enemies and Control*. World Crop Pests Vol. 1B. Elsevier, Amsterdam.

Hölldobler, B. 1971. Communication between ants and their guests. *Sci. Amer.* 224(3):86–93.

Howard, R. W., C. A. McDaniel, and G. J. Blomquist. 1980. Chemical mimicry as an integrating mechanism: cuticular hydrocarbons of a termitophile and its host. *Science* 210:431–433.

Hubbard, S. F., G. Marris, A. Reynolds, and G. W. Rowe. 1987. Adaptive patterns in the avoidance of superparasitism by solitary parasitic wasps. *J. Anim. Ecol.* 56:387–401.

Kaiser, R. 1987. Night-scented flowers, not only attractive to moths. Unpublished paper presented at the EUCHEM 1987 Conference, "Semiochemicals in Plant and Animal Kingdoms," Angers, France, October 12–16, 1987.

Kullenberg, B. and G. Bergström. 1976. The pollination of *Ophrys* orchids. *Bot. Notiser* 129:11–19.

Lambers, H. and P. Dijkstra. 1987. A physiological analysis of genotypic variation in relative growth rate: can growth rate confer ecological advantage? In *Disturbance in Grasslands. Causes, Effects and Processes*, eds. J. van Andel, J. P. Bakker, and R. W. Snaydon, pp. 239–253. Dr. W. Junk Pubs., Dordrecht, The Netherlands.

Lambers, H. and A. Rychter. 1989. The biochemical background of variation in respiration rate: respiratory pathways and chemical composition. In *Causes and Consequences of Variation in Growth Rate and Productivity of Higher Plants*, eds. H. Lambers, M. L. Cambridge, H. Konings, and T. L. Pons, pp. 199–225. SPB Academic Publishing, The Hague.

Lenteren, J. C. van, 1981. Host discrimination by insect parasitoids. In D. A. Nordlund, R. L. *Semiochemicals. Their Role in Pest Control*, eds. D. A. Nordlund, R. L. Jones and W. J. Lewis, pp. 153–179. John Wiley, New York.

Lenteren, J. C. van, H. W. Nell, L. A. Sevenster-van der Lelie, and J. Woets. 1976. The parasite-host relationship between *Encarsia formosa* (Hymenoptera: Aphelinidae) and *Trialeurodes vaporariorum* (Homoptera: Aleyrodidae). III. Discrimination between parasitized and unparasitized hosts by the parasite. *Z. Ang. Ent.* 81:377–380.

Lenteren, J. C. van, H. W. Nell, and L. A. Sevenster-van der Lelie. 1980. The parasite-host relationship between *Encarsia formosa* (Hymenoptera: Aphelinidae) and *Trialeurodes vaporariorum* (Homoptera: Aleyrodidae). IV. Oviposition behaviour of the parasite, with aspects of host selection, host discrimination and host feeding. *Z. Ang. Ent.* 89:442–454.

Malmqvist, B. 1987. Drift of mayflies in response to disturbance by predacious stonefly nymphs. *Proc. 3rd Eur. Congr. Entomol.* 1:107–110.

Mangel, M. and B. D. Roitberg. 1989. Dynamic information and host acceptance by a tephritid fruit fly. *Ecol. Entomol.* 14:181–189.

Marris, G. C., S. F. Hubbard, and C. M. Scrimgeour. 1989. The discrimination of genetic similarity in the oviposition behaviour of solitary insect parasitoids. (submitted).

Matile, P. and R. Altenburger. 1988. Rhythms of fragrance emission in flowers. *Planta* 174:242–247.

Maurer, B., A. Hauser, and J. C. Froidevaux. 1986. (E)-4,8-dimethyl-1,3,7-nonatriene and (E, E)-4,8,12-trimethyl-1,3,7,11-tridecatetraene, two unusual hydrocarbons from cardamom oil. *Tetrahedron. Lett.* 27:2111–2112.

McNeil, J. N. and J. Delisle. 1989. Host plant pollen influences calling behavior and ovarian development of the sunflower moth, *Homoeosoma electellum*. *Oecologia* 80:201–205.

Mihaliak, C. A., J. Gershenzon, and R. Croteau. 1991. Lack of rapid monoterpene turnover in rooted plants—Implications for theories of plant chemical defense. *Oecologia* 87:373–376.

Nault, L. R. and P. L. Phelan. 1984. Alarm pheromones and sociality in pre-social insects. In *Chemical Ecology of Insects*, eds. W. J. Bell & R. T. Cardé, pp. 237–256. Chapman and Hall, London.

Nault, L. R., M. E. Montgomery, and W. S. Bowers. 1976. Ant-aphid association: role of aphid alarm pheromone. *Science* 192:1349–1351.

Nguyen, R. and R. I. Sailer. 1987. Facultative hyperparasitism and sex determination of *Encarsia smithi* (Silvestri)(Hymenoptera: Aphelinidae). *Ann. Entomol. Soc. Am.* 80:713–719.

Noldus, L. P. J. J. 1989. Chemical Espionage by Parasitic Wasps: How *Trichogramma* species Exploit Moth Sex Pheromone Systems. Ph. D. dissertation, Agricultural University, Wageningen, The Netherlands.

Nordlund, D. A. 1981. Semiochemicals: a review of the terminology. In *Semiochemicals. Their Role in Pest Control*. eds. D. A. Nordlund, R. L. Jones, and W. J. Lewis, pp. 13–28. Plenum, New York.

Nordlund, D. A. and W. J. Lewis. 1976. Terminology of chemical-releasing stimuli in intraspecific and interspecific interactions. *J. Chem. Ecol.* 2:211–220.

Nordlund, D. A., W. J. Lewis, and M. A. Altieri. 1988. Influences of plant-produced allelochemics on the host and prey selection behavior of entomophagous insects. In *Novel Aspects of Insect-Plant Interactions.* eds. P. Barbosa and D. K. Letourneau pp. 65–90. John Wiley, New York.

Price, P. W., C. E. Bouton, P. Gross, B. A. McPheron, J. N. Thompson, and A. E. Weis. 1980. Interactions among three trophic levels: influence of plant interactions between insect herbivores and natural enemies. *Annu. Rev. Ecol. Syst.* 11:41–65.

Prokopy, R. J. 1981. Epideictic pheromones that influence spacing patterns of phytophagous insects. In *Semiochemicals. Their Role in Pest Control*, eds. D. A. Nordlund, R. L. Jones and W. J. Lewis, pp. 183–213. John Wiley, New York.

Prokopy, R. J., B. D. Roitberg, and A. L. Averill. 1984. Resource partitioning. In *Chemical Ecology of Insects*, eds. W. J. Bell and R. T. Cardé, pp. 301–330. Chapman and Hall, London.

Raffa, K. F. and A. A. Berryman. 1983. The role of host plant resistance in the colonization behavior and ecology of bark beetles (Coleoptera: Scolytidae). *Ecol. Monogr.* 53:27–49.

Raina, A. K. 1988. Selected factors influencing neurohormonal regulation of sex pheromone production in *Heliothis* species. *J. Chem. Ecol.* 14:2063–2069.

Rhoades, D. F. 1983. Responses of alder and willow to attack by tent caterpillars and webworms: evidence for pheromonal sensitivity of willows. In *Plant Resistance to Insects*, ed. P. A. Hedin, pp. 55–68. American Chemical Society Symposium Series 208, Washington, DC.

Rhoades, D. F. 1985. Pheromonal communication between plants. *Chemically Mediated Interactions Between Plants and Other Organisms*, eds. G. A. Cooper-Driver, T. Swain, and E. C. Conn. *Recent Advances in Phytochemistry* 19:195–218. Plenum Press, New York.

Riddiford, L. M. 1967. *Trans*-2-hexenal: mating stimulant for polyphemus moths. *Science* 158:139–141.

Riddiford, L. M. and C. M. Williams. 1967. Volatile principle from oak leaves: role in sex life of the polyphemus moth. *Science* 155:589–590.

Roitberg, B. D. 1992. Why an Evolutionary Perspective? (this volume).

Roitberg, B. D. and M. Mangel, M. 1989. On the evolutionary ecology of marking pheromones. *Evol. Ecol.* 2:289–315.

Roughgarden, J. 1979. *Theory of Population Genetics and Evolutionary Ecology: An Introduction.* MacMillan, New York.

Rutowski, R. L., 1981. The function of pheromones. *J. Chem. Ecol.* 7:481–484.

Sabelis, M. W. 1981. Biological control of two-spotted spider mites using phytoseiid predators. Part I. Modelling the predator-prey interaction at the individual level. *Agric. Res. Rep.* 910, Pudoc, Wageningen, The Netherlands.

Sabelis M. W. and B. P. Afman, 1984. Factors initiating or suppressing aerial dispersal of the predatory mite *Phytoseiulus persimilis*. *Abstr. 17th Int. Cong. Entomol., Hamburg, August 1984*, p. 445.

Sabelis, M. W. and H. E. van de Baan. 1983. Location of distant spider mite colonies by phytoseiid predators: demonstration of specific kairomones emitted by *Tetranychus urticae* and *Panonychus ulmi*. *Entomol. Exp. Appl.* 33:303–314.

Sabelis, M. W. and M. Dicke. 1985. Long-range dispersal and searching behaviour. In *Spider Mites. Their Biology, Natural Enemies and Control. World Crop Pests, 1B*, eds. W. Helle and M. W. Sabelis, pp. 141–160. Elsevier, Amsterdam.

Sabelis, M. W. and M. C. M. de Jong. 1988. Should all plants recruit bodyguards? Conditions for a polymorphic ESS of synomone production in plants. *Oikos*, 53:247–252.

Sabelis, M. W. and A. Janssen. 1991. Optimal food selection by passive dispersers (in prep.).

Sabelis, M. W. and J. van der Meer. 1986. Local dynamics of the interaction between predatory mites and two-spotted spider mites. In *Dynamics of Physiologically Structured Populations*, eds. J. A. J. Metz and O. Diekmann, pp. 322–343. Lecture Notes in Biomathematics, Springer-Verlag, Berlin.

Sabelis, M. W., B. P. Afman, and P. J. Slim. 1984a. Location of distant spider mite colonies by *Phytoseiulus persimilis*: localization and extraction of a kairomone. *Acarology VI*, 1:431–440.

Sabelis, M. W., J. E. Vermaat, and A. Groeneveld. 1984b. Arrestment responses of the predatory mite, *Phytoseiulus persimilis*, to steep odour gradients of a kairomone. *Physiol. Entomol.* 9:437–446.

Schneider, D. 1987. The strange fate of pyrrolizidine alkaloids. In *Proceedings in Life Sciences: Perspectives in Chemoreception and Behavior*. eds. R. F. Chapman, E. A. Bernays, and J. G. Stoffolano, Jr. pp. 123–142. Springer-Verlag, New York.

Stowe, M. K., J. H. Tumlinson, and R. R. Heath. 1987. Chemical mimicry: bolas spider emits components of moth prey species sex pheromones. *Science* 236:964–967.

Sullivan, T. P. and D. R. Crump. 1986. Feeding responses of snowshoe hares (*Lepus americanus*) to volatile constituents of red fox (*Vulpes vulpes*) urine. *J. Chem. Ecol.* 12:729–735.

Takabayashi, J. and M. Dicke. 1992. Response of predatory mites with different rearing histories to volatiles of uninfested plants. *Entomol. Exp. Appl.* (in press).

Takabayashi, J. and S. Takahashi. 1986. Effect of kairomones in the searching behavior of *Apanteles kariyai* Watanabe (Hymenoptera: Braconidae), a parasitoid of the common armyworm, *Pseudaletia separata* Walker (Lepidoptera: Noctuidae). III. Synthesis and bioassay of arrestants and related compounds. *Appl. Ent. Zool.* 21:519–524.

Takabayashi, J. and S. Takahashi. 1990. An allelochemical elicits arrestment in *Apanteles kariyai* in feces of non-host larvae *Acantholeucania loreyi*. *J. Chem. Ecol.* 16:2009–2017.

Takabayshi, J., M. Dicke, J. Kemerink, and T. Veldhuizen. 1990. Environmental effects on production of plant synomone that attracts predatory mites. In *Insect-Plant Interactions*. eds. T. Jeremy and A. Szentesi. Akademiai Kiado, Budapest, Hungary, *Symp. Biol. Hung.* 39:541–542.

Takabayashi, J., M. Dicke, and M. A. Posthumus. 1991. Variation in composition of predator-attracting allelochemicals emitted by herbivore-infested plants: relative influence of plant and herbivore. *Chemoecology* 2:1–6.

Takabayshi, J., T. Noda, and S. Takahashi. 1985. Effect of kairomones in the searching behavior of *Apanteles kariyai* Watanabe (Hymenoptera: Braconidae), a parasitoid of the common armyworm, *Pseudaletia separata* Walker (Lepidoptera: Noctuidae). I. Presence of arresting stimulants produced by the host larvae. *Appl. Ent. Zool.* 20:484–489.

Tamaki, Y. 1985. Sex pheromones. In *Comprehensive Insect Physiology, Biochemistry and Pharmacology*, eds. G. A. Kerkut and L. I. Gilbert, Vol. 9, pp. 145–190. Pergamon, Oxford.

Tietjen, W. J. and J. S. Rovner. 1982. Chemical communication in lycosids and other spiders. In *Spider Communication: Mechanisms and Ecological Significance*, eds. P. N. Witt and J. S. Rovner, pp. 249–279. Princeton University Press, Princeton, NJ.

Tomczyk, A. and Kropczynska, D. 1985. Effects on the host plant. In *Spider Mites. Their Biology, Natural Enemies and Control. World Crop Pests*, Vol. 1A, eds. W. Helle and M. W. Sabelis, pp. 317–329. Elsevier, Amsterdam.

VanderMeer, R. K. and D. P. Wojcik. 1982. Chemical mimicry in the Myrmecophilous beetle *Myrmecaphodius excavaticollis*. *Science* 218:806–808.

Visser, J. H. 1986. Host odor perception by phytophagous insects. *Annu. Rev. Entomol.* 31:121–144.

Vité, J. P. and G. B. Pitman, 1968. Bark beetle aggregation: effects of feeding on the release of pheromones in *Dendroctonus* and *Ips*. *Nature* 218:169–170.

Whitman, D. W. 1988. Allelochemical interactions among plants, herbivores, and their predators. In *Novel Aspects of Insect-Plant Interactions*, eds. P. Barbosa and D. K. Letourneau, pp. 11–64. John Wiley, New York.

Whittaker, R. H. and P. P. Feeny. 1971. Allelochemics: chemical interactions between species. *Science* 171:757–770.

Wood, D. L. 1982. The role of pheromones, kairomones, and allomones in the host selection and colonization behavior of bark beetles. *Annu. Rev. Entomol.* 27:411–446.

Woodhead, S. and E. A. Bernays. 1977. Changes in release rates of cyanide in relation to palatability of *Sorghum* to insects. *Nature* 270:235–236.

Yeargan, K. V. 1988. Ecology of a bolas spider, *Mastophora hutchinsoni*: phenology, hunting tactics, and evidence for aggressive chemical mimicry. *Oecologia* 74:524–530.

Zeringue, H. J. Jr. 1987. Changes in cotton leaf chemistry induced by volatile elicitors. *Phytochemistry* 26:1357–1360.

5

A Physiological Perspective

Murray B. Isman

University of British Columbia

5.1. Introduction

5.1.1. Chemicals in the Environment as Mediators

There remains little doubt that for many insect species, chemicals in the environment are instrumental mediators of both intraspecific and interspecific interactions. Although other sensory modalities are more significant for many aspects of the life history of particular species, it is likely that for every insect species, some exogenous chemicals act as natural ecological effectors.

With over a million species described to date (primarily based on morphology) and perhaps several million yet undescribed, insects represent the most species-rich taxon on earth. In parallel with this staggering diversity, one might expect an equally overwhelming diversity of chemicals produced by insects. However, relative to the plant kingdom, insects appear to be conservative with respect to the diversity of natural products they produce (that is, pheromones and allomones). On the other hand, because insects are the masters of adaptive radiation and occupy the broadest range of ecological niches, we should expect that insects collectively possess the capacity to respond to what appears as an almost limitless array of natural substances in their environments which serve as cues to hosts for both feeding and oviposition. Such evolutionary "exposure" to environmental chemicals demands both an expansive receptor breadth and effective filters in insect sensilla with which to receive and interpret (make sense of) these signals.

In reality, only a small subset of all the potentially information-conveying chemicals in any particular environment will be ecologically relevant to a particular species of insect, perhaps resulting in economy in sensillar function to focus on those chemicals which are of greatest importance to individuals of that species. In simpler terms, insects must have evolved mechanisms to optimize the "signal-to-noise" ratio for the detection of chemical mediators which enhance genetic fitness (via growth, development, survival, and fecundity) by acting as "positional and behavioral indicators" (they permit the insect to optimize its position in the environment and act appropriately).

In addition to nutrients, insects (particularly plant-feeders) frequently must ingest substances which are non-nutritional, or in the extreme, potentially deleterious to fitness. In this context, insects must have the capacity to cope with potentially toxic substances, especially in cases where the "toxins" fail to elicit a behavioral response (avoidance). Detailed study of certain insect species which specialize on "toxic" plants have uncovered some remarkable adaptations for dealing with putative dietary toxins, which, in the most extreme cases, result in idiosyncratic utilization of toxins as serendipitous nutrients or as sequestered defenses against natural enemies. Again, because insects are faced with such a diversity of putative toxins in nature, the mechanisms for dealing with xenobiotics, at least initially in evolutionary time, must have been substrate nonspecific. Where an insect species has evolved to specialize on a limited diet, we might expect to see increasing specificity of the mechanisms (for example, detoxicative enzymes) in parallel to the reduced potential breadth of toxins encountered.

5.1.2. The Physiological Filter

Insects face physiological limitations in their abilities to perceive, act upon or otherwise deal with chemical mediators in the environment. In the case of olfaction, many insects have the biological "need" for a finely-tuned sensory system capable of perceiving minute quantities of pheromones emitted by potential mates. At the same time, though, the insect may need a more broadly-tuned chemosensory system allowing for discrimination of acceptable from unacceptable host plants with respect to oviposition. Although these "goals" seem diametrically opposed, many insects appear remarkably adept at responding to both types of chemical signals in a manner which enhances their survival. Generally speaking, location of a mate is a more specific challenge than the location of hosts for oviposition and feeding (or subsequent feeding and development of the offspring), and therefore, in some species, evolution may have favored olfactory systems which optimize the perception of conspecific pheromones, perhaps at the expense of a more discriminatory olfactory system aimed at hosts.

In acquiring nutrients, insects run the disk of exposure to dietary toxins. Insects have evolved a number of physiological strategies and adaptations to avoid a toxic outcome, and although many of these adaptations in theory entail a cost to the particular individuals possessing them, demonstrations of such costs have been most difficult. For example, an insect depending on metabolic enzymes to detoxify a dietary constituent may have to invest energy in the production of the enzyme; in the absence of the toxin in the diet, that investment could be used more effectively for structural proteins or other enzymes with the end result of faster development (see Caprio and Tabashnik, this volume).

Another means of avoiding toxicity is to limit the passage of deleterious chemicals into the hemolymph from the digestive tract. However, the insect midgut is relatively non-selective as a physiological filter, and thus, reduced nutrient availability may be an undesired result (Berenbaum and Isman, 1989).

The fact that insects are observed to suffer from ingested allelochemicals, fail to perceive conspecific pheromones under certain conditions, and make ovipositional "mistakes" is evidence that there are real limitations on insects for the utilization of chemicals as ecological mediators. There are many examples were reproductive isolation between species is maintained by differences in behavioral response to sex pheromones, or by host plant shifts, presumably with concomitant differences in response to respective host chemistries. The objectives of this chapter are to explore facets of insect physiology that influence the adaptiveness of insects to environmental chemicals and to consider physiological targets on which natural selection might act to guide the evolution of chemically-mediated systems.

5.2. Sensory Perception of Chemicals

5.2.1. Receptor Specificity and Peripheral Response

Olfactory sensilla of insects are located exclusively on the antennae, but contact (gustatory) sensilla can be found on the mouthparts, in the oral cavity, on the tarsi, on antennae, and even on the ovipositor. The chemosensory cells of the sensilla can be broadly categorized as either generalists or specialists, depending on the specificity of receptors on the dendritic membrane (of the cell) (see Stadler, 1984, and Mustaparta, 1984, for reviews). Specialist cells, by definition, show extreme sensitivity to one (or a very few) substance(s). A classic example is the "pheromone receptor" on the antennae of lepidopterans. Such cells can discriminate between positional, geometric, and stereochemical (enantiomeric) isomers (Ritter, 1989). However, the specificity of these cells is not absolute; closely related compounds can often be stimulatory, albeit at concentrations far greater (orders of magnitude) than that required of the key stimulant. Some cases have been described wherein closely related compounds (e.g., stereoisomers) actually inhibit the behavioral response to an attractant pheromone (Mustaparta, 1984), although it is unclear as to whether this is attributable to inhibition at the receptor site or to stimulation of a second cell which is interpreted by the central nervous system in an opposite manner.

At the behavioral level, a simplistic system would be one in which each insect species produces a unique pheromonal substance, and individuals of the attracted sex would have olfactory receptors specifically tuned to that substance alone. In reality, at least based on our current understanding of lepidopteran and coleopteran species, there is considerable economy with respect to the synthesis of different pheromonal substances. As a result, reproductive isolation frequently depends on blends of two or more constituents, with different species responding most strongly to blends of characteristic proportions. It has been well-documented that many species of moths have sensilla containing neurones (cells) of two or more kinds, based on the specific pheromonal constituents they respond to. However, it is important to note that interactions between pheromonal components occur at the level of integration (in the brain) and not at the peripheral level. An extreme example would be in the case of a two-component pheromone in which one component is stimulatory, based on the insect's behavioral response (e.g., upwind flight), and the other inhibitory. The observed behavior would be a lack of response, yet at the peripheral level (the antennae), both types of neurons would be stimulated and sending their respective signals to the brain.

Generalist cells are well-exemplified by the gustatory receptors found on the mouthparts of insects. Given the chemical complexity of food, whether the insect is phytophagous or predatory (in the broadest sense, that is, including blood-feeders), it would seem logical that the sensory organs charged with discriminating acceptable from unacceptable foods would consist of a large number of relatively nonspecific cells with broadly overlapping spectra of sensitivity. In this way, a chemical search image or "template" can be used as a basis for food acceptance. This is not to say that individual host chemicals cannot be pivotal in food selection. Although some insect species enjoy a wide host range, the majority (at least among phytophagous species) have limited host ranges (they are oligophagous or monophagous). Two hypotheses exist which seek to explain host range limitation. The first is that the limitation of host range results from the insects' response to characteristic (= token) feeding cues in their hosts. Perhaps surprisingly, no such host-unique chemicals have been found for some well-studied monophagous plant feeders such as the Colorado potato beetle, *Leptinotarsa decemlineata*, the tobacco hornworm, *Manduca sexta*, and the monarch butterfly, *Danaus plexippus*. However, Mitchell et al. (1990) recently found that adult *L. decemlineata* show a highly characteristic electrophysiological response to expressed leaf sap of potato, their normal host, and highly variable responses to plant saps of non-hosts. This response, though, is not likely related to the characteristic glycoalkaloids of potato and related solanaceous plants, but to more widely distributed nutrient chemicals instead (Mitchell, 1988). An alternate hypothesis is that most insects avoid non-hosts owing to feeding deterrents in these plants (Jermy, 1966; Chapman and Bernays, 1989). This latter hypothesis failed to receive support from a recent study which indicates that *M. sexta* is insensitive to a wide range of allelochemicals from non-host plants when presented on host (*Nicotiana tabacum*) leaf discs (Wrubel and Bernays, 1990, and discussion therein). Thus, the mechanism accounting for limitation in host range for many phytophagous insects remains a point of controversy.

Sources of variation in chemoreceptor activity have been of interest to a number of investigators. There is evidence for desensitization of sensilla at the peripheral level. For example, the triterpenoid antifeedant azadirachtin inhibits sensilla which normally respond to phagostimulants such as sugars (Simmonds and Blaney, 1984). Internal inputs may also influence the receptivity of peripheral receptors. A striking example of this is seen in locusts: injection of amino acid solutions into the hemolymph significantly reduces the sensitivity of amino acid receptors on the maxillary palps (Abisgold and Simpson, 1988).

5.2.2. Neural Integration and Central Inputs

Before an insect can respond in an appropriate manner (that is, with a specific behavior) to an ecologically important chemical cue, the chemical signal must first be translated into an electrical signal, the "currency of exchange" of the nervous system. That translation from a chemical signal into a pattern of electrical events is the process of neural coding, that is to say, the message (quality and quantity of the stimulus) is encoded. Two general models for neural coding have been described (see Stadler, 1984, for a concise review). Where a high degree of specificity is involved (e.g., pheromone perception), neurones primarily sensitive to a single substance may have evolved. Such dedicated neural pathways, in which the message is encoded largely by a single class of neural input, are termed "labeled lines." Primary neurons (those with dendrites in the chemosensory organs) involved in pheromone perception are thought to act as labeled lines, but there is also some evidence for labeled lines in the mouthparts of grasshoppers and locusts which respond to specific feeding deterrents (Chapman, 1988).

Where the chemical signal is a complex one, with several to many stimulatory components, the insect may rely on a large number of cells with broadly overlapping sensitivity. These transmit information on the quality and quantity of the stimulus by an "across-fiber patterning" and is probably the basis for acceptance of hosts for feeding and oviposition for most species. In this case, a theoretical chemical search image or template exists at the level of integration of the primary neurons. Note, however, that Mitchell et al. (1990) have documented what amounts to a labeled line for a complex stimulus, potato leaf sap, in the Colorado potato beetle.

However, creation of a neural code representative of the stimulus may not always result in a behavioral response. Primary neurons converge in ganglia where the information may be integrated with messages from other external stimuli as well as with internal inputs representing phenomena such as hunger (in the case of feeding) or chronological age (as one modifier of pheromone perception). Among the functions of the integrating ganglia are enhancement of the signal-to-noise ratio and amplification of the signal.

One particularly well-investigated chemosensory system is the pheromone reception system of the male tobacco hornworm moth (*Manduca sexta*) (Homberg et al., 1989). Primary neurons which are stimulated by either of two constituents of the pheromone converge in the macroglomerular complex, a male-specific region of the antennal lobe of the brain. Some of the output interneurons arising in the macroglomerular complex are stimulated by one or both of the pheromonal constituents, whereas other interneurons are inhibited by them.

Through interconnections within this ganglion, a number of different physiological responses (neural codes) can be generated, including excitation to one of the pheromonal components alone, excitation by either component, or a "mixed" response to the complete pheromone (Christensen et al., 1989). The last response may be of special importance in that it could allow the moth to track discontinuities in the pheromonal plume, and therefore optimize sustained upwind flight toward the pheromone source (Baker, 1989). Studies of the neuroanatomy of male hornworm moths provide evidence "that the pheromone-processing subsystem is largely separate from the rest of the olfactory system" (Homberg et al., 1989). However, experimental evidence supports the likelihood that a common motor path is used by the pheromone-specific interneurones of males and the tobacco (host plant)-odor-specific interneurones of females, both of which result in anemotactic behaviors.

A comparison between the activity of chemosensilla during feeding and absolute feeding rate in the cabbageworm (*Pieris brassicae*) allowed the construction of a simple model of a hypothetical "feeding center" in the central nervous system (Schoonhoven and Blom, 1988). Consumption in this insect is highly correlated to chemosensory cell stimulation; nerve impulses signalling the presence of different phagostimulants (sugars, amino acids, glucosinolates) are algebraically summed in the feeding center. However, impulses from phagodeterrent cells counteract the effects of phagostimulants. Such impulses from deterrent receptors take priority in the feeding center, since on average, one impulse from a deterrent neuron "neutralizes" 2.5 impulses from the phagostimulant neurons. Internal inputs (monitoring satiation) are likely also integrated in the feeding center, which is probably located in the suboesophageal ganglion. The authors speculate that the feeding center may consist of only a single "decision-making" interneuron.

It is noteworthy that feeding deterrents for phytophagous insects, based on studies conducted to date, may act both centrally as in the cabbage-worm, or peripherally, as in the case of azadirachtin in the locust. Similarly, internal inputs may exert their influence both centrally (as in the Schoonhoven and Blom model) or peripherally (see 5.2.1). Studies of pheromone interruption between sympatric species suggest that in both lepidopterans and coleopterans (Scolytidae), the site for interspecific inter-ruption is at the level of integration, rather than at the receptor (periph-eral) level (reviewed in Mustaparta, 1984 and Byers, 1989). However, a recent investigation of pheromone analogues in the European corn borer (Schwarz et al., 1990) revealed that a number of these compounds blocked the response of male moths to their normal pheromone in flight tunnel tests; the inhibitory analogues themselves were not attractive. These authors suggest that the mechanism for disruption of attraction may be competitive binding of these analogues to the antennal receptors, with the further possibility that they undergo slower degradation compared to the pheromonally active compounds.

5.2.3. Neural Targets for Selection

As both reception (peripheral) and integration (central) are important in chemically-mediated behaviors of insects, an obvious question is: does selection act predominantly at the peripheral level or the central level? Although there is not an overwhelming body of data that addresses this question, the available evidence suggests that either site may be a proximal target for natural selection. The best means of investigating this issue is by comparisons of closely related taxa, for example conspecific races or sibling species.

One such taxon which has been the subject of this type of investigation is the genus *Yponomeuta*, the ermine moths. This genus consists of eight closely related species, each with narrow, but different, host ranges: speciation may well be the result of host plant shifts from the ancestral host plant family Celastraceae. Larvae of the three species which still utilize hosts in the Celastraceae have gustatory cells sensitive to dulcitol, a characteristic substance of this host family which stimulates feeding. *Yponomeuta* species which feed on rosaceous hosts have cells sensitive to sorbitol, a stereoisomer of dulcitol, although they have retained cells keyed to dulcitol (van Drongelen, 1979); species feeding on non-rosaceous plants are sorbitol-insensitive. In other species which have shifted in evolutionary time away from the Celastraceae, dulcitol sensitivity has been lost. Similar changes in chemosensory physiology have arisen with respect to feeding deterrents. *Y. evonymellus*, which feeds on *Prunus*, possess cells stimulated by the deterrent glycoside phlorizin. *Y. mallinellus*, feeding on *Malus* which is rich in phlorizin, lacks this receptor (van Drongelen, 1979).

A recent investigation of gustatory neurons in larvae of the closely related cabbageworms *Pieris brassicae* and *P. rapae* found significant differences in sensitivity to various amino acids in the two species (van Loon and van Eeuwijk, 1989). In particular, three amino acids which are only weak stimulants for *P. rapae*, histidine, phenylalanine and tryptophan, were among the four most stimulatory for *P. brassicae*. The authors suggest that "such a clear shift in effectiveness between the related species may indicate the presence of separate specific receptor sites for these amino acids on the amino acid chemoreceptor" (van Loon and van Eeuwijk, 1989).

Intraspecific differences (the raw material for evolution by natural selection) of this nature have also been reported. Hovanitz (1969) noted differences in host plant preference between two strains of *P. rapae*, which were correlated to quantitative differences in glucosinolates in the hosts. F_1 hybrids resulting from a cross of the two strains were exactly intermediate in their preference for high versus low glucosinolate content, suggesting a simple genetic basis for preference in this case. Although it is tempting to speculate that the difference in preference is a manifestation of differences in receptor numbers, differences in neural integration cannot be ruled out. A similar, but more detailed, study investigated the differences in sensitivity of deterrent cells in two strains of the noctuid *Mamestra brassicae* (Wieczorek, 1976). The strains showed quantitative differences in their response to two deterrents, which may be a reflection of differences in the ratio of receptors sites on the dendritic membrane (Schoonhoven, 1982).

Intraspecific differences in feeding behavior have also been observed in the silkworm, *Bombyx mori*, and the fruitfly, *Drosophila melanogaster*, where mutants exist for broadened host range and high salt tolerance, respectively. In both cases the peripheral sensory systems do not differ between the mutants and their respective wild types, supporting the conclusion that changes in gustatory perception have arisen at the level of integration (Ishikawa et al., 1963; Falk and Atidia, 1975). In addition, host plant adaptation in geographically distinct populations of the Colorado potato beetle appears to have evolved independently of adaptations to host alkaloids at the sensory level (Harrison and Mitchell, 1988).

That both peripheral and central targets for selection may coexist is suggested by studies on bark beetles in the genus *Ips* (Mustaparta et al., 1980) and the European corn borer, *Ostrinia nubilalis*, a pyralid (Roelofs et al., 1987). Reproductive isolation is maintained in sympatric Californian populations of *I. pini* and *I. paraconfusus* by their attraction to different enantiomers ([−] and [+] respectively) of the pheromone ipsdienol. In both species, the "opposite" enantiomer is behaviorally inhibitory. At the chemosensillar level, both species have neurons receptive to each of the enantiomers; in both species the ratio of neurons for each enantiomer favors the attractant enantiomer (i.e. the pheromone). As mentioned in the previous section, electrophysiological recording experiments demonstrate that the enantiomers are not antagonistic at the receptor level, indicating that inhibition is a central phenomenon. Curiously, an eastern population of *I. pini* from New York State, like its western counterpart, has a majority of its neurons keyed to the (−) enantiomer ([−, +] ratio of 6 : 3), even though females in the eastern population produce a pheromone which is a blend of (−, +) in a ratio of 35 : 65 (Mustaparta et al., 1980)! Though it is totally speculative, one could imagine selection acting first on the CNS to recognize the shift in pheromone constituents, then acting at the peripheral level to "adjust" the ratio of receptors to correspond to the blend being produced by females.

Investigations of *O. nubilalis* indicate that pheromone production, reception, and behavioral response are all under genetic control (Roelofs et al., 1987; Klun and Maini, 1979). However, none of the presumed genes responsible for these actions appear closely linked, and more surprisingly, genes for pheromone reception and pheromone-mediated behavior (attraction) are likely on different chromosomes. In North America, this species has two distinct races. One race emits E and Z isomers of 11-tetradecenyl acetate in an approximate rate of 97 : 3 (the E-strain) whereas the other (Z-strain) emits the same isomers in the inverse ratio (3 : 97, E/Z).

F_1 hybrids between the two races are phenotypically intermediate for pheromone production (emitting a 65 : 35 blend of E : Z) and the presence of neurons keyed to the two enantiomers (Roelofs et al, 1987). Data from F_2 progeny and parental backcrosses led to the conclusion that both traits are controlled by single autosomal genes, each with two alleles. In contrast, the inheritance pattern for upwind flight in response to specific pheromone blends appears to be controlled by a sex-linked gene with two alleles. Assuming then that upwind flight, the culmination of pheromone-mediated behavior, results from neural integration, it may well be that natural selection acts on the total behavior, and therefore on a central target, in the first instance.

A recent study examined inter- and intraspecific differences in host odor recognition based on electroantennogram recordings from the apple maggot fly (*Rhagoletis pomonella*) and the blueberry maggot fly (*R. mendax*) in response to host fruit extracts and volatile fruit odor constituents (Frey and Bush, 1990). Antennal sensitivity appears selectively adapted to species-specific host odors. Both inter- and intraspecific differences were found in response to fruit odor constituents, suggesting, according to the authors, differences in antennal receptor types and/or receptor numbers. They suggest "that antennal sensitivity plays an important role in host shifts and speciation in this genus" (Frey and Bush, 1990).

Far more studies of this type, which seek to unravel the genetic basis of specific behaviors and their physiological basis (at the cellular level), are needed before the question of the primary target for selection can be satisfactorily answered.

5.3. Physiological Responses to Chemicals

5.3.1. Barriers to Toxicity from Host Chemicals

Toxicity, in the strictest sense, is the outcome of an interaction between a toxin and a sensitive target site in an organism, resulting in a physiological lesion. Physiological targets (usually referred to as receptors, but confusion with the term "receptor" as used in sensory systems should be avoided) are most often proteins found on cell (or organelle) membranes. Organ systems in insects which have been reported to be physiological targets for naturally-occurring toxins (e.g., plant allelochemicals) include the nervous system, the endocrine system, the gonads, the heart, and the epidermis. A more liberal interpretation of toxicity would include substances that act as physical disruptants—the physiological lesion results from physical damage which destroys the integrity of membranes.

Insects, like all complex multicellular organisms, can avoid (or more realistically, lessen) toxicity from ingested poisons by three main strategies, all of which revolve around the same outcome, namely interference with or defeat of the toxin-target interaction. The first is to grossly prevent the toxicant from reaching sensitive target sites, which can be called the pharmacokinetic strategy. This could include any aspect of the fate of the toxicant once it enters the insect that prevents the substance from arriving at the target site in appreciable quantity. Examples of the pharmacokinetic strategy are reduced penetration from the alimentary canal into the hemolymph, rapid excretion of the toxicant from the bloodstream, and sequestration of the toxicant away from the target site and into a physiologically inert depot. A second strategy for avoiding toxicity is to metabolize the toxicant into less-toxic or non-toxic products before it can reach sensitive targets in appreciable amounts. This strategy is probably the best documented in both insects and higher animals (see Brattsten, 1980, for a review of this subject). Like the pharmacokinetic strategy, the goal is to prevent toxicant from reaching the target site. The third strategy is target site insensitivity. This last strategy has been recently reviewed by Berenbaum (1986).

Locusts and grasshoppers well exemplify the pharmacokinetic strategy, at least with respect to the digestive tract as a barrier to the uptake of plant toxins. Cottee et al. (1988) compared the toxicity of seven allelochemicals to *Locusta migratoria* and *Schistocerca gregaria* via injection into the hemolymph or cannulation into the gut. Five of the compounds were relatively toxic via injection to both species, but none of these were appreciably toxic to *Schistocerca* via the oral route. In *Locusta*, two of the five were toxic orally, but with LD_{50} values 7-fold (nicotine) and 17-fold (allylisothiocyanate) greater than those obtained via injection. Similarly, the grasshopper *Melanoplus sanguinipes* is able to tolerate oral doses of sesquiterpene lactones far in excess of those which were toxic via injection (Isman, 1985).

Rapid excretion of toxins as a defensive strategy is best known from the tobacco hornworm, *Manduca sexta*. This insect is capable of excreting an ingested dose of the insecticidal alkaloid nicotine almost completely within two hours (Self et al., 1964). However, excretion is just one of several adaptations in this insect to avoid nicotine intoxication (reviewed in Brattsten, 1986). The grasshopper *M. sanguinipes* may also rely partly on this strategy to avoid toxicity from pyrrolizidine alkaloids. Grasshoppers excreted 90% of an injected dose of senecionine-N-oxide with little metabolism (Ehmke et al., 1989). However, oral doses of this alkaloid were subject to considerable metabolism, and metabolism of this alkaloid in the gut was demonstrated *in vitro*. Nonetheless, excretion of alkaloids from the hemolymph may be important in preventing toxicity from the fraction of ingested toxin which eludes metabolism and is absorbed from the gut.

There are numerous cases of detoxicative metabolism of allelochemicals documented in insects; only two examples will be given here. Furanocoumarins are photoactivated toxins from the plant family Apiaceae. One of these compounds, xanthotoxin, is toxic to several insects including the fall armyworm, *Spodoptera frugiperda*, a species with a wide host range. In contrast, the black swallowtail, *Papilio polyxenes* is a specialist feeding on members of the Apiaceae. Both insects are capable of metabolizing xanthotoxin to non-toxic products, but the rate of metabolism in the swallowtail is 33 times faster than that of the armyworm (Ivie et al., 1983; Bull et al., 1984). Precocene II is a chromene which acts as an antihormone in some insect species via destruction of the corpora allata glands which are the site of synthesis of juvenile hormone. The milkweed bug *Oncopeltus fasciatus* is an extremely sensitive species to this compound, whereas the corn earworm, *Heliothis zea*, is not. In this case, differential sensitivity is at least partly attributable to metabolism; the earworm metabolized 90% of a topical dose after 20 hours, while the milkweed bug metabolized only 10% of the dose in the same time period (Haunerland and Bowers, 1985).

Finally, sequestration can be a means of avoiding toxicity (Duffey, 1980), although there are no clear-cut examples of this as an exclusive strategy. Sequestration of toxic plant substances has been best documented in highly-adapted aposematic insects, where the sequestered materials are stored for future defense against the insects' predators (Bowers, this volume). Examples include cardenolides (cardiac glycosides) in danaid butterflies and lygaeid bugs feeding on milkweeds, cyanogenic glycosides in zygaenid moths and quinolizidine alkaloids in lupine-infesting aphids (see Harborne, 1988, for a review).

Target site insensitivity has been well studied in insects with respect to synthetic insecticides (where the target is the nervous system), but has been little studied with respect to natural toxins, particularly those of ecological relevance. However, examples of target site insensitivity in insects to plant toxins include the insensitivity of tobacco hornworm nerve cord to nicotine (Morris, 1984) and that of Na,K-dependent ATPases to cardenolides in the monarch butterfly and milkweed bugs (Scudder et al., 1986).

5.3.2. Evidence for Selection in Handling of Xenobiotics

Although the genetic basis for resistance to putatively toxic allelochemicals has not been characterized in insects (cf. resistance to synthetic insecticides; see Caprio and Tabashnik, this volume), comparisons of allelochemical processing in closely related taxa provide evidence for past selection which has resulted in specific metabolic capabilities in species routinely confronted with dietary toxins. The best evidence of these evolved traits comes from the Lepidoptera, and in particular, from the swallowtail butterflies (Papilionidae; reviewed in Berenbaum, 1991).

As mentioned in the previous section, the black swallowtail, *Papilio polyxenes*, is capable of rapid metabolism of the furanocoumarin xanthotoxin, which is a predictable constituent of its host plants. Metabolism in this case is accomplished by the cytochrome P450 monooxygenase enzyme system. However, this insect possesses a unique P450 isozyme which is both substrate-inducible and highly substrate-specific (Cohen et al., 1989). In addition, this species possesses a high level of antioxidant enzymes capable of rendering activated forms of oxygen, the toxic products of furanocoumarins and other photoactivated toxins, less harmful (Berenbaum, 1990, cite Lee, 1989).

The pattern of enzymatic adaptations to host toxins discovered in *P. polyxenes* is somewhat mirrored in the eastern tiger swallowtail, *P. glaucus*. This insect occurs as at least two well-defined subspecies, the northern *P. glaucus canadensis* and *P. glaucus glaucus* (Scriber, 1988). The subspecies have a zone of sympatry across the northern United States, but utilize different host plant families. In particular, *P. g. canadensis* feeds specifically on host plants in the family Salicaeae, such as balsam poplar and quaking aspen tree species on which survival of *P. g. glaucus* is very low if at all (Scriber, 1988). The ability of *P. g. canadensis* to utilize these hosts is related to specific enzymic adaptation. Both host species contain phenol glycosides which are toxic to *P. g. glaucus*; *P. g. canadensis* is relatively immune from these compounds because it possesses relatively high levels of substrate-inducible midgut esterases which degrade the toxic phenols (Lindroth, 1989). Evidence for the specificity of the esterases for the host plant toxins comes from the observation that the presence of these enzymes does not render *P. g. canadensis* resistant to synthetic insecticides (malathion, permethrin), which are known to be metabolized by esterase-type enzymes.

The interface between zebra butterflies (*Heliconius* species) and their host plants, *Passiflora* species, provides a further example of evidence for selection in the ability of insects to overcome allelochemical defenses of plants. A detailed analysis of host range on 111 *Passiflora* species amongst 51 *Heliconius* species indicated a nonrandom specialization of the butterflies on their hosts (Spencer, 1988). The distribution of butterfly species on hosts is best explained by host chemistry; the hosts are characterized by the production of several types of cyanogenic glycosides. These potential toxins are rendered harmless to *Heliconius* larvae by beta-glucosidase enzymes in the larval gut which block cyanogenesis (Spencer, 1988). The nonrandom host distribution of the butterflies is thought to be a product of selection for insect glucosidases specific to the glucosidases of the particular host plant(s), which under normal circumstances, act to liberate the toxic product, hydrogen cyanide.

5.3.3. Serendipitous Nutrients and Defensive Chemicals

In some cases, insects have evolved not only to avoid toxicity from putatively toxic substances in their diet, but in turn to utilize these dietary substances to their benefit, such that their acquisition results in a net benefit to the individual's fitness. In this section, I review a handful of well-documented examples where "toxins" are utilized as nutrients for highly-adapted insects, are sequestered and stored for defense against predators, or, in one case, both. In each of these species, the ultimate fate of the ingested toxin is (or must be) the product of a series of specific adaptations acting in concert, although the precise details are not completely understood for each.

The best documented case of a dietary toxin being utilized as a nutrient is that of the bruchid beetle *Caryedes brasilienis* (Rosenthal, 1983). The larval stage of this species feeds exclusively on the seeds of the tropical legume *Dioclea megacarpa*, which contain high concentrations (up to 13% by weight) of the non-protein amino acid L-canavanine. This substance acts as a mimic of the structural amino acid L-arginine; ingestion of L-canavanine by a susceptible insect such as the tobacco hornworm leads to counterfeit incorporation of the mimic resulting in the synthesis of non-functional proteins.

C. brasiliensis has two primary lines of defense against intoxication by L-canavanine. The beetle possesses a specific arginyl-t-RNA-synthetase enzyme which discriminates between L-arginine and L-canavanine, preventing counterfeit incorporation. Secondly, the beetle is capable of rapid and complete metabolism of L-canavanine (Rosenthal et al., 1977). L-canavanine is degraded to L-canaline and urea; in the presence of urease, this substrate is further degraded to CO_2 and $2NH_3$, the latter constituting an important source of free nitrogen for the beetle. Rosenthal and co-workers found that *C. brasiliensis* has at least 30,000-fold higher urease activity than any of several species of insects assayed (Rosenthal et al., 1977). As L-canavanine accounts for over 90% of the free amino acid nitrogen in the seeds of *Dioclea*, these physiological adaptations allow the beetle to exploit a nutrient resource which is unavailable to other insects.

Some simple plant phenolics deter feeding or otherwise inhibit growth in a number of insects, although many phenolics are possibly benign in this regard. However, simple catecholic phenols added to the diet of the tree locust *Anacridium melanorrhodon* actually improved the growth of this insect (Bernays and Woodhead, 1982a). Further investigation led to the discovery that dietary phenols were incorporated into the cuticle where they likely function in the stabilization of cuticular proteins (that is, as tanning agents) (Bernays and Woodhead, 1982b). For this species, which normally feeds on plants low in protein but with relatively high concentrations of phenolics, the latter compounds may serve to spare phenylalanine, an amino acid of crucial nutritional significance because of its role in cuticle synthesis (Bernays, 1982).

The chrysomelid beetle *Phratora vitellinae* represents a unique case wherein a phenol glycoside from the host plant provides both nutrition and a precursor for chemical defense. This beetle feeds preferentially on *Salix* species from which it acquires salicin, a glucoside. Once ingested, the salicin is transported to special eversible glands in the integument, in which the phenolic is metabolized by a beta-glucosidase enzyme into glucose and saligenin. The latter compound is then reduced to salicylade-hyde, which is stored in the glands and released when the insect is molested (Pasteels et al., 1983). Experimental evidence supports the conclusion that the glucose moiety, derived from salicin, enhances larval growth in this species (Rowell-Rahier and Pasteels, 1986). Further, salicyl-aldehyde is more deterrent to ants in laboratory bioassays than salicin or saligenin. The duel advantages (nutrition, defense) of salicin acquisition in *P. vitellinae* contrasts with the situation in related chrysomelids (e.g., *P. tibialis*) where the production of autogenous defensive compounds entails a measurable cost (Rowell-Rahier and Pasteels, 1986).

The large milkweed bug *Oncopeltus fasciatus* is just one example of many insect species in which the sequestration and storage of plant toxins for defense against predators is the end result of a series of physiological adaptations (Scudder et al., 1986). These adaptations include enzymes that metabolically alter the plant cardenolides, a specialized epithelial layer that concentrates and stores the compounds, a system of active transport in the Malpighian tubules to clear excess quantities from the hemolymph, and target site insensitivity. Although the evolutionary order of appear-ance of these adaptations is unclear, it is highly likely that the morphologi-cal and physiological changes in the integument to permit the storage of cardenolides in high concentrations were the most recent.

5.3.4. Biochemical Targets for Selection

Among the biochemical targets upon which selection might act to provide "resistance" to natural toxins in insects are metabolic enzymes (Brattsten, 1988), often membrane-bound, and carrier proteins circulating in the hemolymph (Shapiro 1989; Haunerland and Bowers, 1986). It is likely that the latter materials are involved in cases of sequestration of dietary toxins for use as a defense against predation, rather than as a primary means of protection from intoxication. The best documented cases of sequestration of xenobiotics usually involve transport across the gut epithelium or into specialized glands against a concentration gra-dient, indicating the need for specific carrier molecules (see Wink and Schneider, 1988, for an example). In this context, it is interesting to note that the ability to sequester exogenous chemicals for defense may be less specific than first thought, based on elegant experimental evidence from the lubber grasshopper *Romalea guttata*. Jones et al. (1989) have demon-strated that restricting this polyphagous insect on wild onion plants results in sequestration of sulfur volatiles from the host into the defensive glands of the grasshopper; the defensive secretions resulting from enforced monophagy are more deterrent to ants than those from wild conspecifics feeding on a mixed diet including wild onion. Similar forced monophagy on an exotic host, catnip, led to the sequestration of terpenoid lactones from the host into the defensive glands (Blum et al., 1990).

As sequestration of toxins represents an uncommon phenomenon, relative to the number of phytophagous insect species, we should focus our attention on enzymes as the key target for natural selection from allelochemicals. For most species, the biological endpoint of avoiding toxicity from a dietary constituent need require little more than an efficient enzyme(s) to degrade the toxin. In some cases, even simple metabolic transformations result in dramatic reductions in toxicity of the parent molecule (e.g., oxidation of chromenes; Isman, 1989). In a previous section, I highlighted some lepidopteran species where adaptation to putatively toxic host plants is characterized by the evolution of highly inducible, substrate-specific enzymes. Although much of the earlier literature emphasizes the role of the cytochrome P450 family of monooxygenases in defense against allelochemicals (Brattsten, 1980), some of the studies I have highlighted indicate that esterases, and especially beta-glucosidases, may be pivotal in the adaptation of insects to toxic plants. Berenbaum (1990) postulates that feeding specialists are characterized by possessing very specific isozymes capable of degrading potential toxins predictably encountered by those monophagous insects. Such specialization may arise at the cost of a more broad-spectrum enzymatic system.

In the case of L-canavanine-consuming coleopterans, extremely high urease activity is a clear adaptation for the utilization of ammonium as a breakdown product of L-canavanine as a nutrient pool. However, many phytophagous insects have sufficient arginase to cleave L-canavanine into L-canaline and urea (Bleiler et al., 1988). Instead, the authors suggest that it is the discriminatory arginyl-tRNA-synthetase of *Caryedes*, preventing the formation of deleterious canavanyl proteins, that allowed radiation of the beetle onto host plants containing high concentrations of L-canavanine. Therefore, as will likely be discovered in other systems where toxins are serendipitously utilized as nutrients, the initial step was avoidance of toxicity, which in turn permitted the exploitation of the toxin to the insect's benefit.

Studies of the mechanisms of resistance to synthetic insecticides may provide some clues as to the likely targets for selection from natural toxins. Increased metabolism appears to be the predominant mechanism of resistance, followed by target site insensitivity, and finally reduced penetration (Oppenoorth and Welling, 1976). However, a single selecting agent, such as DDT, can give rise to four or more different mechanisms of resistance. According to Brattsten (1988), resistance based on enhanced metabolism is probably the first kind to develop, which may create an opportunity for target site insensitivity or other kinds of resistance to develop.

One characteristic that separates most toxic allelochemicals from synthetic insecticides is that the former frequently occur at sublethal levels in nature, and thus, selection pressure from them may be diffuse. Enzyme induction is non-hereditary, but many allelochemicals are known to induce metabolic enzymes. Brattsten (1988) suggests that induction may permit the survival and reproduction of insects with unexpressed resistance mechanisms, such that the frequency of their resistance genes increases in following generations.

What this chapter hopefully has done is point out the obvious need for detailed, long-term studies of the physiological bases of adaptation to exogenous chemicals in closely-related taxa. With such knowledge in hand, we can then turn to investigations of the genetic basis for those adaptations, through both artificial selection experiments and hybridization studies utilizing insect strains showing differential responses to chemical mediators (e.g. Roelofs et al., 1987). Only when we have a much better comprehension of insect physiology and its genetic basis in the ecological context will we be in the reasonable position to make predictions or models for the direction of evolution to semiochemicals.

Acknowledgements

I especially thank B. K. Mitchell for his helpful advice on the section of this chapter dealing with chemosensory physiology, and G. G. E. Scudder, M. R. Berenbaum, N. Haunerland, B. D. Roitberg, and an anonymous reviewer for their constructive comments on the manuscript. I also acknowledge the continuing support for my research program from the Natural Sciences and Engineering Research Council of Canada (NSERC).

References

Abisgold, J. D. and S. J. Simpson. 1988. The effect of dietary protein levels and hemolymph composition on the sensitivity of the maxillary palp chemoreceptors of locusts. *J. Exp. Biol.* 135:215–229.

Baker, T. C. 1989. Sex pheromone communication in the Lepidoptera: new research progress. *Experientia* 45:248–262.

Berenbaum, M. R. 1986. Target site insensitivity in insect-plant interactions. In *Molecular Aspects of Insect-Plant Associations*, eds. L. B. Brattsten and S. Ahmad, pp. 257–272. Plenum Press, New York.

Berenbaum, M. R. 1991. Comparative allelochemical processing in the Papilionidae (Lepidoptera). *Arch. Insect Biochem. Physiol.* 17:213–221.

Berenbaum, M. R. and M. B. Isman. 1989. Herbivory in holometabolous and hemimetabolous insects: contrasts between Orthoptera and Lepidoptera. *Experientia* 45:229–235.

Bernays, E. A. 1982. The insect on the plant—a closer look. *Proc. 5th Intl. Symp. Insect-Plant Relationships*, eds. J. H. Visser and A. K. Minks, pp. 3–17. Pudoc, Wageningen, The Netherlands.

Bernays, E. A. and S. Woodhead, 1982a. Plant phenols utilized as nutrients by a phytophagous insect. *Science* 216:201–203.

Bernays, E. A. and S. Woodhead. 1982b. Incorporation of dietary phenols in the cuticle in the tree locust *Anacridium melanorrhodon. J. Insect. Physiol.* 28:601–606.

Bleiler, J. A., G. A. Rosenthal, and D. H. Janzen, 1988. Biochemical ecology of canavanine-eating seed predators, *Ecology* 69:427–433.

Blum, M. S., R. F. Severson, R. F. Arrendale, D. W. Whitman, P. Escoubas, O. Adeyeye, and C. G. Jones. 1990. A generalist herbivore in a specialist mode. Metabolic, sequestrative, and defensive consequences. *J. Chem. Ecol.* 16:223–244.

Brattsten, L. B. 1980. Biochemical defense mechanisms in herbivores against plant allelochemicals. In *Herbivores. Their Interaction with Secondary Plant Metabolites*, eds. G. A. Rosenthal and D. H. Janzen, pp. 199–270. Academic Press, New York.

Brattsten, L. B. 1986. Fate of ingested plant allelochemicals in herbivorous insects. In *Molecular Aspects of Insect-Plant Associations*, eds. L. B. Brattsten and S. Ahman, pp. 211–255. Plenum Press, New York.

Brattsten, L. B. 1988. Potential role of plant allelochemicals in the development of insecticide resistance. In *Novel Aspects of Insect-Plant Interactions*, eds. P. Barbosa and D. K. Letourneau, pp. 313–348. Wiley-Interscience, New York.

Bull, D. L., G. W. Ivie, R. C. Beier, N. W. Pryor, and E. H. Oertli. 1984. Fate of photosensitizing furanocoumarins in tolerant and sensitive insects. *J. Chem. Ecol.* 10:893–911.

Byers, J. A. 1989. Chemical ecology of bark beetles. *Experientia* 45:271–283.

Chapman, R. F. 1988. Sensory aspects of host-plant recognition by Acridodea: questions associated with the multiplicity of receptors and variability of response. *J. Insect Physiol.* 34:167–174.

Chapman, R. F. and E. A. Bernays. 1989. Insect behavior at the leaf surface and learning as aspects of host plant selection. *Experientia* 45:215–222.

Christensen, T. A., H. Mustaparta, and J. G. Hildebrand. 1989. Discrimination of sex pheromone blends in the olfactory system of the moth. *Chemical Senses* 14:463–477.

Cohen, M. B., M. R. Berenbaum, and M. A. Schuler. 1989. Induction of cytochrome P450-mediated detoxification of xanthotoxin in the black swallowtail. *J. Chem. Ecol.* 15:2347–2356.

Cottee, P. K., E. A. Bernays, and A. J. Mordue. 1988. Comparisons of deterrency and toxicity of selected secondary plant compounds to an oligophagous and a polyphagous acridid. *Entomol. Exp. Appl.* 46:241–247.

Duffey, S. S. 1980. Sequestration of plant natural products by insects. *Annu. Rev. Entomol.* 25:447–477.

Ehmke, A., P. Proksch, L. Witte, T. Hartmann, and M. B. Isman. 1989. Fate of ingested pyrrolizidine alkaloid N-oxide in the grasshopper *Melanoplus sanguinipes. Naturwiss.* 76:27–29.

Falk, R. and J. Atidia. 1975. Mutation affecting taste perception in *Drosophila melanogaster, Nature* 254:325–326.

Frey, J. E. and G. L. Bush. 1990. *Rhagoletis* sibling species and host races differ in host odor recognition. *Entomol. Exp. Appl.* 57:123–131.

Harborne, J. B. 1988. *Introduction to Ecological Biochemistry*, 3rd ed. Academic Press, London.

Harrison, G. D. and B. K. Mitchell. 1988. Host-plant acceptance by geographic populations of the Colorado potato beetle, *Leptinotarsa decemlineata*. Role of solanaceous alkaloids as sensory deterrents. *J. Chem. Ecol.* 14:777–788.

Haunerland, N. H. and W. S. Bowers. 1985. Comparative studies on pharmacokinetics and metabolism of the anti-juvenile hormone precocene II. *Arch. Insect Biochem. Physiol.* 2:55–63.

Haunerland, N. H. and W. S. Bowers. 1986. Binding of insecticides to lipophorin and arylphorin, two hemolymph proteins of *Heliothis zea. Arch. Insect Biochem. Physiol.* 3:87–96.

Homberg, U., T. A. Christensen, and J. G. Hildebrand. 1989. Structure and function of the deutocerebrum in insects. *Annu. Rev. Entomol.* 34:477–501.

Hovanitz, W. 1969. Inherited and/or conditioned changes in host-plant preference in *Pieris. Entomol. Exp. Appl.* 12:729–735.

Ishikawa, S., Y. Tazima, and T. Hirao. 1963. Responses of the chemoreceptors of maxillary sensory hairs in a "non-preference" mutant of the silkworm. *J. Sericult. Sci. Jpn.* 32:125–129.

Isman, M. B. 1985. Toxicity and tolerance of sesquiterpene lactones in the migratory grasshopper, *Melanoplus sanguinipes* (Acrididae). *Pestic. Biochem. Physiol.* 24:348–354.

Isman, M. B. 1989. Toxicity and fate of acetylchromenes in pest insects. In *Insecticides of Plant Origin*, eds., J. T. Arnason, B. J. R. Philogene, and P. Morand, pp. 44–58. Amer. Chem. Soc. Symp. Ser. 387.

Ivie, G. W., D. L. Bull, R. C. Beier, N. W. Pryor, and E. H. Oertli. 1983. Metabolic detoxification: mechanism of insect resistance to plant psoralens. *Science* 221:374–376.

Jermy, T. 1966. Feeding inhibitors and food preference in chewing phytophagous insects. *Entomol. Exp. Appl.* 9:1–12.

Jones, C. G., D. W. Whitman, S. J. Compton, P. J. Silk, and M. S. Blum. 1989. Reduction in diet breadth results in sequestration of plant chemicals and increases efficacy of chemical defense in a generalist grasshopper. *J. Chem. Ecol.* 15:1811–1822.

Klun, J. A. and S. Maini. 1979. Genetic basis of an insect chemical communication system: the European cornborer. *Environ. Entomol.* 8:423–426.

Lindroth, R. L. 1989. Differential esterase activity in *Papilio glaucus* subspecies: absence of cross-resistance between allelochemicals and insecticides. *Pestic. Biochem. Physiol.* 35:185–191.

Mitchell, B. K. 1988. Adult leaf beetles as models for exploring the chemical basis of host plant recognition. *J. Insect Physiol.* 34:213–225.

Mitchell, B. K. B. M. Rolseth, and B. G. McCashin. 1990. Differential responses of galeal gustatory sensilla of the adult Colorado potato beetle, *Leptinotarsa decemlineata* (Say), to leaf saps from host and non-host plants. *Physiol. Entomol.* 15:61–72.

Morris, C. E. 1983. Uptake and metabolism of nicotine by the CNS of a nicotine-resistant insect, the tobacco hornworm (*Manduca sexta*). *J. Insect Physiol.* 29:807–817.

Mustaparta, H. 1984. Olfaction. In *Chemical Ecology of Insects*, eds. W. J. Bell and R. T. Carde, pp. 37–70. Chapman and Hall, London.

Mustaparta, H., M. E. Angst, and G. N. Lanier. 1980. Receptor discrimination of enantiomers of the aggregation pheromone ipsdienol, in two species of *Ips*. *J. Chem. Ecol.* 6:689–701.

Oppenoorth, F. J. and W. Welling. 1976. Biochemistry and physiology of resistance. In *Insecticide Biochemistry and Physiology*, ed. C. F. Wilkinson, pp. 507–551. Plenum Press, New York.

Pasteels, J. M., M. Rowell-Rahier, J. C. Braekman, and D. Daloze. 1983. Salicin from host plant as precursor of salicylaldehyde in defensive secretion of chrysomeline larvae. *Physiol. Entomol.* 8:307–314.

Ritter, F. J. 1988. Steric factors in pheromonal pest control. In *Stereoselectivity of Pesticides. Biological and Chemical Problems*, eds. E. J. Ariens, J. J. S. van Rensen, and W. Welling, pp. 327–356. Elsevier, Amsterdam.

Roelofs, W., T. Glover, X.-H. Tang, I. Sreng, P. Robbins, C. Eckenrode, C. Lofstedt, B. S. Hansson, and B. O. Bengtsson. 1987. Sex pheromone production and perception in European cornborer moths is determined by both autosomal and sex-linked genes. *Proc. Natl. Acad. Sci. USA* 84:7585–7589.

Rosenthal, G. A. 1983. The adaptation of a beetle to a poisonous plant. *Sci. Amer.* 249:164–171.

Rosenthal, G. A., D. L. Dahlman, and D. H. Janzen. 1977. Degradation and detoxification of canaline by a specialized seed predator. *Science* 202:528–529.

Rowell-Rahier, M. and J. M. Pasteels. 1986. Economics of chemical defense in Chrysomelinae. *J. Chem. Ecol.* 12:1189–1203.

Schoonhoven, L. M. 1982. Biological aspects of antifeedants. *Entomol. Exp. Appl.* 31:57–69.

Schoonhoven, L. M. and F. Blom. 1988. Chemoreception and feeding behaviour in a caterpillar: towards a model of brain functioning in insects. *Entomol. Exp. Appl.* 49:123–129.

Schwarz, M., J. A. Klun, and E. C. Uebel. 1990. European corn borer sex pheromone. Inhibition and elicitation of behavioral response by analogs. *J. Chem. Ecol.* 16:1591–1604.

Scriber, J. M. 1988. Tale of the tiger: Beringial biogeography, binomial classification, and breakfast choices in the *Papilio glaucus* complex of butterflies. In *Chemical Mediation of Coevolution*, ed. K. C. Spencer, pp. 241–301. Academic Press, New York.

Scudder, G. G. E., L. V. Moore, and M. B. Isman. 1986. Sequestration of cardenolides in *Oncopeltus fasciatus*: morphological and physiological adaptations. *J. Chem. Ecol.* 12:1171–1187.

Self, L. S., F. Guthrie, and E. Hodgson. 1964. Adaptation of tobacco hornworms to the ingestion of nicotine. *J. Insect Physiol.* 12:224–230.

Shapiro, J. P. 1989. Xenobiotic absorption and binding by proteins in hemolymph of the weevil *Diaprepes abbreviatus*. *Arch. Insect Biochem. Physiol.* 11:65–78.

Simmonds, M. S. J. and W. M. Blaney. 1984. Some neurophysiological effects of azadirachtin on lepidopterous larvae and their feeding response. In *Natural Pesticides From the Neem Tree and Other Tropical Plants*, eds. H. Schmutterer and K. R. S. Ascher, pp. 163–179. Schriftenreihe der GTZ, No. 161, Eschborn.

Spencer, K. C. 1988. Chemical mediation of coevolution in the *Passiflora-Heliconius* interaction. In *Chemical Mediation of Coevolution*, ed. K. C. Spencer, pp. 167–240. Academic Press, New York.

Stadler, E. Contact chemoreception. In *Chemical Ecology of Insects*, eds. W. J. Bell and R. T. Carde, pp. 3–35. Sinauer, Sunderland, MA.

van Drongelen, W. 1979. Contact chemoreception of host plant specific chemicals in larvae of various *Yponomeuta* species (Lepidoptera). *J. Comp. Physiol.* 134A:265–279.

van Loon, J. J. A. and F. A. van Eeuwijk. 1989. Chemoreception of amino acids in larvae of two species of *Pieris*. *Physiol. Entomol.* 14:459–469.

Wieczorek, H. 1976. The glycoside receptor of the larvae of *Mamestra brassicae* L. (Lepidoptera, Noctuidae). *J. Comp. Physiol.* 106A:153–176.

Wink, M. and D. Schneider. 1988. Carrier-mediated uptake of pyrrolizidine alkaloids in larvae of the aposematic and alkaloid-exploiting moth *Creatonotos*. *Naturwiss.* 75:524–525.

Wrubel, R. P. and E. A. Bernays. 1990. The relative insensitivity of *Manduca sexta* larvae to non-host plant secondary compounds. *Entomol. Exp. Appl.* 54:117–124.

Part II

Specific Problems

Section Overview

The first part of this volume dealt with problems of a general nature. In this second part, specific problems within chemical ecology of insects are addressed. These problems are quite diverse, yet there is a common theme upon which the respective authors develop their theses, namely, evolutionary processes. Each contribution fits into the mandate of the volume as follows.

Caprio and Tabashnik (Chapter 6) address the problem of how and why insects develop resistance to toxic compounds produced by their host plants. They suggest that development of such resistance depends upon factors at both the molecular and populational level, and that a complimentary approach that considers both levels of analysis is essential in order to predict both rates and directions of evolution of resistance. Further, they suggest that, using molecular techniques, we should be prepared to find that the genome can respond in a number of different ways to cope with plant toxins and that the genetic basis of resistance found in a given population will be a function of both stochastic processes and population structure.

In a related contribution, Bowers reviews the evolution of chemical defenses of insects. She considers how unpalatability to predators might evolve through two different mechanisms, sequestration and *de novo* synthesis. Using both phylogenetic and mechanistic approaches, she concludes that unpalatability has evolved independently numerous times across a variety of taxa, but that certain physiological preconditions may be necessary. Bowers completes Chapter 6 with a consideration of the putative costs of chemical defense in insects and suggests how such costs might be demonstrated.

In Chapter 8, Jaenike and Papaj address the problem of how phenotypic plasticity, as a function of learning and motivation, can affect the evolution of host choice. Using elegant, biologically-reasonable computer simulations, they find that even modest effects of learning can have substantive outcomes on patterns of host use by insects. Interestingly, their models suggest that learning can facilitate formation of host races within populations, particularly when the effects of learning are moderate.

Phelan considers the use of chemical communication by male and female insects with respect to mating and reproduction in Chapter 9.

In Chapter 10, Winston considers the evolution of sociality in different insect taxa and draws the conclusion that chemical signals (particularly pheromones) are important facilitators of social organization. They are, in his words, "the glue that holds insect colonies together." He begins by discussing kin selection theory and then demonstrates how semiochemicals can facilitate kin recognition, thus providing a mechanism for kin recognition and asymmetric altruism. Winston extends his discussion to aspects of reproduction and again shows how semiochemicals can facilitate both the timing and extent of reproduction at the colony and individual levels. His discussion illustrates the great complexities involved in chemical ecology and the need for unifying theories in light of recent dramatic advances in analytical chemistry.

In the final chapter, McNeil asks how an evolutionary perspective might be employed within pest management systems, and whether such an approach would be advantageous. He finds that there are several areas (for example, correlation of pheromone trap captures with pest populations) for which an evolutionary perspective could help to clarify otherwise enigmatic patterns. First, he proposes several reasons why trap captures may not always be representative of the density of "matable" males present at a specific location. Second, using an evolutionary perspective, he points to conditions under which resistance to mating disruption strategies might evolve. He concludes, like many of the authors, with a plea for complementarity of approaches.

Taken together, the final six chapters of this volume provide a broad perspective on problems in insect chemical ecology, and suggest how such problems might be better approached by applying the concepts and methods of evolutionary biologists as outlined in the first part.

6

Evolution of Resistance to Plant Defensive Chemicals in Insects

Michael A. Caprio and *Bruce E. Tabashnik*
University of Hawaii

The remarkable diversity of plant defensive chemicals is matched only by the ability of insects to overcome them. Evolution of resistance requires genetic changes in insect populations that reduce the negative effects of plant defensive chemicals. In principle, such resistance can be demonstrated by documenting evolutionary change within populations through time. In practice, however, one typically infers evolution of resistance when populations differ in their success in coping with one or more defensive chemicals.

Resistance of insects to plant defensive chemicals has theoretical as well as practical significance. Knowledge of insect adaptation to plant defenses has played a key role in development and testing of evolutionary theory, particularly in shaping ideas about coevolution (Ehrlich and Raven, 1964; Feeny, 1976; Gilbert, 1979; Futuyma and Slatkin, 1983) and speciation (Diehl and Bush, 1984; Futuyma, 1987). The ability of insects to resist plant defenses has enabled them to remain major crop pests throughout the world; insects have also been used successfully as biological control agents against weeds. Understanding how insects evolve resistance to plant defenses can facilitate development of crop systems with more effective and durable protection against insects.

Recent reviews and books focusing on various aspects of this topic include those on ecology and behavior (Wiens, 1976; Hanson, 1983; Miller and Strickler, 1984), population genetics (Felsenstein, 1976; Hedrick et al., 1976; Levins, 1976; Futuyma, 1983; Gould, 1983; Mitter and Futuyma, 1983; Diehl and Bush, 1984; Futuyma and Peterson, 1985; Hedrick, 1986), chemical ecology (Feeny et al., 1983; Scriber, 1984; Barbosa and Saunders, 1985; Slansky and Rodriguez, 1987; Spencer, 1988), and plant-insect interactions (Denno and McClure, 1983; Futuyma and Slatkin, 1983; Boethel and Eikenbary, 1986; Green and Hedin, 1986; Brattsten and Ahmad, 1986; Barbosa and Letourneau, 1988).

In this chapter, we focus on the evolutionary genetics of insect adaptation to plant defensive chemicals. We consider evolution of resistance in the context of population genetic theory to assess the influences of interactions among loci (Wright's shifting balance theory), population structure, gene flow, population dynamics, and niche breadth. We also consider the molecular genetics of insect resistance to insecticides and prokaryote adaptation to novel substrates. We caution that findings from these systems may not always be applicable to insect-plant interactions. Our aim is to use knowledge of these systems to stimulate development and testing of hypotheses about the molecular basis of evolutionary adaptation to plant defenses in insects.

6.1. Population Biology and Insect Resistance to Toxins

In this section, we examine how genomic changes that confer resistance to plant defenses arise and behave in insect populations. After mutations conferring resistance to host plant toxins have arisen in one or more individuals, they must increase in frequency before they can have a significant effect on the population. Our emphasis is on how new alleles are maintained or become common in populations in which they increase fitness. We discuss Wright's shifting balance theory as a general theoretical framework, review ecological factors affecting new mutations, and consider some of the population genetic models which underscore the role of population structure in maintaining genetic polymorphisms.

6.1.1. Alleles in Populations: Shifting Balance Theory

How do rare alleles that confer resistance to plant toxins become common in insect populations? If alleles at a single locus determined the phenotype, or if the effects of alleles at more than one locus had only additive effects on the phenotype, then directional selection would always increase the frequency of rare alleles that conferred higher fitness. According to the shifting balance theory, however, alleles in populations exist within an adaptive topography that depends on their effect on fitness in combination with alleles from other loci (Wright, 1969). In other words, non-additive interactions among loci are important. Because of these non-additive interactions, certain combinations of alleles may be particularly fit, forming a coadapted gene complex, a peak on the adaptive topography. The peak occupied by the population need not be the only peak, nor the highest. The population is prevented, under normal conditions, from shifting to other peaks by the maladaptive troughs in between. Small changes in any allele frequency will move the population away from the peak, and selection will tend to return the population to its original state.

For example, consider an insect with two types of host plants X and Y (X and Y could be different species or different chemotypes within a population). Assume that two loci interact to control the insect's response to these host plants. The first locus affects the insect's ability to detoxify plant defensive chemicals, the second affects its host preference. At the first locus, alleles *A* and *a* confer effective detoxification of toxins from plants X and Y, respectively. Alleles *B* and *b* at the second locus confer preference for plants X and Y, respectively. Genotype *AABB* will prefer and perform best on plant X; genotype *aabb* will prefer and be most fit on plant Y. Thus, genotypes *AABB* and *aabb* form coadapted gene complexes, that is, "particularly happy combinations of genes" (Crow and Kimura, 1970). Conversely, the least fit combinations of genes are *AAbb* and *aaBB*. Individuals with these genotypes tend to feed on plants to which they are poorly adapted. Assuming no dominance at either locus, heterozygotes have intermediate fitness.

Assume that the fitness of *aabb* individuals is greater than that of *AABB* individuals, but the populations is nearly fixed for both the *A* and *B* alleles. Because alleles *a* and *b* are rare, they will occur primarily in heterozygotes. Selection will favor *AABB* homozygotes over all heterozygotes, thus preventing the population from evolving towards *aabb*, the most fit genotype. In Wrightian terms, selection prevents the population from shifting to a higher adaptive peak.

Shifting balance theory (Wright, 1977) is concerned with mechanisms that allow populations to cross maladaptive troughs. Wright emphasized that the importance of genetic drift increases in small populations and at some point overcomes the constricting force of selection, allowing a population to randomly meander to an area under the influence of a different adaptive peak. As population size increases, selection regains its dominance over genetic drift and draws the population up to the new peak. Wright also realized that temporary environmental changes could alter the topography, allowing selection to move genotype frequencies into a range which, when the original environment was restored, could lie under the influence of a different peak. Wright termed this mass selection in response to changing conditions.

The probability of a peak shift is inversely related to the strength of selection on the trait (Charlesworth and Rouhani, 1988). The probability of peak shifts increases if the heterozygote fitness is equal to that of the most fit homozygote in each habitat. In the previous example, this occurs if an *AaBb* heterozygote has the same fitness of the *AABB* homozygote in a habitat containing plant X, while the same individual would have the same fitness of an *aabb* homozygote in a habitat containing plant Y. This type of interaction increases the potential for maintenance of polymorphisms (Gillespie, 1976; Hoekstra et al., 1985).

6.1.2. Ecology and Population Genetics

Ecological factors can affect the potential of an insect to adapt to host plant toxins. This section considers how habitat patchiness, host longevity, herbivore niche breadth, migration, and population growth rates can affect insect adaptation to host plant toxins.

6.1.2.1. Habitat Patchiness

Habitat patchiness can lead to reduced gene flow between populations, increasing the effects of local selection and adaptation to local variation in host plant defenses. The concept of environmental "grain" has been used to describe the degree of patchiness relative to the dispersal abilities of an individual (Levins, 1976; Wiens, 1976). In a coarse-grained environment, an individual experiences only one habitat type during its lifetime. Non-random mating results because most of the potential mates for the insect are derived from the same patch type. As long as migration is infrequent, the genetic makeup of the population in the patch responds to local selection, even when the genetic changes in one population would be maladaptive in other patches. In species where migration is more frequent, selection in other patches becomes important, and local adaptation may not be possible. Rather than acting as a force opposing local adaptation, gene flow determines the relative significance of global versus local selection (Spieth, 1979).

When host plant patches are small relative to insect dispersal, an individual will experience most of the available patch types during its lifetime, and the environment is considered fine-grained. Under these conditions, adaptation will occur in response to global selection pressures, that is, selection pressures averaged over all patches. No adaptation to a novel host plant defense will occur unless the allele(s) responsible for the adaptation lead to a global increase in fitness for the insects, at least as heterozygotes. In this case, the fitness of the new allele will depend upon the frequency of the novel host plant type to which it provides resistance. A new allele that increases fitness to a rare host plant phenotype may be globally maladaptive and will remain rare while gene flow is high and the novel host plant phenotype is rare. Only when that host plant phenotype becomes sufficiently common will the new allele become selectively advantageous globally and increase in frequency.

Thus, insect species that experience their environment as fine-grained will evolve resistance only when the resistance allele becomes globally advantageous. If the resistance allele has no fitness costs associated with it, the resistance allele will be globally advantageous as soon as the new host plant phenotype appears. However, if there are fitness costs associated with the resistance allele in the absence of the novel host plant phenotype, then the resistance allele will not become globally advantageous until the novel host plant phenotype has become common enough so that selection for the resistance allele on the novel host plant phenotype is stronger than the selection against the resistance phenotype on other host plants.

In contrast, species that experience their environment as coarse-grained patches can evolve resistance in local populations even though the resistance allele is not favored globally. The rate at which resistance will evolve in a subpopulation will depend upon the frequency of the novel host plant in the patch and the rate of immigration of maladapted alleles from other populations. A special case occurs when resistance alleles are rare and occur initially only in a few subpopulations (Caprio, 1990). When gene flow is very low, those subpopulations that contain the rare resistance allele will evolve resistance fast because there is no influx of maladapted alleles from neighboring subpopulations. Adaptation to the novel plant phenotype in other subpopulations, however, will be delayed because of the slow spread of the resistance allele. The fastest overall rate of resistance evolution will occur at an intermediate gene flow rate where sufficient gene flow occurs to move rare alleles between populations, but not so much gene flow that the immigration of maladapted alleles from other subpopulations delays resistance evolution in those subpopulations where, by chance, resistance alleles initially occurred.

The effects of habitat patchiness caused by differences in selection regimes between patches (coarse-grain) will be greatest when the differences between patches are large and patches are of similar size (Gillespie, 1974). There will be a tendency toward a fine-grained response when patch differences become smaller, patch size variation increases, or inter-patch distance decreases.

In the special case in which the adaptive allele confers a behavioral preference for a different host plant phenotype (rather than a physiological adaptation), a frequency-dependent equilibrium can be established. If one host plant phenotype is more common, the corresponding insect phenotype will increase in frequency. This may result in increased herbivore pressure on that host plant phenotype, eventually lowering is frequency and thereby decreasing the selective advantage for the insect phenotype adapted to feeding on that host plant type (Gould, 1983).

Environmental grain is not fixed for an insect population; it can vary with the particular trait examined and with population density. An insect may experience its environment as coarse-grained with respect to some traits, such as oviposition, while other traits, such as feeding sites, may be fine-grained. This occurs because habitat patchiness is not only an inter-play between patch distribution and insect mobility, but it is also a function of the behavioral response of the insect. For example, in some populations of *Colias* butterflies, individual females obtain nectar from plants of numerous species, yet they show strong oviposition preference for one or a few of many potential species of larval host plants (Watt et al., 1974, 1977; Tabashnik et al., 1981). Thus, nectar feeding can be considered fine-grained whereas oviposition may be coarse-grained for the same population.

6.1.2.2. Herbivore Longevity

The longevity of herbivores relative to patch persistence can affect the potential for local adaptation and the maintenance of genetic variation (Wiens, 1976). Short herbivore generation time relative to patch persistence promotes local adaptation. In this case, many generations of selection can operate within a patch before the patch deteriorates and the insects move to another patch. The high proportion of examples of host race formation which come from insects feeding on trees (Via, 1984) may be a result of such longevity differences. Alternatively, when herbivore longevity is greater than patch persistence, the insect will experience a more fine-grained habitat (because it must move more often), lowering the potential for formation of locally-adapted host plant races.

6.1.2.3. Population Density

Population density may also affect the behavioral responses of insects to habitat patchiness. Fretwell and Lucas (1969) suggest that at low densities, only the most optimal habitats would be occupied, and the environment would be viewed as relatively coarse-grained. As population densities increased, preference for the optimal habitats would be decreased because of overcrowding, and habitats with lower initial rankings would become more preferred. Eventually, all available habitats would be filled, and the densities in the different habitats would reflect the initial preference rankings. At this point, there would be essentially no habitat preference, and the insect would view its environment as fine-grained.

6.1.2.4. Gene Flow

Ecological factors can affect the level of gene flow between populations. High gene flow shifts the balance towards global adaptation. Conversely, factors limiting gene flow increase the occurrence of local adaptation and the potential for adaptation to plants with novel chemical defenses. Slatkin (1985, 1987) and Slatkin and Barton (1989) reviewed the measurement and implications of migration and gene flow in natural populations.

Movement of individuals does not necessarily cause gene flow (Ehrlich et al., 1975, Ehrlich and Raven, 1969; Slatkin, 1987). Gene flow requires reproduction in the new environment by migrants. Gene flow estimates based on migration of individuals are generally overestimates unless mortality and fecundity curves are considered (Endler, 1979).

The age of dispersing individuals can greatly affect their reproductive success in the new habitat (Endler, 1979). In many insect species, females lay a large portion of their eggs in early adulthood. Females dispersing as older adults will lay fewer eggs and can be expected to be under-represented in the next generation in the new population. This can be especially important when females migrate into stable populations, where strong competition may reduce the reproductive success of older dispersers, lowering the transfer of genes between populations. In areas of high mortality, the average migration distance will be less, decreasing the neighborhood size and effective population size (Endler, 1979). This might be important in peripheral populations, since local adaptation would be enhanced with smaller neighborhoods, potentially allowing the population to adapt to suboptimal habitats.

6.1.2.5. Reproduction and Population Growth

In addition to the age of the dispersing females, reproductive status can also affect gene flow between populations. Females that mate before dispersing carry to the new habitat the genetic complement of at least two individuals (herself and her mate). In contrast, females that mate after reaching the new habitat bring only their own genes and thus lower their contribution to gene flow by 50%. These examples are only extremes in a continuum. Females may mate periodically, so that the eggs they lay initially will have a genetic complement derived solely from the original population. Eggs that are laid later, following a mating with a male from the new population, will derive only 50% of their genes from the original population. Increased mating frequency will therefore tend to lower the female contribution to gene flow.

When populations undergo frequent extinctions and recolonizations or bottlenecks, innate growth rate and compensation mechanisms can influence the amount of gene flow into the underutilized habitat. Insecticide resistance models (Comins, 1977; May and Dobson, 1986; Tabashnik, 1990) have demonstrated that following insecticide treatment (a population bottleneck), evolution of resistance in the population (local adaptation) occurred more rapidly when growth rates were higher. These models demonstrated that the effects of migration by susceptible individuals can be strongly influenced by density-dependent reproduction. When growth rates were lower (undercompensating density dependence), the population remained at a low level longer, allowing susceptible migrants to form a relatively larger portion of the population. In contrast, when the innate growth rate was high (overcompensating density dependence), the few insects remaining in the habitat following the bottleneck reproduced rapidly before many migrants arrived, thus increasing their genotypic representation in the re-established population. Since the survivors following the insecticide treatment were more likely to be adapted to the local selection pressures favoring insecticide resistance, this led to rapid local adaptation. Taylor et al. (1983) examined the evolution of resistance in caged housefly (*M. domestica*) populations. Susceptible immigrants were added three times per week into each cage. Two levels of pesticide persistence were simulated by adjusting dieldrin concentrations in the rearing medium. Immigration slowed resistance evolution more at the shorter level of persistence because fewer immigrants were killed by pesticide residues.

6.1.3. Polymorphism and the Multiple-Niche Hypothesis

The ability to maintain polymorphisms is important in the evolution of resistance because it promotes persistence of adapted phenotypes (races) for some or all of the plant phenotypes. In the simplest case, envision a single locus with two alleles in which one allele confers high fitness on one host plant phenotype, while the second allele confers high fitness on another host plant phenotype. If the species were not able to maintain the polymorphism, perhaps because of high gene flow, global selection would prevail. Selection would favor the allele that conferred highest overall fitness. Selection for adaptation to a new plant phenotype would be hindered until that phenotype became common enough to confer a global fitness increase. On the other hand, in a highly polymorphic species, selection for adaptation to a new host plant phenotype could occur even though that phenotype still accounted for a relatively small proportion of the overall plant population. This would probably occur in local concentrations of the novel plant phenotype, as might be found when a new resistant allele is first spreading though a plant population. In addition, polymorphism may serve to reduce the potential pleiotropic and epistatic costs associated with the new adaptive alleles because there is a greater number of genomic combinations that can be formed, some of which may by chance reduce fitness costs.

Several authors (DaCunha et al., 1950; DaCunha and Dobzhansky, 1954; Van Halen, 1965; Levene, 1953) have suggested that niche breadth may be correlated with the maintenance of genetic variability. They reasoned that insects with broader niches will encounter greater environmental heterogeneity and thus retain greater genetic variation. Many models describe conditions under which such genetic polymorphisms can be maintained in populations. An excellent review of genetic polymorphism with respect to host plant adaptation was provided by Mitter and Futuyma (1983). More general reviews of this topic can be found in Felsenstein (1976), Hedrick et al. (1976), and Hedrick (1986).

Levene (1953) modeled the case in which offspring are deposited into niches where selection occurs, while adults from all niches form a single, randomly-mating pool. The niches where selection occurs can be thought of as host plant phenotypes on which eggs are laid. The fitness of a larva depends upon whether its alleles are adapted to that plant phenotype. For example, the niches might be resistant and susceptible host plant phenotypes. The relevant locus in the herbivore would have alleles which provide high fitness on susceptible and resistant phenotypes, respectively. Levene found that both alleles could be maintained only when the harmonic mean fitnesses of the homozygotes, over all habitats, are both less than the fitness of the heterozygote (harmonic mean overdominance). Wallace (1968) noted that if the contribution in offspring of a niche to the total population depends on the genetic makeup of the subpopulations (hard selection), maintenance of stable polymorphism requires arithmetic mean overdominance. This condition is more restrictive than harmonic mean overdominance. Hence, independent population regulation in subpopulations will foster local adaptations.

Arnold and Anderson (1983) used a linear logistic form of selection in each niche, so that the selection in each niche depended upon the number and ecological fitness of the individuals in that particular habitat. Gametes are contributed in proportion to the number of individuals that lived in and survived selection in that niche. Their results indicate that a sufficient condition for stable polymorphism is the weighted harmonic mean overdominance of a fitness parameter, the adjusted carrying capacity.

If habitat preference is added to the above model in the form of habitat fidelity (adults tend to lay eggs in the habitat from which they came), polymorphisms become easier to maintain (Christiansen, 1974). However, this model has been criticized for a lack of robustness (Hoekstra et al., 1985). If niche preference is correlated with fitness and genetically controlled, robustness greatly increases (Garcia-Dorado, 1986). This model is sensitive to linkage between loci controlling behavior and those controlling fitness.

The correlation between preference behavior and fitness on various hosts has been examined for several insects. Taylor and Condra (1983) examined resource partitioning (habitat selection) by *Drosophila pseudoobscura* in the lab in fine-grained environments. They found significant (though small) genotype x environment interactions in several phenological traits, indicating that selection for the various genotypes differed in the different habitats. They also found that strains chose the habitat in which they fared best and that if two strains were mixed, individuals from each of the strains tended to choose different habitats. Hence, strains partitioned, at least to a degree, the available resources, and there were selective differences in the different habitats. Similarly, Via (1986) found that leafminers (*Liriomyza sativae*) oviposited preferentially on the host on which their female offspring were likely to gain their greatest pupal weight (a measure of fitness). Host preference and larval fitness were also correlated between two populations of *Heliothis virescens*, one collected from native hosts in the Virgin Islands, the other from cotton in Mississippi (Schneider and Roush, 1986). Singer et al. (1988) found significant correlations between female oviposition and larval performance within a single population of *Euphydryas editha*. However, many other studies have demonstrated a lack of correspondence between host preferences and larval fitness (Tabashnik, 1983; Williams, 1983; Singer, 1983). Tabashnik and Slansky (1987) reviewed potential ecological and evolutionary explanations for the lack of such correlations.

Local adaptation to host plant phenotypes is easier when populations are substructured so that all adults do not form a randomly-mating pool. Models of substructured populations emphasize the importance of migration and gene flow between populations. In island models, each subpopulation is equally likely to exchange migrants with every other subpopulation. Stepping stone models assume that each population exchanges migrants only with its nearest neighbors. Both types of models have consistently demonstrated that polymorphisms are more likely when migration is below some critical value (Haldane, 1930; Hanson, 1966; Maynard Smith, 1966; Pollak, 1974; Spieth, 1974; Gillespie, 1975). Kimura and Maruyama (1971) showed that more than one migrant per generation between populations prevents significant local differentiation in island models.

There is considerable confusion over the exact meaning of local differentiation, however. Allendorf and Phelps (1981) suggest that one migrant per generation will ensure that the same alleles will be shared among populations over long periods of time, but does not ensure identical allele frequencies between populations. Using Monte Carlo simulation, they found significant genetic divergence among 20 subpopulations in 60% of the trials when 25 migrants were exchanged per generation ($\alpha = 0.05$). Spieth (1979) demonstrated that low levels of gene flow could not overcome local selection. He suggested that the rule of thumb of one migrant per generation to prevent local differentiation is too low by an order of magnitude. In Monte Carlo simulations, MacCluer (1974) also found that one migrant per generation was insufficient to deter local differentiation. Gould (1983) has suggested that the spatial arrangement of habitat patches can also affect the maintenance of polymorphisms, particularly when considering adaptation to several intermixed host plant types. Clumping of habitats would lower gene flow between alternate habitats.

Much like reduced migration between substructured populations, non-random mating between genotypes or selective oviposition by a genotype in the habitat in which it was raised can also promote local adaptation (Mitter and Futuyma, 1983). If the environment is coarse-grained (an insect usually spends its entire life in one habitat patch), then females will most likely mate with males from the same habitat. This type of non-random mating eases requirements necessary for local adaptation. Sympatric speciation in the treehopper species complex, *Echenopa binotata*, is mediated by coarse-grained use of habitats by both adults and larvae. Limited movement of adults increases differentiation, and gene flow is further reduced because adults mate on the host plant they emerged on (Wood, 1980; Wood and Guttman, 1982; Wood and Keese, 1990).

Coarse-grained use of habitats and the consequent formation of host-associated populations has also been observed in the apple maggot, *Rhagoletis pomonella* (Bush, 1966; Prokopy et al., 1982; Diehl, 1984). This species was endemic to North America and was found on native species of hawthorn (*Crataegus* spp.). In the mid-nineteenth century, the fly expanded its host range to include apple (*Malus pumila*), an introduced species. Subsequent studies have demonstrated genetic variation between local populations found on the two different hosts, as well as clinal variations in allele frequencies (Feder et al., 1988, 1990a, 1990b; McPheron et al., 1988).

Seasonal asynchrony in host phenotype and differences in host plant acceptance behavior have been suggested as mechanisms that isolate the sympatric populations. Smith (1988) found heritable variation in emergence times of flies collected from three different hosts and suggested that seasonal asynchrony in host plant phenology could reduce matings and gene flow between sympatric populations. Divergence in host plant acceptance behavior has also been noted for *R. pomonella* (Prokopy et al., 1988). Flies collected from apple and hawthorn did not differ in their oviposition on hawthorn, but flies collected from apple were more likely to lay eggs on apple than were flies collected from hawthorn. No difference in larval survivorship between populations was found.

Electrophoretic examination of 13 polymorphic loci showed moderate genetic differentiation between apple and hawthorn populations of *R. pomonella*. Significant genetic differences between local pairs of apple and hawthorn populations were observed consistently at six polymorphic loci, while at the seven other loci, no consistent differences were found (Feder et al., 1990a, 1990b). The six polymorphic loci for which inter-populational differences were observed to map consistently to three distinct regions in the genome (Feder et al., 1989). These loci may be subject to selection, either directly or via epistasis, or may be hitch-hiking via linkage with other selected loci. The weak differentiation among the other seven loci suggests that gene flow occurs between populations on the different hosts. The presence of gene flow was also indicated by the relatively high frequencies of otherwise rare alleles in several paired hawthorn and apple populations. Gene flow and differentiation between sympatric populations may vary with the quantity and quality of fruit set by each of the respective host plants (Prokopy et al., 1988).

In summary, the results suggest that while some level of gene flow between host-associated populations of the apple maggot does occur, the selective forces acting on each host are sufficient to maintain, at least for parts of the genome, host-associated polymorphisms. Thus, evolution of resistance to host-specific toxins could occur in *R. pomonella*, though this does not appear to be the case because larval survival did not differ between populations collected from the two host plants.

Parthenogenic genotypes of the fall cankerworm, *Alsophila pometaria*, associated with maple (*Acer rubrum*) and oak (*Quercus* spp.), were examined for physiological or behavioral differences in response to the two host plants (Futuyma et al., 1984). The genotype associated with maple had the highest growth and efficiency of conversion of ingested food on both host plants, suggesting that trade-offs in physiological adaptation did not exist between the populations. Genetic correlations in performance among rare genotypes also provided little evidence of pleiotropic effects associated with performance on different host plants (Futuyma and Philippi, 1987). In behavioral tests, there was no difference between the genotypes in their tendency to disperse from oak and dogwood, but the maple-associated genotype was much less likely to disperse from maple leaves than the oak-associated genotype (Futuyma et al., 1984). These results explain the prevalence of the maple-associated genotype in maple stands, but they do not explain its scarcity in oak stands. The oak-associated genotype may survive in oak stands because of greater synchrony between hatch and bud break in that genotype (Mitter et al., 1979), or because of other ecological factors such as soil type, which might affect mortality of the pupae (Futuyma et al., 1984).

The multiple-niche hypothesis predicts that species with multiple or broader niches will have more genetic variation than species with fewer or narrower niches. Species with wider niches, therefore, would be more likely to have the genetic variance necessary to adapt to a novel host plant phenotype. Experimental tests of the correlation between environmental and genetic variance have been inconclusive. Lacy (1982) examined a mycophagous guild of *Drosophila* spp. with included niches (the niche of the species with the narrowest niche was included in the species with the next broadest niche, etc.). The species with the broadest niche had the most electrophoretic variation, supporting the niche breadth hypothesis. Likewise, Mark (1982) found that bruchid beetles adapted behavior to spatial and temporal variation in habitat (when temporal variation was predictable). In a laboratory study of *Drosophila melanogaster*, Mackay (1980, 1981) found that heterogeneity increased variance in some characters. Surprisingly, temporal variance was equally if not more effective in the maintenance of polymorphisms than was spatial variance. Initial studies of one strain indicated that the environmental heterogeneity did not increase the genetic variance, but rather retained the initial variance better than the controls. Correlations between environmental and genetic variance have also been reported by Van Halen (1965), Bernstein (1979), Nevo et al. (1983), Shugart and Blaylock (1973), Bryant (1974), Steiner (1977), Bell (1982), and Lynch (1987).

However, results from other studies have contradicted the multiple-niche hypothesis. For instance, Patterson (1983) examined morphological variation within populations of different subfamilies of grasshoppers (Acrididae) with different niche breadths. No correlation in ranking between niche breadth and mandible size variation was found. The data suggest that selection favors one overall best compromise phenotype rather than maintaining phenotypic polymorphism. Caution in interpretation of the results should be exercised, since the comparisons were made between species in different subfamilies that had substantially different diets. Lister (1976), Soule and Stewart (1970), Soule and Yang (1973), Sabath (1974), Mitter and Futuyma (1979) and Smith (1981) also present evidence against the niche breadth hypothesis.

The lack of consistent findings with regard to the niche breadth hypothesis may be a reflection of the fact that no one answer is suitable for every organism. Genetic responses are constrained by factors such as epistatic interactions between genes, linkage, and evolutionary history. Thus, insects with the same niche breadth cannot always be expected to respond in the same manner. Evolutionary paths depend upon the spectrum of genetic variability present in finite populations, which in turn depends upon chance mutations and the history of each population. Perhaps the best that can be said is that factors such as low gene flow will increase the probability of polymorphism, but they are not directly the cause of it. If genetic variability is absent in the populations, no differentiation of populations will occur no matter how low the gene flow.

6.2. The Genetic Basis of Resistance to Plant Defensive Chemicals in Insects

In this section, we examine the kinds of changes within the genome that promote adaptation to plant toxins. We ask whether resistance is most likely to result from mutations in structural or regulatory sites in the genome. We also ask whether resistance to host plant toxins evolves primarily as a result of changes at a single locus or many loci.

6.2.1. Molecular Adaptation

Increased understanding of the molecular nature of the genome has revealed considerable flexibility in its ability to respond to stresses such as novel host plant toxins. To understand how insects can adapt to plant toxins, it is necessary to examine the molecular mechanisms involved in adaptation before one can suggest which pathways are most likely to be followed in resistance evolution. Unfortunately, little is known about the molecular basis of adaptation to plant toxins in insects. We shall therefore draw examples from a somewhat larger spectrum, mostly from prokaryotes. We also consider evolution of insect resistance to other toxins, such as synthetic insecticides. We focus on evolution of resistance to insecticides that are similar to plant compounds, such as pyrethroids, nicotine, and rotenone, because such examples are most likely to provide insights about insect response to plant toxins. We examine gene organization, the role of structural (qualitative) versus regulatory (quantitative) changes in enzymes in adaptation, and adaptation by prokaryotes in response to novel nutritional substrates and viral pathogens. Although an understanding of prokaryote adaptation and evolution of resistance to synthetic insecticides may stimulate development and testing of hypotheses, we caution that conclusions from these areas may not always be applicable to evolution of resistance to plant defensive chemicals in insects.

6.2.1.1. Gene Organization

Gene structure, the physical arrangement of regulatory and structural genes on the chromosomes, affects evolution in several ways. First, the regulatory elements, which control when and where genes are expressed, are small and can move about in the genome. When exposed to stress, cells may be capable of responding by increasing regulatory mutations (McClintock, 1984), allowing rapid adaptation merely by altering the pattern of expression of existing genes. Second, many regulatory elements are closely linked with structural genes. Selection can then act on the entire gene complex without recombination, potentially allowing for co-adaptation between the structural genes and the regulatory elements surrounding them.

Regulation of a protein product within the eukaryotic cell can occur at three different levels: gene activation, transcription, and protein synthesis/activation (Macintyre, 1983; Paigen, 1986). Chromatin surrounding the DNA must be activated before transcription can be initiated. Once transcription has occurred, the primary transcript may be modified via diverse

Figure 6.1. Hypothetical promoter region upstream of a gene. Mutations within boxed areas greatly reduce transcription of the structural gene; mutations in other areas have little effect on transcription.

processes before mature mRNA is produced. Finally, this mature mRNA is used to synthesize polypeptides, which may be further processed to yield an active protein. At each step, degradation or inactivation of the product will also be important in determining the ultimate rate of synthesis.

Initial studies of prokaryotic organisms indicated that regulation occurred mainly at the transcriptional level. Eukaryotic organisms exhibit more diverse methods of regulation (Paigen, 1986). Eukaryotic DNA contains much nontranscribed, or "junk," DNA (Li, 1983; Edgell et al., 1983; Dover, 1982). Even within a region encoding a structural protein, there may be large nontranscribed spacers or areas that are spliced out of the primary transcript when it is processed to mature mRNA. Sequences that control gene regulation may be embedded within these nontranscribed areas. Upstream (towards the 5' end of the DNA where transcription is initiated) of the start of the structural gene lie promoters and enhancers, the regulatory elements of the gene (Fig. 6.1). Promoters determine the site of transcript initiation, while enhancer elements determine the rate of transcription.

The promoter is usually a small element, containing an AT-rich "box," such as the TATA or CCAAT boxes, which lie 20–100 base pairs upstream of the gene (Arnheim, 1986). These boxes are highly conserved. Mutations in the surrounding DNA will have little effect on transcription, but alteration of the promoter will severely inhibit transcription (Maniatis et al., 1987). Since these promoter sequences are common to many genes, it is unlikely that they play an important role in gene-specific regulation, except for repression. Farther upstream, about 50–300 base pairs are cis-acting upstream promoter elements. These small elements (8–10 base pairs), closely linked to the structural gene, bind proteins and are necessary for maximal transcription.

Enhancer elements are similar to upstream promoter elements in structure, but are not tightly linked with the structural gene. Enhancers may be located several thousand base pairs away, or even on other chromosomes. These trans-acting enhancers generally act through intermediate transmitters and are therefore likely to be dominant in their effect (Paigen, 1986). Enhancers and upstream promoter elements are more likely to provide differential regulation of structural genes (Maniatis et al., 1987). Since most of these elements are small, they may be easily moved about in the genome by transposons (McClintock, 1984; Paigen, 1986).

The number of regulatory elements linked to a gene may vary. For example, regulatory elements involved in transcription of the major late promoter of adenovirus by human RNA polymerase II included a TATA box 28 bases upstream and an upstream promoter element 58 bases upstream (Van Dyke et al., 1988). Three transcription factors (ancillary proteins) and RNA polymerase II were required for transcription, while maximal activity required a fourth gene-specific transcription factor which bound to the upstream element.

6.2.1.2. Structural versus Regulatory Changes

At the molecular level, structural or regulatory changes can promote adaptation to plant toxins and novel substrates. Structural changes result from mutations that alter the amino acid sequence of a protein. The altered protein interacts with the toxin differently, perhaps metabolizing the toxin more rapidly or, in the case of target site insensitivity, reducing the affinity of the protein for the toxin. Alternatively, changes in gene regulation that alter the amount of a protein without concomitant change in structure can also lead to adaptation. Gene duplication results in increased production of unmodified protein. We therefore regard gene duplication as a regulatory change.

Resistance of insects to toxins has been shown in several cases to be the result of structural changes. L-canavanine, which functions as a defensive chemical agent against non-adapted herbivores, is a structural analog of the amino acid arginine. For example, in the tobacco hornworm (*Manduca sexta*), L-canavanine is incorporated into polypeptides, resulting in non-functional proteins. Larvae of the beetle, *Caryedes brasiliensis*, eat seeds of the legume *Dioclea megacarpa*, which contain large concentrations of L-canavanine (Rosenthal et al., 1977). *C. brasiliensis* avoids incorporation of L-canavanine because the t-RNA synthetase of the insect discriminates between the analogs, resulting in the incorporation predominantly of arginine into polypeptides (Rosenthal et al., 1976; Rosenthal, 1983).

Structural changes have also been implicated in adaptation of the black swallowtail butterfly (*Papilio polyxenes*) to plants in the Umbelliferae and Rutaceae families. The larval host range of this butterfly includes species with high concentrations of the plant defensive compound, xanthotoxin and related furanocoumarins (Berenbaum and Feeny, 1981). Ivie et al. (1983) demonstrated that, compared with a generalist (*Spodoptera frugiperda*), *P. polyxenes* rapidly metabolized the toxins, particularly in the midgut area. Bull et al. (1986) suggested that the increased metabolism was due to a higher specific activity of the mixed-function oxidase system (P450) for substrates typical of the insect's diet. A small fraction of the total P450 content was responsible for the increased activity, suggesting that there are multiple forms of P450, only some of which are induced by the presence of furanocoumarins in the diet (Cohen et al., 1989).

Many structural changes in the genome are known from studies of insecticide resistance (Oppenoorth and Welling, 1976; Soderlund and Bloomquist, 1990; Wood, 1981). For example, qualitative differences in induced cytochrome P450, an enzyme responsible for increased metabolism of insecticides, were found in the housefly, *Musca domestica* (Yu and Terriere, 1979; Vincent et al., 1985). Of particular interest is the response of insects to synthetic pyrethroid insecticides, which are based on pyrethrins, natural constituents of the plant *Chrysanthemum cinerafolis*. The pyrethrins are thought to act by interfering with nerve sodium channels, keeping them open and leading to repetitive nerve discharges (Vijverberg et al., 1982; Osborne and Smallcombe, 1983; Narahashi, 1983).

Farnham (1971, 1973) reported that resistance to natural pyrethrins was associated with cross-resistance to synthetic pyrethroids in the housefly, *Musca domestica*. These results show that the same mechanisms can provide resistance to both pyrethrins and pyrethroids. However, the housefly strains were artificially selected with pyrethrins, and experienced different selection regimes than would be typical among field populations of flies. Thus, results of this experiment may not be indicative of responses to natural selection in the field.

Resistance to synthetic pyrethroids as a result of target site insensitivity and decreased cuticular penetration has been reported in the housefly (DeVries and Georghiou, 1981a, 1981b) and mosquitoes (Omer et al., 1980). In contrast, increased metabolism of pyrethroids led to resistance in the cattle tick (*Boophilus microplus*) (De Jersey et al., 1985), the tobacco budworm (*Heliothis virescens*) (Dowd et al., 1987), and the mite *Amblyseius fallacis* (Chang and Whalon, 1986). There were both qualitative and quantitative changes leading to resistance in the tobacco budworm (Dowd et al., 1987). The cattle tick and mite studies were not detailed enough to allow separation of the two alternatives.

Changes in gene regulation can also lead to adaptation to novel toxins. Adaptation by regulatory gene changes are exemplified best by response of the bacterium *Escherichia coli* challenged with novel sugar substrates. *E. coli* strains adapted to novel sugar substrates by increasing the level of preexisting enzymes via derepression (Mortlock, 1982). The production of ribitol dehydrogenase was normally induced by the sugar ribitol, for which it had high affinity. When the wild strain was placed on a medium containing only the novel sugar xylitol, a mutation in the regulatory pathway for ribitol dehydrogenase caused constitutive (continuous) production of the enzyme, which had a limited capability to metabolize xylitol. Further mutations in the borrowed structural gene for ribitol dehydrogenase then increased the specificity of the enzyme to its new task, while changes at other loci increased uptake of the new substrate via derepression. Mutations acquired in adaptation to the new substrate may reduce fitness in the presence of the original substrate (Hall, 1983).

Gene regulation has also been suggested as a mechanism by which species of *Drosophila* have adapted to the ethanol-rich environments of rotting fruits. Variation in alcohol dehydrogenase (ADH) affected ethanol tolerance (McKenzie and Parsons, 1972; David et al., 1976), and use of ethanol-rich habitats was correlated with ADH activity across *Drosophila* spp. (McDonald and Avise, 1976). In *D. melanogaster*, expression of the ADH gene was regulated in part by trans-acting regulatory genes on the third chromosome; the ADH gene is on the second chromosome (McDonald et al., 1977; McDonald and Ayala, 1978). Alleles on the third chromosome coding for high ADH activity were either partially or fully dominant over genes that coded for lower activity, as would be predicted for trans-acting regulatory genes. Among the Hawaiian *Drosophila*, many of the differences in tissue-specific ADH expression have been traced to cis-acting regulatory elements closely linked to the structural locus (Rabinow and Dickinson, 1981; Dickinson et al., 1984). Genetic variation in enzyme activity, the raw material for selection, certainly exists. The activities of six of seven enzymes examined in *D. melanogaster* showed genetic variation, and in five out of the seven, the regulatory elements were trans-acting, that is, not closely linked to the structural gene (Laurie-Ahlberg et al., 1980).

Gene duplication followed by independent regulatory adjustment of each gene allows one copy of the gene to adapt to a new function, while the original copy of the gene continues to function in its former role, reducing the costs associated with adaptation (Markert et al., 1975). Analysis of DNA sequences and the discovery of repetitive sequences provides evidence for the frequency of gene duplication and its importance in genomic adaptation (Li, 1983; Paigen, 1986). Alternatively, adjustment of both regulatory and structural elements of the original gene complex could reduce fitness costs while retaining the ability of an enzyme to function on both the novel and original substrates (Uyenoyama, 1986).

Batterham et al. (1984) documented a duplication of the ADH structural locus in the *mulleri* subgroup of the genus *Drosophila*. Species with the duplication did not show higher levels of ADH activity, which suggested that the duplication did not cause increased production of ADH. In three species examined, however, expression of the two ADH genes had differentiated. Expression of the first ADH gene occurred throughout the larval and pupal instars. Expression of the second ADH gene started during the second larval instar and continued throughout adulthood (Batterham et al. 1983, 1984). These results demonstrate that differentiation can occur between duplicated genes, resulting in stage-specific variation in expression of the genes. Since significant evolutionary time had passed since the duplication, the lack of increased ADH production associated with the duplication may have been caused by additional changes at regulatory sites.

Repeated duplication of structural genes, also called gene amplification, is responsible for several cases of insecticide resistance. Resistance to organophosphate insecticides in the green peach aphid, *Myzus persicae*, is due to duplication of a structural gene for carboxylesterase (Devonshire and Sawicki, 1979; Devonshire 1977, 1980, 1989). The carboxylesterase is relatively inefficient at metabolizing insecticides, but it binds them with high affinity. Gene amplification increases the production of carboxylesterase so much that 3% of the insect's total protein is accounted for by carboxylesterase. The enormous quantity of enzyme and its affinity for insecticides enables *M. persicae* to sequester insecticides and thus reduce or avoid their toxic effects (Devonshire and Moores, 1982). Resistance to organophosphate insecticides in *Culex quinquefasciatus* is also attributed to amplification of an esterase gene (Mouches et al., 1986). In both cases, the duplication led to an increase in the enzyme produced, and in neither case is there evidence of modifications of the amplified sequences.

Hedrick and McDonald (1980) used a hierarchial model of gene regulation, in which a regulatory gene controlled the amount of gene product at two or more loci, to examine the types of environmental change that would lead to regulatory or structural adaptation. A critical assumption of the model is that regulatory changes are likely to have much larger phenotypic effect than are changes in structural genes. The model also assumes that phenotypic values result from several structural loci, each having a small effect. Under these conditions, adaptation via gene regulation occurred more rapidly and was likely to be a response to large changes in the environment. The polygenic structural gene system was better able to track small environmental changes because less phenotypic variation occurred around the optimum phenotype. Thus, rapid environmental changes were more likely to result in changes at regulatory loci, while slower, gradual environmental changes resulted in changes at structural loci.

6.2.2. Pleiotropy

Pleiotropy refers to the tendency of a gene to affect more than one trait simultaneously, including seemingly unrelated traits (Hartl, 1987). Thus, adaptation to a new host plant can affect fitness on the original host plant. Pleiotropy can be measured as the change in fitness experienced by the newly-evolved genotype compared to the original genotype when exposed to the original selection regime (Lenski, 1988a). Whether the changes involved in adaptation to host plant toxins are qualitative or quantitative, pleiotropic costs may be incurred. Given time and sufficient isolation from susceptible genotypes, selection may act on fitness variation within resistant phenotypes to reduce pleiotropic costs.

To evaluate the potential for pleiotropy in host plant adaptation, Gould (1979) examined host range evolution in the spider mite *Tetranychus urticae* by isolating two populations, one on the preferred host (*Phaseolus vulgaris*), bean, and another on a marginal host, mite-resistant cucumber (*Cucumis sativus*). After 21 months (~ 50 generations), the population isolated on the cucumber survived significantly better on that host than did the bean population. The cucumber population showed a slightly lowered fitness on the bean plants, indicating a cost associated with adaptation to cucumber. Surprisingly, fitness on potato and tobacco, two other marginal hosts, was significantly higher for the selected cucumber line than for the unselected bean line. These positively correlated changes in fitness demonstrate that pleiotropy need not always entail reductions in fitness.

Tabashnik (1983) found that adaptation by populations of *Colias philodice* (Edwards) to a new host (alfalfa) reduced fitness to the natural hosts. Two populations of the butterfly *Euphydras editha*, collected from different hosts, were also shown to have higher fitness on their natural host than on the alternate host plant (Rausher 1982). Via (1984a) found that fitness of genotypes of the leafminer *Liriomyza sativa* changed ranks on different host plants, but the genetic variation was contained within all populations, with no specialization occurring between populations from different habitats. High gene flow between populations may have reduced population-level differentiation. In the tortoise beetle, *Deloyala guttata*, three traits were found to be positively correlated across two different hosts, while a fourth trait, fecundity, was suggestive of negative correlation and trade-offs in performance between the two hosts (Rausher 1984). The presence of positive genetic correlations among some traits would not necessarily interfere with adaption to local host plants as long as at least one trait exhibited negative correlations (Rausher, 1984). These negative correlations would result in selection favoring alternate genotypes on different host plants, despite other positive genetic correlations.

Pashley (1988) found genotype-host plant interactions among host-associated strains of the fall armyworm (*Spodoptera frugiperda*) collected from corn and rice. The corn-associated strain had higher fitness on both hosts than the rice-associated strain. The rice-associated strain performed poorly on corn, but did not differ significantly from the corn-associated strain when reared on rice. Because there was no change in the ranks of fitness on the two hosts, these results do not fully explain the presence of the host-associated strains. Low gene flow (Pashley, 1986) and reproductive incompatibilities between strains (Pashley and Martin, 1987) allow the two strains to coexist presently, but shed little light on how the two strains evolved. Ovipositional preferences have not been measured in these strains, and genetic correlations between oviposition behavior and fitness traits may help to explain evolution of the two host-associated strains.

Laboratory studies estimating genetic correlations should be interpreted with care. Selection in the field will tend to fix genes contributing to positive genetic correlations so that the genes contribute little to the covariance of the traits. The frequencies of genes responsible for negative genetic correlations are less responsive to selection (Falconer, 1981). Thus, these genes will continue to contribute to covariance. Genetic correlations measured in a novel environment, where selection has not had time to fix alleles with positive pleiotropic effects, may tend to be more positive than genetic correlations measured in normal environments. Service and Rose (1985) found that the additive genetic correlations between early-life fecundity and starvation resistance in *D. melanogaster* increased from -0.913 to -0.453 when measured in a novel laboratory environment. This example demonstrates both the presence of negative pleiotropic interactions between two fitness-associated traits and the potential bias towards more positive genetic correlations in novel environments.

Lenski (1988a, b) examined pleiotropic costs associated with resistance to the T4 virus in *E. coli*. Resistance to T4 was due to mutations in the structural gene encoding for the lipopolysaccharide core of the cell envelope (Lenski, 1988a; Wright et al., 1980). In the absence of the virus, all 20 resistant strains examined were less fit than the parent susceptible strain, and fitness varied significantly among resistant strains. Fitness in five resistant and six susceptible strains was monitored over 400 generations. Compared to a control T4-sensitive strain with a fitness of 1.0, the susceptible strains increased on average from 1.00 to 1.10, whereas the average fitness of the resistant strains increased from 0.66 to 1.03. Thus, genetic changes in the resistant strains reduced the pleiotropic costs associated with T4 resistance.

We do not know if results from the evolution of resistance to T4 in *E. coli* apply to resistance to host plant toxins in insects. Nevertheless, if different fitnesses are associated with resistance alleles, then selection among these alleles will be important. Thus, the same resistance-conferring mutation may become fixed in many populations, even in the absence of gene flow. Such convergent evolution at the molecular level occurs even between widely separated taxa. For example, amino acid sequencing of lysozymes from cows and lemurs showed similar mutations in response to a ruminant lifestyle (Stewart et al., 1987; Brown, 1987).

Evolution of insecticide resistance often reduces fitness in the absence of the insecticide, as evidenced by declines in resistance when selection with insecticides is stopped. Such fitness costs, however, are sometimes minor or absent (Roush and McKenzie, 1987). Reduction of pleiotropic costs associated with resistance can occur by formation of coadapted gene complexes or by further mutations at the same locus. Continued selection with insecticides eventually reduced the differences in fitness associated with the resistant and susceptible alleles (Abedi and Brown, 1960; White, 1978; Roush and McKenzie, 1987). In some cases, fitness costs associated with resistant homozygotes decreased after repeated backcrossing with susceptible strains (Amin and White, 1984; Helle 1965), indicating the potential for the evolution of fitness modifiers. Selection cannot favor such modifiers until the resistant allele has reached an appreciable frequency. In the field, selection for resistance is usually withdrawn when resistance becomes common due to failure of the insecticide (Roush and McKenzie, 1987). In other studies, repeated backcrosses of resistant lines with susceptible lines failed to demonstrate evidence of fitness modifiers (Whitehead et al., 1985; Beeman and Nanis, 1986). Failure of backcrosses to demonstrate negative pleiotropic costs does not rule out the possibility of continued evolution at a single locus to reduce fitness costs. As evidenced by *E. coli* resistance to the T4 virus (Lenski, 1988a, b), reduction of negative pleiotropic costs could occur by successive substitutions of increasingly competitive resistant alleles.

6.2.3. Mode of Inheritance

The response of an insect species to a new plant toxin will vary depending upon the insect, the strength of selection applied by the new toxin, the genetic variability present in the species, mutation rates, and so on. In this section, we consider conditions that favor adaptation to novel plant toxins by monogenic versus polygenic changes. Throughout, it must be remembered that evolution is affected by stochastic processes. Evolution uses the genetic variance available in populations, which is in turn dependent on chance mutations (Wood and Bishop, 1981). Thus, different populations of the same species may find alternate solutions, even when faced with similar conditions, because of differences in genetic variation between populations. For example, the housefly, *Musca domestica*, became resistant to DDT via one gene (kdr) in Denmark (Barbesgaard and Keiding, 1955), while in the U.S., resistance developed by means of another gene (deh) (Georghiou, 1972).

Resistance to insecticides is thought to be primarily monogenic (Roush and McKenzie, 1987). Some cases of resistance, however, including many examples from the field, show patterns that differ significantly from monogenic inheritance (Keiding, 1977; Wood, 1981; Halliday and Georghiou, 1985; Roush et al., 1986; Pree, 1987). The apparent prevalence of monogenic resistance to insecticides may be a result of the strong selection for rare phenotypes. Thus, caution should be used in extending these results to adaptation to host plant toxins. Since insecticides are generally withdrawn once they become ineffective, selection for loci which modify the action of the major gene rarely proceeds long enough for coadaptation to occur (Roush and McKenzie, 1987).

The nature of resistance evolution in response to host plant toxins may depend upon the manner in which an insect species is exposed to the toxin. Lande (1983) noted that mutations with major effects (with monogenic inheritance) occur less frequently and are likely to have greater deleterious effects than mutations with minor effects (and polygenic inheritance). Modeling adaptation via monogenic versus polygenic inheritance, he found that evolution of a new phenotype occurred most readily via monogenic inheritance when small changes in phenotype would be selected against (as when crossing a Wrightian trough, see Section 6.1.1), and when selection is sufficiently strong to overcome the greater negative pleiotropic costs usually associated with major mutations. Coevolution, where an insect species is gradually exposed to a new toxin as it evolves in the host plant, may tend to result in polygenic resistance to the new toxin. On the other hand, sudden exposure to a new toxin, as might happen in a host plant shift, might tend to result in monogenic resistance.

Monogenic resistance to host plant cultivars was found in resistance in wheat to the Hessian fly, *Mayetiola destructor*. Resistance in wheat is controlled by thirteen dominant or partially dominant genes (Hatchett et al., 1981). Ten races of the Hessian fly vary in their viability on different cultivars of wheat, and it is believed that there is a "gene-for-gene" correspondence between the resistance-breaking genes in the fly and those providing resistance in the wheat (Hatchett and Gallun, 1970; Gallun, 1978), that is, each resistance gene in the fly provides resistance to one wheat genotype (see Diehl and Bush [1984] for alternative interpretations of the data).

Huettel and Bush (1972) examined host preference in the tephritid fly, *Procecidochares australis*, which forms galls on the asters *Heterotheca subaxillaris* and *H. latifolia*, and an undescribed sibling species which parasitized the aster *Machaerantera phyllocephala*. Ovipositional preferences of offspring from hybrid adults backcrossed to adults of the parental lines suggested that ovipositional preference was controlled by a single major gene, while survival on each of the two hosts was affected by one to two loci (Diehl and Bush, 1984).

In contrast, the development of tolerance to resistant rice cultivars in *Nilaparvata lugens*, the rice brown planthopper, occurred by changes at many loci (Den Hollander and Pathak, 1981). Four major genes provide resistance in rice (Sogawa, 1982). Differences in host plant preference may help to isolate planthopper populations, increasing the potential for development of polygenic resistance to local rice cultivars.

6.3. Conclusions

The evolution of resistance to host plant toxins in insects depends on factors ranging from the molecular structure of the insect's genome to its population structure. Plants contain an array of compounds that affect herbivore fitness, which vary in structure from relatively simple ketides to large, complex proteins and tannins. Insects, however, have shown an equally diverse ability to evolve resistance to these chemicals. Few plants escape insect herbivory. That lesson has been repeated with the evolution of resistance to synthetic insecticides. No compounds have proven to be fail-safe, not even those that mimic insect growth and molting hormones.

Little information exists on the types of genomic changes that provide resistance to host plant toxins. Recent advances in molecular techniques that allow rapid sequencing of DNA from individual insects may help to clarify this issue in the future. We should be prepared to find that the genome can respond in number of different ways, including modification of structural genes, gene duplication, and alteration of gene regulation. The genetic basis of resistance in a particular case may be decided stochastically, depending on which mutations happen to be present in the population at the time selection for resistance occurs. As the insect adapts to the plant toxin over many generations, multiple mutations, both structural and regulatory, may be incorporated which collectively fine-tune the genome to its new function.

By examining mutations associated with resistance, molecular studies can help us understand whether resistance originates in a relatively small number of populations and is spread throughout other populations by migration or whether there is sufficient genetic variability in most populations to allow development of resistance locally. If resistant alleles are rare, one would expect migration to spread the same allele to many populations. If resistant alleles are initially common, or population sizes are large, resistance will develop locally, and different alleles in most populations would be expected as a consequence. The strategies for management of host plant resistance in each case are quite different. In the first case, increased migration (such as that caused by crop rotation) would hasten the spread of the resistance allele, while in the latter, increases in migration would increase the effects of global versus local selection, retarding selection for resistance in insects until the frequency of the resistant host plant phenotype confers a fitness advantage to insects with resistant genotypes.

Population structure is important in determining how mutations are spread and maintained. If a mutation is globally advantageous, high gene flow will rapidly spread the new allele to all populations. If, however, resistance to a toxin confers some disadvantage to insects feeding on host plants without the toxin, then high gene flow will hinder adaptation to a new host plant phenotype. Gene flow will be reduced when there is: (1) low migration between habitats and/or low reproductive success of migrants; (2) assortative mating such that the females in one habitat tend to mate with males from the same habitat, due either to behavioral or spatial characteristics; (3) density-dependent reproductive response (the more undercompensating the response, the greater the potential for gene flow); and (4) host plant/habitat preference (especially when there is a preference/fitness correlation). In addition, if population regulation is independent of genotypes (i.e., the number of offspring produced in a habitat is not affected by the genotype) or there is dominance switching so that the fitness of the heterozygotes approaches the most fit homozygote in that habitat, adaptation to host plant toxins may be easier.

We encourage the continued study of host plant adaptation via quantitative genetics. Such studies to determine genotype x environment interactions and correlations between host preference and offspring fitness are an important step in understanding the evolution of host plant adaptation. Without further studies on the population ecology of the species involved, however, there is little hope of understanding why one species will readily adapt to new host phenotypes while others will not. Such studies should therefore attempt to include, either quantitatively or at the very least qualitatively, information on the level of gene flow between populations within and between habitats, the expression and dominance of resistant alleles, and population regulation, especially with regard to genotype. The latter issue is best addressed in field studies, as population regulation may be affected not only by plant toxins, but also predator/prey complexes (Bernays, 1988) and abiotic factors.

Knowledge of resistance to natural host plant toxins in insects may be applied to prolong the efficacy of genetically engineered plants (Gould, 1988). If we understand how insects overcome plant defenses at both the molecular and population levels, then we may be able to engineer more durable defenses in transgenic plants and implement cultural practices that slow the spread of resistant alleles.

References

Abedi, Z. H. and A. W. A. Brown. 1960. Development and reversion of DDT-resistance in *Aedes aegypti*. *Can. J. Genet. Cytol.* 2:252–261.

Allendorf, F. W. and S. R. Phelps. 1981. Use of allelic frequencies to describe population structure. *Can. J. Fish. Aquat. Sci.* 38:1507–1514.

Amin, A. M. and G. B. White. 1984. Relative fitness of organophosphate-resistant and susceptible strains of *Culex quinquefasciatus* Say (Diptera: Culicidae). *Bull. Entomol. Res.* 74:91–98.

Arnheim, N. 1986. The evolution of transcriptional control signals: coevolution of ribosomal gene promoter sequences and transcription factors. In *Evolutionary Processes and Theory*, eds. S. Karlin and E. Nevo, pp. 37–52. Academic Press, Orlando.

Arnold, J. and W. W. Anderson. 1983. Density-regulated selection in a heterogeneous environment. *Am. Nat.* 121:656–668.

Barbesgaard, P. and J. Keiding. 1955. Crossing experiments with insecticide resistant houseflies (*Musca domestica* L.). *Vidensk. Meddr. dansk natur. Foren.* 117:84–116.

Barbosa, P. and D. K. Letourneau. 1988. *Novel Aspects of Insect Plant Interactions.* John Wiley, New York.

Barbosa, P. and J. A. Saunders. 1985. Plant allelochemicals: linkages between herbivores and their natural enemies. In *Chemically-Mediated Interactions Between Plants and Other Organisms*, Vol. 19. *Recent advances in phytochemistry.* Plenum Press, New York.

Batterham, P., J. A. Lovett, W. T. Starmer, and D. T. Sullivan. 1983. Differential regulation of duplicate alcohol dehydrogenase genes in *Drosophila mojavensis.* *Dev. Biol.* 96:346–354.

Batterham, P., G. K. Chambers, W. T. Starmer, and D. T. Sullivan. 1984. Origin and expression of an alcohol dehydrogenase gene duplication in the genus *Drosophila.* *Evolution* 38:644–657.

Beeman, R. W. and S. M. Nanis. 1986. Malathion resistance alleles and their fitness in the red flour beetle (Coleoptera: Tenebrionidae). *J. Econ. Entomol.* 79:580–587.

Bell, M. A. 1982. Differentiation of adjacent stream populations of three-spined sticklebacks. *Evolution.* 36:189–199.

Berebaum, M. and P. Feeny. 1981. Toxicity of angular furanocoumarins to swallowtails: escalation in the coevolutionary arms race. *Science* 212:927–929.

Bernays, E. 1988. Host specificity in phytophagous insects: selection pressure from generalist predators. *Entomol. Exp. Appl.* 49:131–140.

Bernstein, R. A. 1979. Evolution of niche breadth in populations of ants. *Am. Nat.* 114:533–544.

Boethel, D. J. and R. D. Eikenbary. 1986. *Interactions of Plant Resistance and Parasitoids and Predators of Insects.* Ellis Horwood, Chichester, England.

Brattsten, L. B. and S. Ahmad. 1986. *Molecular aspects of insect-plant associations.* Plenum Press, New York.

Brown, A. L. 1987. Gut reactions of lysozyme. *Nature* 330:315–316.

Bryant, E. H. 1974. On the adaptive significance of enzyme polymorphisms in relation to environmental variability. *Am. Nat.* 108:1–19.

Bull, D. L., G. W. Ivie, R. C. Beier, and N. W. Pryor. 1986. In vitro metabolism of a linear furanocoumarin (8-methoxypsor len, xanthotoxin) by mixed-function oxidases of larvae of black swallowtail butterfly and fall armyworm. *J. Chem. Ecol.* 12:885–892.

Bush, G. L. 1966. Taxonomy, cytology and evolution of the genus *Rhagoletis* in North America. *Bull. Harvard Mus. Comp. Zool.* 134:431–562.

Caprio, M. A. 1990. Gene flow as a factor in the evolution of insecticide resistance. Ph. D. dissertation, Univ. of Hawaii at Manoa, Honolulu.

Chang, C. K. and M. E. Whalon. 1986. Hydrolysis of permethrin by pyrethroid esterases from resistant and susceptible strains of *Amblyseius fallacis* (Acari: Phytoseiidae). *Pestic. Biochem. Physiol.* 25:446–452.

Charlesworth, B. and S. Rouhani. 1988. The probability of peak shifts in a founder population II. An additive polygenic trait. *Evolution* 42:1129–1145.

Christiansen, F. B. 1974. Sufficient conditions for protected polymorphism in a subdivided population. *Am. Nat.* 108:157–166.

Cohen, M. B., M. R. Berenbaum, and M. A. Schuler. 1989. Induction of cytochrome P450-mediated detoxification of xanthotoxin in the black swallowtail. *J. Chem. Ecol.* 15:2347–2355.

Comins, H. N. 1977. The development of insecticide resistance in the presence of migration. *J. Theor. Biol.* 64:177–197.

Crow, J. F. and M. Kimura. 1970. *An Introduction to Population Genetics Theory.* Harper & Row, New York.

DaCunha, A. B. and T. Dobzhansky. 1954. A further study of chromosomal polymorphism in *Drosophila willistoni* in its relation to the environment. *Evolution* 8:119–134.

DaCunha, A. B., H. Burla, and T. Dobzhansky. 1950. Adaptive chromosomal polymorphism in *Drosophila willistoni. Evolution* 4:212–235.

David, J. R., C. Bocquet, M. F. Arens, and P. Fouillet. 1976. Biological role of alcohol dehydrogenase in the tolerance of *Drosophia melanogaster* to aliphatic alcohols: utilization of an ADH-null mutant. *Biochem. Genet.* 14:989–997.

De Jersey, J., J. Nolan, P. A. Davey, and P. W. Riddles. 1985. Separation and characterization of the pyrethroid-hydrolyzing esterases of the cattle tick, *Boophilus microplus*. *Pestic. Biochem. Physiol.* 23:349–357.

Den Hollander, J. and P. K. Pathak. 1981. The genetics of the 'biotypes' of the rice brown planthopper, *Nilaparvata lugens*. *Entomol. Exp. Appl.* 29:76–86.

Denno, R. F. and M. S. McClure. 1983. *Variable Plants and Herbivores in Natural and Managed Ecosystems*. Academic Press, London.

Devonshire, A. L. 1977. The properties of a carboxylesterase from the peach-potato aphid, *Myzus persicae* (Sulz.), and its role in conferring insecticide resistance. *Biochem. J.* 167:675–683.

Devonshire, A. L. 1980. Evolution by gene duplication in insecticide-resistant *Myzus persicae*. *Nature* 284:578–xxx.

Devonshire, A. L. 1989. Insecticide resistance in *Myzus persicae*: from field to gene and back again. *Pest. Sci.* 26:375–382.

Devonshire, A. L. and G. D. Moores. 1982. A carboxylesterase with broad substrate specificity causes organophosphorus, carbamate and pyrethroid resistance in peach-potato aphids (*Myzus persicae*). *Pestic. Biochem. Physiol.* 18:235–246.

Devonshire, A. L. and R. M. Sawicki. 1979. Insecticide-resistant *Myzus persicae* as an example of evolution by gene duplication. *Nature* 280:140–141.

DeVries, D. H. and G. P. Georghiou. 1981a. Absence of enhanced detoxication of permethrin in pyrethroid-resistant houseflies. *Pestic. Biochem. Physiol.* 15:242–252.

DeVries, D. H. and G. P. Georghiou. 1981b. Decreased nerve sensitivity and decreased cuticular penetration as mechanisms of resistance to pyrethroids in a (1R)-trans-permethrin-selected strain of the housefly. *Pestic. Biochem. Physiol.* 15:234–241.

Dickinson, W. J., R. G. Rowan, and M. D. Brennan. 1984. Regulatory gene evolution: adaptive differences in expression of alcohol dehydrogenase in *Drosophila melanogaster* and *Drosophila simulans*. *Heredity* 52:215–225.

Diehl, S. R. 1984. The role of host plant shifts in the ecology and speciation of *Rhagoletis* flies. Ph. D. dissertation, University of Texas, Austin.

Diehl, S. R. and G. L. Bush. 1984. An evolutionary and applied perspective of insect biotypes. *Annu. Rev. Entomol.* 29:471–504.

Dover, G. 1982. Molecular drive: a cohesive mode of species evolution. *Nature* 299:111–117.

Dowd, P. F., C. C. Gagne, and T. C. Sparks. 1987. Enhanced pyrethroid hydrolysis in pyrethroid-resistant larvae of the tobacco budworm, *Heliothis virescens* (F.). *Pestic. Biochem. Physiol.* 28:9–16.

Edgell, M. H., S. C. Hardies, B. Brown, C. Voliva, A. Hill, S. Phillips, M. Comer, F. Burton, S. Weaver, and C. A. Hutchison. 1983. Evolution of the mouse beta globin complex locus. In *Evolution of Genes and Proteins*, eds. M. Nei and R. K. Koehn, pp. 1–13. Sinauer, Sunderland, MA.

Ehrlich, P. R. and P. H. Raven. 1964. Butterflies and plants: a study in coevolution. *Evolution* 18:586–608.

Ehrlich, P. R. and P. H. Raven. 1969. Differentiation of populations. *Science* 165:1228–1232.

Ehrlich, P. R., R. R. White, M. C. Singer, S. W. McKechnie, and L. E. Gilbert. 1975. Checkerspot butterflies: a historical perspective. *Science* 188:221–228.

Endler, J. A. 1979. Gene flow and life history patterns. *Genetics* 93:263–284.

Falconer, D. S. 1981. *Introduction to Quantitative Genetics*, 2nd ed. Longman, London.

Farnham, A. W. 1971. Changes in cross-resistance patterns of houseflies selected with natural pyrethrins or resmethrin (5-benzyl-3-furylmethyl [+ −]-cis-trans-chyrysanthemate). *Pestic. Sci.* 2:138–143.

Farnham, A. W. 1973. Genetics of resistance of pyrethroid-selected houseflies, *Musca domestica* L.. *Pestic. Sci.* 4:513–520.

Feder, J. L., C. A. Chilcote, and G. L. Bush. 1988. Genetic differentiation between sympatric host races of the apple maggot fly *Rhagoletic pomonella*. *Nature* 336:61–64.

Feder, J. L., C. A. Chilcote, and G. L. Bush. 1989. Inheritance and linkage relationships of allozymes in the apple maggot fly. *J. Hered.* 80:277–283.

Feder, J. F., C. A. Chilcote, and G. L. Bush. 1990a. The geographic pattern of genetic differentiation between host-associated populations of *Rhagoletis pomonella* (Diptera: Tephritidae) in the eastern United States and Canada. *Evolution* 44:570–594.

Feder, J. L., C. A. Chilcote, and G. L. Bush. 1990b. Regional, local and microgeographic allele frequency variation between apple and hawthorn populations of *Rhagoletis pomonella*. *Evolution* 44:595–608.

Feeny, P. P. 1976. Plant apparency and chemical defense. In *Biochemical Interactions Between Plants and Insects. Rec. Adv. Phytochem*, Vol. 10, eds. J. Wallace and R. Mansell, pp. 1–40. Plenum Press, New York.

Feeny, P. L., L. Rosenberry, and M. Carter. 1983. Chemical aspects of oviposition behavior in butterflies. In *Herbivorous Insects: Host-Seeking Behavior and Mechanisms*, ed. S. Ahmad, pp. 27–75. Academic Press, New York.

Felsenstein, J. 1976. The theoretical population genetics of variable selection and migration. *Annu. Rev. Genet.* 10:253–280.

Fretwell, S. D. and H. L. Lucas. 1969. On territorial behavior and other factors influencing habitat distribution in birds. I. Theoretical developments. *Acta Biotheor.* 19:16–36.

Futuyma, D. J. 1983. Selective factors in the evolution of host choice by phytophagous insects. In *Herbivorous Insects: Host-Seeking Behavior and Mechanisms*, ed. S. Ahmad, pp. 227–244. Academic Press, New York.

Futuyma, D. J. 1987. The role of behavior in host-associated divergence in herbivorous insects. In *Evolutionary Genetics of Invertebrate Behavior*, ed. M. Huettel, pp. 283–294. Plenum Press, New York.

Futuyma, D. J. and S. C. Peterson. 1985. Genetic variation in the use of resources by insects. *Annu. Rev. Entomol.* 30:217–238.

Futuyma, D. J. and T. E. Philippi. 1987. Genetic variation and covariation in responses to host plants by *Alsophila pometaria* (Lepidoptrea: Geometridae). *Evolution* 41:269–279.

Futuyma, D. J. and M. Slatkin. 1983. *Coevolution*. Sinauer, Sunderland, MA.

Futuyma, D. J., R. P. Cort, and I. V. Noordwijk. 1984. Adaptation to host plants in the fall cankerworm (*Alsophila pometaria*) and its bearing on the evolution of host affiliation in phytophagous insects. *Am. Nat.* 123:287–296.

Gallun, R. L. 1978. Genetics of biotypes B and C of the Hessian fly. *Ann. Entomol. Soc. Amer.* 71:481–486.

Garcia-Dorado, A. 1986. The effect of niche preference on polymorphism protection in a heterogeneous environment. *Evolution* 40:936–945.

Georghiou, G. P. 1972. The evolution of resistance to insecticides. *Annu. Rev. Ecol. and Syst.* 3:133–168.

Gilbert, L. E. 1979. Development of theory in the analysis of insect-plant interactions. In *Analysis of ecological systems*, eds. D. Horn, R. Mitchell, and G. Stairs, pp. 117–154. Ohio State University Press, Columbus.

Gillespie, J. H. 1974. Polymorphism in patchy environments. *Am. Nat.* 108:145–151.

Gillespie, J. H. 1975. The role of migration in the genetic structure of populations in the temporarily and spatially varying environments I. Conditions for polymorphism. *Am. Nat.* 109:127–135.

Gillespie, J. H. 1976. A general model to account for enzyme variation in natural populations. II. Characterization of the fitness functions. *Am. Nat.* 110:809–821.

Gould, F. 1979. Rapid host range evolution in a population of the phytophagous mite *Tetranychus urticae* Koch. *Evolution* 33:791–802.

Gould, F. 1983. Genetics of plant-herbivore systems: interactions between applied and basic study. In *Variable Plants and Herbivores in Natural and Managed Ecosystems*, eds. R. F. Denno and M. S. Mclure, pp. 599–653. Academic Press, London.

Gould, F. 1988. Genetic engineering, integrated pest management and the evolution of pests. *TREE 3/Bitech* 6:S15–S19.

Green, M. B. and P. A. Hedin. 1986. *Natural Resistance of Plants to Pests*. Amer. Chem. Soc., New York.

Haldane, J. B. 1930. A mathematical theory of natural and artificial selection. VI. Isolation. *Proc. Cambridge Philos. Soc.* 26:220–230.

Hall, B. G. 1983. Evolution in a petri dish: the evolved β-galactosidase system as a model for studying acquisitive evolution in the laboratory. *Evol. Biol.* 15:85–150.

Halliday, W. R. and G. P. Georghiou. 1985. Inheritance of resistance to permethrin and DDT in the southern house mosquito (Diptera: Culicidae). *J. Econ. Entomol.* 78:762–767.

Hanson, W. D. 1966. Effects of partial isolation (distance), migration, and different fitness requirements among environmental pockets upon steady state gene frequencies. *Biometrics* 22:453–468.

Hanson, F. E. 1983. The behavioral and neurophysiological basis of food-plant selection by lepidopterous larvae. In *Herbivorous Insects: Host-Seeking Behavior and Mechanisms*, ed. S. Ahmad, pp. 3–23. Academic Press, New York.

Hartl. D. L. 1987. *A Primer of Population Genetics*. Sinauer, Sunderland, MA.

Hatchett, J. H. and R. L. Gallun. 1970. Genetics of the ability of the Hessian fly, *Mayetiola destructor*, to survive on wheats having different genes for resistance. *Ann. Entomol. Soc. Amer.* 63:1400–1407.

Hatchett, J. H., T. J. Martin, and R. W. Livers. 1981. Expression and inheritance of resistance to Hessian fly in synthetic hexaploid wheats derived from *Triticum tauschii* (Coss) Schmal. *Crop Sci.* 21:731–734.

Hedrick, P. W. 1986. Genetic polymorphism in heterogeneous environments: a decade later. *Annu. Rev. Ecol. Syst.* 17:535–566.

Hedrick, P. W. and J. F. McDonald. 1980. Regulatory gene adaptation: an evolutionary model. *Heredity* 45:83–97.

Hedrick, P. W., M. E. Ginevan, and E. P. Ewing. 1976. Genetic polymorphism in heterogeneous environments. *Annu. Rev. Ecol. Syst.* 7:1–32.

Helle, W. 1965. Resistance in the acarina: mites. *Adv. Acarol.* 2:71–93.

Hoekstra, R. F., R. Bijlsma, and A. J. Dolman. 1985. Polymorphism from environmental heterogeneity: models are only robust if the heterozygote is close in fitness to the favored homozygote in each environment. *Genet. Res.* 45:299–314.

Huettel, M. D. and G. L. Bush. 1972. The genetics of host selection and its bearing on sympatric speciation in *Procecidochares* (Diptrea: Tephritidae). *Entomol. Exp. Appl.* 15:465–480.

Ivie, G. W., D. L. Bull, R. C. Beier, N. W. Pryor, and E. H. Oertli. 1983. Metabolic detoxification: Mechanism of insect resistance to plant psoralens. *Science* 221:374–376.

Keiding, J. 1977. Resistance in the housefly in Denmark and elsewhere. In *Pesticide Management and Insecticide Resistance*, eds. D. L. Watson and A. W. A. Brown, pp. 261–302. Academic Press, New York.

Kimura, M. and T. Maruyama. 1971. Pattern of neutral polymorphism in a geographically structured population. *Genet. Res.* 18:125–131.

Lacy, R. C. 1982. Niche breadth and abundance as determinants of genetic variation in populations of mycophagous drosophilid flies (Diptera: Drosophilidae). *Evolution* 36:1265–1275.

Lande, R. 1983. The response to selection on major and minor mutations affecting a metrical trait. *Heredity* 50:47–65.

Laurie-Ahlberg, C. C., G. Maroni, G. C. Bewley, J. C. Lucchesi, and B. S. Weir. 1980. Quantitative genetic variation of enzyme activities in natural populations of *Drosophila melanogaster*. *Proc. Natl. Acad. Sci. USA* 77:1073–1077.

Lenski, R. E. 1988a. Experimental studies of pleiotropy and epistasis in *Escherichia coli*. I. Variation in competitive fitness among mutants resistant to virus T4. *Evolution* 42:425–432.

Lenski, R. E. 1988b. Experimental studies of pleiotropy and epistasis in *Escherichia coli*. II. Compensation for maladaptive effects associated with resistance to virus T4. *Evolution* 42:433–440.

Levene, H. 1953. Genetic equilibrium when more than one ecological niche is available. *Am. Nat.* 87:331–333.

Levins, R. 1976. *Evolution in Changing Environments*. Princeton Univ. Press, Princeton, NJ.

Li, W. H. 1983. Evolution of duplicate genes and pseudogenes. In *Evolution of Genes and Proteins*, eds. M. Nei and R. K. Koehn, pp. 14–37. Sinauer, Sunderland, MA.

Lister, B. C. 1976. The nature of niche expansion in West Indian Anolis lizards. II. Evolutionary components. *Evolution* 30:677–692.

Lynch, M. 1987. The consequences of fluctuating selection for isozyme polymorphisms in *Daphnia*. *Genetics* 115:657–669.

MacCluer, J. W. 1974. Monte Carlo simulation: the effects of migration on some measures of genetic distance. In *Genetic Distance*, eds. J. Crow and C. Denniston, pp. 77–95. Plenum Press, New York.

Macintyre, R. J. 1983. Regulatory genes and adaptation. *Evol. Biol.* 15:247–285.

Mackay, T. F. 1980. Genetic variance, fitness, and homeostasis in varying environments: an experimental check of the theory. *Evolution* 34:1219–1222.

Mackay, T. F. 1981. Genetic variation in varying environments. *Genet. Res.* 37:79–93.

Maniatis, T., S. Goodbourn, and J. A. Fischer. 1987. Regulation of inducible and tissue-specific gene expression. *Science* 236:1237–1244.

Mark, G. A. 1982. An experimental study of evolution in heterogeneous environments: Phenological adaptation by a bruchid beetle. *Evolution* 1982:984–997.

Markert, C. L. and J. B. W. G. Shaklee. 1975. Evolution of a gene. *Science* 189:102–114.

May, R. M. and A. P. Dobson. 1986. Population dynamics and the evolution of pesticide resistance. In *Pesticide Resistance: Strategies and Tactics for Management*, pp. 170–193. National Academy Press, Washington, DC.

Maynard Smith, J. 1966. Sympatric speciation. *Am. Nat.* 100:637–650.

McClintock, B. 1984. The significance of responses of the genome to challenge. *Science* 226:792–801.

McDonald, J. F. and J. C. Avise. 1976. Evidence for the adaptive significance of enzyme activity levels: interspecific variation in alpha-GPDH and ADH in *Drosophila*. *Biochem. Genetics* 14:347–355.

McDonald, J. F. and F. J. Ayala. 1978. Genetic and biochemical basis of enzyme activity variation in natural populations. I. Alcohol dehydrogenase in *Drosophila melanogaster*. *Genetics* 89:371–388.

McDonald, J. F., G. K. Chambers, J. David, and F. J. Ayala. 1977. Adaptive response due to changes in gene regulation: a study with *Drosophila*. *Proc. Natl. Acad. Sci. USA* 74:4562–4566.

McKenzie, J. A. and P. A. Parsons. 1972. Alcohol tolerance: an ecological parameter in the relative success of *Drosophila melanogaster* and *Drosophila simulans*. *Oecologia* 10:373–388.

McPheron, B. A., D. C. Smith, and S. H. Berlocher. 1988. Genetic differences between host races of *Rhagoletis pomonella*. *Nature* 336:64–66.

Miller, J. M. and K. L. Strickler. 1984. Finding and accepting host plants. In *Chemical ecology of insects*, eds. W. J. Bell and R. T. Carde, pp. 127–158. Sinauer, Sunderland, MA.

Mitter, C. and D. J. Futuyma. 1979. Population genetic consequences of feeding habits in some forest *Lepidoptera*. *Genetics* 92:1005–1021.

Mitter, C. and D. J. Futuyma. 1983. An evolutionary-genetic view of host-plant utilization by insects. In *Variable Plants and Herbivores in Natural and Managed Ecosystems*, eds. R. F. Denno and M. S. Mclure, pp. 427–459. Academic Press, London.

Mitter, C., D. J. Futuyma, J. C. Schneider, and J. D. Hare. 1979. Genetic variation and host plant relations in a parthenogenetic moth. *Evolution* 36:777–790.

Mortlock, R. P. 1982. Regulatory mutations and the development of new metabolic pathways by bacteria. *Evol. Biol.* 14:205–268.

Mouches, C., N. Pasteur, J. B. Berge, O. Hyrien, M. Raymond, B. R. Saint Vincent, M. Silvestri, and G. P. Georghiou. 1986. Amplification of an esterase gene is responsible for insecticide resistance in a California *Culex* mosquito. *Science* 233:778–780.

Narahashi, T. 1983. Nerve membrane sodium channels as the major target site of pyrethroids and DDT. In *Pesticide Chemistry: Human Welfare and the Environment*, eds. J. Miyamoto and P. C. Kearney, pp. 109–114. Pergamon Press, New York.

Nevo, E., C. Bar-El and Z. Bar. 1983. Genetic diversity, climatic selection and speciation of *Sphincterochila* landsnails in Israel. *Biol. J. Linn.* 19:339–373.

Omer, S. M., G. P. Georghiou, and S. N. Irving. 1980. DDT/pyrethroid resistance inter-relationships in *Anopheles stephensi*. *Mosq. News* 40:200–209.

Oppenoorth, F. J. and W. Welling. 1976. In *Insecticide Biochemistry and Physiology*, ed. C. F. Wilkinson, pp. 507–551. Heydon, London.

Osborne, M. P. and A. Smallcombe. 1983. Site of action of pyrethroid insecticides in neuronal membranes as revealed by the kdr resistance factor. In *Pesticide Chemistry: Human Welfare and the Environment*, eds. J. Miyamoto and P. C. Kearney, pp. 103–107. Pergamon Press, New York.

Paigen, K. 1986. Gene regulation and its role in evolutionary processes. In *Evolutionary Processes and Theory*, eds. S. Karlin and E. Nevo, pp. 3–36. Academic Press, Orlando.

Pashley, D. P. 1986. Host-associated genetic differentiation in fall armyworm (Lepidoptera: Noctuidae): a sibling species complex? *Ann. Entomol. Soc. Amer.* 79:756–762.

Pashley, D. P. 1988. Quantitative genetics, development, and physiological adaptation in host strains of fall armyworm. *Evolution* 42:93–102.

Pashley, D. P. and J. A. Martin. 1987. Reproductive incompatibility between host strains of fall armyworm (Lepidoptera: Noctuidae). *Ann. Entomol. Soc. Amer.* 80:509–578.

Patterson, B. D. 1983. Grasshopper mandibles and the niche variation hypothesis. *Evolution* 37:375–388.

Pollak, E. 1974. The survival of a mutant gene and maintenance of polymorphism in subdivided populations. *Am. Nat.* 108:20–28.

Pree, D. J. 1987. Inheritance and management of cyhexatine and dicofol resistance in the European red mite (Acari: Tetranychidae). *J. Econ. Entomol.* 80:1106–1112.

Prokopy, R. J., A. L. Averill, S. S. Cooley, S. L. Roitberg, and C. Kallet. 1982. Variation in host acceptance pattern in apple maggot flies. In *Insect-plant relationships*, eds. J. H. Visser and A. K. Minks, pp. 123–129. Pudoc, Wageningen, The Netherlands.

Prokopy, R. J., S. R. Diehl, and S. S. Cooley. 1988. Behavioral evidence for host races in *Rhagoletis pomonella* flies. *Oecologia* 76:138–147.

Rabinow, L. and W. J. Dickinson. 1981. A cis-acting regulator of enzyme tissue specificity in *Drosophila* is expressed at the RNA level. *Mil. Gen. Genet.* 183:264–269.

Rausher, M. D. 1982. Population differentiation in *Euphydryas editha* butterflies: larval adaptation to different hosts. *Evolution* 36:381–390.

Rausher, M. D. 1984. Trade-offs in performance on different hosts: evidence from within- and between-site variation in the beetle *Deloyala guttata*. *Evolution* 38:582–595.

Rosenthal, G. A. 1983. L-canavanine and L-canaline: protective allelochemicals of certain leguminous plants. In *Plant Resistance to Insects*, ed. P. A. Hedin, pp. 279–290. Amer. Chem. Soc., Washington, D. C.

Rosenthal, G. A., D. L. Dahlman, and D. H. Janzen. 1976. A novel means for dealing with L-canavanine, a toxic metabolite. *Science* 192:256–257.

Rosenthal, G. A., D. H. Janzen, and D. L. Dahlman. 1977. Degradation and detoxification of canavanine by a specialized seed predator. *Science* 196:658–660.

Roush, R. T. and J. A. McKenzie. 1987. Ecological genetics of insecticide and acaricide resistance. *Annu. Rev. Entomol.* 32:361–380.

Roush, R. T., R. L. Combs, T. C. Randolph, and J. A. Hawkins. 1986. Inheritance and effective dominance of pyrethroid resistance in the horn fly (Diptera: Muscidae). *J. Econ. Entomol.* 79:1178–1182.

Sabath, M. D. 1974. Niche breadth and genetic variability in sympatric natural populations of *Drosophilid* flies. *Am. Nat.* 108:533–540.

Schneider, J. C. and R. T. Roush. 1986. Genetic differences in oviposition preference between two populations of *Heliothis virescens*. In *Evolutionary Genetics of Invertebrate Behavior*, ed. M. D. Huettel, pp. 163–171. Plenum Press, New York.

Scriber, J. M. 1984. Host-plant suitability. In *Chemical Ecology of Insects*, eds. W. J. Bell and R. T. Carde, pp. 159–202. Sinauer, Sunderland, MA.

Service, P. M. and M. R. Rose. 1985. Genetic covariation among life-history components: the effect of novel environments. *Evolution* 39:943–945.

Shugart, H. H. and G. B. Blaylock. 1973. The niche-variation hypothesis: an experimental study with *Drosophila* populations. *Am. Nat.* 107:575–579.

Singer, M. C. 1983. Determinants of multiple host use by a phytophagous insect population. *Evolution* 37:389–403.

Singer, M. C., D. Ng, and C. D. Thomas. 1988. Heritability of oviposition preference and its relationship to offspring performance within a single insect population. *Evolution* 42:977–985.

Slansky, F. and J. C. Rodriguez. 1987. *Nutritional Ecology of Insects, Mites, Spiders, and Related Invertebrates*. John Wiley, New York.

Slatkin, M. 1985. Rare alleles as indicators of gene flow. *Evolution* 39:53–65.

Slatkin, M. 1987. Gene flow and the geographic structure of natural populations. *Science* 236:787–792.

Slatkin, M. and N. H. Barton. 1989. A comparison of three indirect methods for estimating average levels of gene flow. *Evolution* 43:1349–1368.

Smith, M. F. 1981. Relationship between genetic variability and niche dimensions among coexisting species of *Peromyscus*. *J. Mammal.* 62:273–285.

Smith, D. C. 1988. Heritable divergence of *Rhagoletis pomonella* host races by seasonal asynchrony. *Nature* 336:66–67.

Soderlund, D. M. and J. R. Bloomquist. 1990. Molecular mechanisms of insecticide resistance. In *Pesticide Resistance in Arthropods*, eds. R. T. Roush and B. E. Tabashnik, pp. 58–96. Chapman and Hall, New York.

Sogawa, K. 1982. The rice brown planthopper: feeding physiology and host plant interactions. *Annu. Rev. Entomol.* 27:49–73.

Soule, M. and B. R. Stewart. 1970. The 'niche-variation' hypothesis: a test and alternatives. *Am. Nat.* 104:85–97.

Soule, M. and S. Y. Yang. 1973. Genetic variation in side-blotched lizards in islands in the Gulf of California. *Evolution* 27:593–600.

Spencer, K. C. 1988. *Chemical Mediation of Coevolution*. Academic Press, New York.

Spieth, P. T. 1974. Gene flow and genetic differentiation. *Genetics* 78:961–965.

Spieth, P. T. 1979. Environmental heterogeneity: a problem of contradictory selection pressures, gene flow, and local polymorphism. *Am. Nat.* 113:247–260.

Steiner, W. W. M. 1977. Niche width and genetic variation in Hawaiian *Drosophila*. *Am. Nat.* 111:1037–1046.

Stewart, C. B., J. W. Schilling, and A. Wilson. 1987. Adaptive evolution in the stomach lysozymes of foregut fermenters. *Nature* 330:401–404.

Tabashnik, B. E. 1983. Host range evolution: the shift from native legume hosts to alfalfa by the butterfly, *Colias philodice eriphyle*. *Evolution* 37:150–162.

Tabashnik, B. E. 1990. Modeling and evaluation of resistance management tactics. In *Pesticide Resistance in Arthropods*, eds. R. T. Roush and B. E. Tabashnik, pp. 153–182. Chapman and Hall, New York.

Tabashnik, B. E. and F. Slansky. 1987. Nutrional ecology of forb foliage-chewing insects. In *Nutrional Ecology of Insects, Mites, and Spiders*, eds. F. Slansky and J. G. Rodriguez, pp. 71–103. John Wiley, New York.

Tabashnik, B. E., H. Wheelock, J. D. Rainbolt, and W. B. Watt. 1981. Individual variation in oviposition preference in the butterfly, *Colias eurytheme*. *Oecologia* 50:225–230.

Taylor, C. E. and C. Condra. 1983. Resource partitioning among genotypes of *Drosophila pseudoobscura*. *Evolution* 37:135–149.

Taylor, C. E., F. Quaglia, and G. P. Georghiou. 1983. Evolution of resistance to insecticides: a case study on the influence of migration and insecticide decay rates. *J. Econ. Entomol.* 76:704–707.

Uyenoyama, M. K. 1986. Pleiotropy and the evolution of genetic systems conferring resistance. In *Pesticide Resistance: Strategies and Tactics for Management*, pp. 207–221. National Academy Press, Washington, DC.

Van Dyke, M. W., R. G. Roeder, and M. Sawadogo. 1988. Physical analysis of transcription preinitiation complex assembly on a class II gene promoter. *Science* 241:1335–1338.

Van Halen, L. 1965. Morphological variation and width of ecological niche. *Am. Nat.* 99:377–390.

Via, S. 1984. The quantitative genetics of polyphagy in an insect herbivore. I. Genotype-environment interaction in larval performance in different host plant species. *Evolution* 38:881–895.

Via, S. 1986. Genetic covariance between oviposition reference and larval performance in an insect herbivore. *Evolution* 40:778–785.

Vijverberg, H. P. M., G. S. F. Ringt, and J. van den Berken. 1982. Structure-related effects of the pyrethroid insecticides on the lateral-line sense organ and on peripheral nerves of the clawed frogs, *Xenopus laevis*. *Pestic. Biochem. Physiol.* 18:315–324.

Vincent, D. R., A. F. Moldenke, D. E. Farnsworth, and L. C. Terriere. 1985. Cytochrome P-450 in insects. 6. Age dependency and phenobarbitol induction of cytochrome P-450, P-450 reductase, and monooxygenase activities in susceptible and resistant strains of *Musca domestica. Pestic. Biochem. Physiol.* 23:182–189.

Wallace, B. 1968. Polymorphism, population size, and genetic load. In *Population Biology and Evolution*, ed. R. C. Lewontin, pp. 87–108. Syracuse University Press, Syracuse, NY.

Watt, W. B., P. C. Hoch, and S. G. Mills. 1974. Nectar resource use by *Colias* butterflies: chemical and visual aspects. *Oecologia* 14:353–374.

Watt, W. B., F. S. Chew, L. R. G. Snyder, A. G. Watt, and D. E. Rothschild. 1977. Population structure of pierid butterflies. 1. Numbers and movements of some montane *Colias* species. *Oecologia* 27:1–22.

White, R. M. 1978. Ecological and population genetics of insecticide resistance in the Australian sheep blowfly, *Lucilia cuprina*. PhD. dissertation, University of Liverpool, Liverpool, England.

Whitehead, J. R., N. B. R. Roush, and B. R. Norment. 1985. Resistance stability and coadaptation in diazinon-resistant houseflies (Diptera: Muscidae). *J. Econ. Entomol.* 78:25–29.

Wiens, J. A. 1976. Population responses to patchy environments. *Annu. Rev. Ecol. Syst.* 7:81–120.

Williams, K. S. 1983. The coevolution of *Euphydras chalcedona* butterflies and their larval host plants. III. Oviposition behavior and host plant quality. *Oecologia* (Berl.) 56:336–340.

Wood, T. K. 1980. Divergence in the *Enchenopa binotata* Say complex (Homoptera: Membracidae) effected by host plant adaptation. *Evolution* 34:147–160.

Wood, R. J. 1981. Strategies for conserving susceptibility to insecticides. *Parasitology* 82:69–80.

Wood, R. J. and J. A. Bishop. 1981. Insecticide resistance: populations and evolution. In *Genetic Consequences of Man-Made Change*, eds. J. A. Bishop and L. M. Cook, pp. 97–127. Academic Press, London.

Wood, T. K. and S. I. Guttman. 1982. Ecological and behavioral basis of reproductive isolation in the sympatric *Echenopa binotata* complex (Homoptera: Membracidae). *Evolution* 36:233–242.

Wood, T. K. and M. C. Keese. 1990. Host-plant-induced assortative mating in *Enchenopa* treehoppers. *Evolution* 44:619–628.

Wright, S. 1969. *Evolution and the Genetics of Populations, Vol. 2. The Theory of Gene Frequencies*. Univ. of Chicago Press, Chicago.

Wright, S. 1977. *Evolution and Genetics of Populations, Vol. 3. Experimental Results and Evolutionary Deductions*. Univ. of Chicago Press, Chicago.

Wright, A., M. McConnell, and S. Kanegasaki. 1980. Lipopolysaccharide as a bacteriophage receptor. In *Virus receptors, Part 1. Bacterial viruses*, eds. L. L. Randall and L. Philipson, pp. 27–57. Chapman and Hall, London.

Yu, S. J. and L. C. Terriere. 1979. Cytochrome P-450 in insects. 1. Differences in the forms present in insecticide resistant and susceptible houseflies. *Pestic. Biochem. Physiol.* 12:239–248.

7

The Evolution of Unpalatability and the Cost of Chemical Defense in Insects

M. Deane Bowers

University of Colorado, Boulder

7.1. Introduction

In addition to the well-known and potent venoms of many Hymenoptera, insect species in several orders exploit chemical defenses for protection from predators and parasitoids (Eisner, 1970; Rothschild, 1972, 1985; Blum, 1981; Pasteels and Gregoire, 1983; Brower, 1984). These defensive chemicals may be synthesized by the insects (*de novo* synthesis) or acquired from the larval or adult food (sequestration). They are of many chemical classes, and their modes of action range from volatile irritation to physical entanglement, distastefulness, and physiological toxicity (Blum, 1981).

The study of chemical defenses of insects, whether they are due to *de novo* synthesis or to sequestration, has been approached from several, very productive directions. From a chemical perspective, techniques for the isolation and identification of defensive chemicals have been developed and refined over the past several years. From the ecological perspective, the efficacy of a diversity of putative insect chemical defenses has been tested against both vertebrate and invertebrate predators (Eisner, 1970; Berenbaum and Milickzy, 1984; Brower, 1984; McLain and Shure, 1985; Pasteels et al., 1986, 1988b; Witz, 1989). In addition, the physiological mechanisms involved in biosynthesis and sequestration of insect chemical defenses are beginning to be understood (Franzl et al., 1986; Moore and Scudder, 1986; Wink, 1988). Finally, the evolution of warning coloration and gregariousness and the diversification of chemical defenses has been studied phylogentically (Sillen-Tullberg, 1988; Pasteels et al., 1984, 1989, 1990).

In this chapter, I discuss (1) the characteristics of unpalatable insects that influence their interactions with predators; (2) the consequences of qualitative and quantitative variation in chemical defense for both prey and predator; (3) the evolution of unpalatability through natural selection and the importance of complete metamorphosis (holometaboly) in the evolution of unpalatability; and (4) the determination of the physiological and evolutionary cost of chemical defense, whether by *de novo* synthesis or sequestration of defensive chemicals.

7.2. Unpalatability as a Defense

7.2.1. Characteristics of Unpalatable Insects

Unpalatable insects often exhibit a suite of associated characteristics that advertise their noxious qualities. They are often conspicuously ("warningly") colored; such colors include black, red, orange, yellow and white. They are also often gregarious and exhibit other behaviors that enhance their conspicuousness, such as slow flight or a warning display (Bates, 1862; Cott, 1940; Brower, 1984).

Blest (1963) suggested that, in addition, warningly-colored, unpalatable insects will be long-lived relative to cryptic, palatable insects. Vertebrate predators, such as birds, learn to find cryptic prey or to avoid aposematic prey with experience. Long-lived cryptic individuals would enhance the learning of these predators, making it easier to find other similarly-colored individuals. In contrast, long-lived aposematic individuals would help "train" predators to avoid other similarly-colored individuals, and therefore would increase the chances of survival of those individuals. Blest (1963) invoked kin selection to explain the evolution of these different life history traits in cryptic versus aposematic species. However, there are few data to support this hypothesis.

Bates (1862), in his classic work on unpalatability and mimicry, observed that the unpalatable models in mimicry complexes had very tough bodies. He suggested that this feature allowed these insects to survive handling and tasting by predators. Anecdotal support for this idea was provided by feeding experiments earlier in this century (Carpenter, 1932; Jones, 1932, 1934). Recent experimental evidence showed that unpalatable insects of several taxa (butterflies, beetles, bugs) may survive handling by bird predators better than palatable insects (Boyden, 1976; Jarvi et al., 1981; Wiklund and Jarvi, 1982).

Warning coloration itself may enhance survival of an individual insect. For example, birds may be more hesitant in attacking warningly-colored or novel prey (Coppinger, 1970; Smith, 1975; Schuler and Hesse, 1985) or may learn to avoid unpalatable prey more effectively when they are conspicuously-colored (Gittleman and Harvey, 1980, Gittleman et al., 1980; Roper and Wistow, 1986). Sillen-Tullberg (1985) found that a warningly-colored, red morph of a distasteful lygaeid bug had a higher survival rate after attack by birds than did a cryptic, gray morph, even though the two morphs were equally distasteful. She attributed this difference in survival to three features of predator behavior in response to the two morphs: (1) an initial reluctance to attack the warningly-colored morph; (2) a faster rate of avoidance-learning with the aposematic morph; and (3) less damage to the aposematic prey when attacked (Sillen-Tullberg, 1985). Warning coloration may also reduce the chance that an experienced predator will make an inappropriate attack (Guilford, 1986). Because warning colors are conspicuous, they may be detected at a greater distance from the predator. This could, in turn, increase the amount of time that the predator spends viewing the prey before attacking, resulting in less chance of a mistaken attack (Guilford 1986).

Gregariousness often accompanies unpalatability (Fisher, 1930; Cott, 1940; Stamp, 1980). Aggregation may be advantageous for unpalatable insects because it increases the magnitude of the warning signal, it potentially increases the efficacy of a chemical defense, and it may immediately reinforce predator avoidance-learning (Cott, 1940; Edmunds, 1974; Aldrich and Blum, 1978; Kidd, 1982; Sillen-Tullberg and Leimar, 1988; Guilford, 1990).

There are, however, exceptions to these generalizations about characteristics of unpalatable insects. For example, some cryptic insects are unpalatable (Jarvi et al., 1981; Turner, 1984), and some species of unpalatable insects, such as larvae of many danaid butterflies, are solitary (Vane-Wright and Ackery, 1986). In addition, putatively warningly-colored insects may, in fact, be palatable mimics (Bates, 1862).

7.2.2. Variation in Unpalatability

Within any population or species of unpalatable insect, there may be variation in the degree to which individuals are unpalatable (Brower et al., 1968; Bowers, 1980; Gardner and Stermitz, 1988). For those insects that sequester their defensive compounds from the host plant, the chemical content of the host plant(s) on which the insect feeds may be a critical determinant of the degree of protection conferred on the insect (Brower et al., 1968; Bowers, 1988a). Substantial information has accumulated demonstrating variation in kinds and amounts of secondary metabolites among plant populations, individuals and organs within an individual plant (Dolinger et al., 1973; McKey et al., 1978; McKey, 1979; Lincoln and Langenheim, 1979, 1981; Lincoln and Mooney, 1984; Lincoln et al., 1986; Louda and Rodman, 1983a, b; Brower et al., 1982, 1984, 1986; Nishio et al., 1983), as well as the consequences of this variation for insects feeding on those plants (Brower et al., 1968; Brower et al., 1982, 1984; Jones et al., 1987, 1989; Belofsky et al., 1989). The result of this host plant chemical variation is that, within a population of "unpalatable" insects, there is likely to be a spectrum of palatability (Brower et al., 1968). This phenomenon is known as automimicry (Brower et al., 1970) or Browerian mimicry (Pasteur 1982). In automimicry, the more palatable individuals in the population benefit from association with the more distasteful individuals. Because the model and mimic are of the same species, the mimicry is exact. Automimicry is probably very common, if not ubiquitous, in unpalatable insects.

In addition, the efficiency with which insects sequester defensive compounds can contribute to their degree of unpalatability. For example, the cardenolide content of different insect species feeding on the same milkweed species may vary substantially, apparently due to differences in the ability of the different insects to sequester those compounds (Isman et al., 1977; Cohen, 1985; Nishio, 1983; Nishio et al., 1983). The queen butterfly, *Danaus gilippus* (Nymphalidae) appears to be much less efficient at sequestering cardenolides than the monarch, *Danaus plexippus* (Cohen, 1985). Differences in sequestrative ability may occur at the subspecies level as well. The south american *D. plexippus megalippe* was more emetic than *D. p. erippus* where both were reared on the same milkweed, *Asclepias curassavica* (Rothschild and Marsh, 1978).

For insects that produce their own defensive compounds, such as repellent droplets, sprays, or volatile irritants, variation in the amount and thus the effectiveness of the defense may depend on a variety of factors. These include: (1) the strength of the stimulus initiating defense production and the consequent amount of the discharge; (2) the number of recent discharges (Eisner, 1965); (3) the time since the last use of the defense (Eisner, 1965; Abushama, 1972; Bjorkman and Larsson, unpublished data); and (4) the energy available to replenish the defensive compounds, which may depend on a variety of factors such as the insect's age or reproductive state.

Whether insects sequester defensive chemicals from host plants or product defensive chemicals *de novo*, the amount of defensive compound that they contain or produce may determine whether or not they survive a predator's attack. Although unpalatable insect species may be more likely to survive the attack of a predator due to their tougher bodies (Wiklund and Jarvi, 1982), some predators may taste several potential prey individuals and reject the least palatable (Boyden, 1976; Brower and Calvert, 1985). In such cases, variation in the amount of defensive chemicals may determine which individuals are killed and eaten and which are rejected. Tasted and rejected individuals (the least palatable) may survive with little or no ill effect (Jarvi et al., 1981; Wiklund and Jarvi, 1982). Insects that have exhausted their supply of a defensive spray or have recently responded to a predator by spray discharge, may also be less able to repel a subsequent attack.

With the increasing sophistication of analytical chemical techniques used by chemical ecologists, we can now determine the degree to which chemical defenses vary among individual insects within a single population or species. For example, Monarch butterflies (*Danaus plexippus*, Nymphalidae) may contain from 0 to over 1,200 μg of milkweed-derived cardenolides per butterfly (just over 1% dry weight) (Brower, 1984). *Oncopeltus fasciatus* (Lygaeidae) also sequesters cardenolides and may contain seven to ten times more cardenlides by weight than monarchs (Scudder et al., 1986). Butterflies in the genus *Euphydryas* (Nymphalidae) may contain from less than 1% to over 10% dry weight iridoid glycosides, compounds sequestered from host plants which render these insects unpalatable to potential predators (Bowers, 1980; Gardner and Stermitz, 1988; Bowers, 1988; Bowers and Farley, 1990). Millipedes, which produce hydrogen cyanide as a defense, may generate from 0 to 600 μg per individual (weighing from 0.6 to 1.4 g) (Eisner et al., 1967). In each of these arthropods, some individuals contain or produce relatively large amounts of defensive compounds, while other individuals may be devoid of chemical defense. Predators may act as selective agents on these chemically-variable individuals, and those individuals with the most potent defense may have the highest fitness (Fink and Brower, 1981; Malcolm and Brower, 1989; Denno et al., 1990).

7.3. The Evolution of Unpalatability

The evolution of unpalatability can be approached from both phylogenetic (Sillen-Tullberg, 1988; Pasteels et al., 1990) and mechanistic (Duffey, 1980) perspectives. Using a phylogenetic approach, Sillen-Tullberg (1988) asked how many times unpalatability evolved in butterflies. By constructing a cladogram based on morphological characters and superimposing on that cladogram the character states of warning coloration in each taxon, she found that warning coloration in butterfly larvae, which she used as an indicator of unpalatability, has evolved 12 times in the families Papilionidae (four times), Pieridae (four times) and Nymphalidae (four times). Although she did not include the Lycaenidae because of complications due to the larvae of many species being ant-tended, unplatability has evolved at least once in the Lycaenidae, in the genus *Eumaeus* (Rothschild et al., 1986; Bowers and Larin, 1989; Bowers and Farley, 1990). Larvae of members of this genus are warningly-colored red and yellow, and feed on cycads. *Eumaeus atala* sequesters cycasin from its cycad host plants

(Rothschild et al., 1986; Bowers and Larin, 1989) and is unpalatable to ants and birds (Bowers and Larin, 1989; Bowers and Farley, 1990). Thus, unpalatability appears to have evolved independently 13 different times in the butterflies. Such a phylogenetic approach has great potential for the analysis of the evolution of other characteristics associated with unpalatability.

One can also test alternative hypotheses about the evolution of certain traits. To determine whether unpalatability evolved before gregariousness in butterflies, Sillen-Tullberg (1988) used existing phylogenies of various taxa of butterflies and superimposed on those phylogenies data on gregariousness and unpalatability (as indicated by warning coloration). The results of this phylogenetic analysis suggested that unpalatability more often evolved before gregariousness, in contrast to Fisher's (1930) and Harvey and Greenwood's (1978) hypotheses that gregariousness evolved before unpalatability.

With a mechanistic approach, one can examine the physiological and biochemical adaptations necessary for the evolution of unpalatability. For example, Scudder and Meredith (1982) found that the midgut of *Oncopeltus fasciatus* (Lygaeidae) was permeable to cardenolides, which it sequesters from its milkweed host plants. However, the midguts of cockroaches (*Periplaneta americana*) and locusts (*Schistocerca americana*), which do not normally ingest cardenolides, were impermeable to cardenolides (Scudder and Meredith, 1982). They suggested that impermeability of the gut may be responsible for the lack of cardenolide sequestration in insects which feed on milkweeds, but do not sequester those compounds (Isman et al., 1977). Thus, the evolution of unpalatability due to sequestration of chemical compounds from the host plant requires a biochemical mechanism that allows passage of potential defensive compounds, such as cardenolides, from the gut and to sites (such as hemolymph) where they may be accumulated. Exceptions to this are certain unpalatable insects, such as sawflies, which accumulate noxious plant compounds in the gut or a foregut diverticulum and do not transport them to other parts of the body (Eisner et al., 1974; Morrow et al., 1976; Larsson et al., 1986).

A combination of phylogenetic and mechanistic approaches could be extremely illuminating in understanding the evolution of unpalatability. For example, a phylogenetic analysis of a group in which unpalatability occurs may show how taxa within that group are related. Examination of members of those taxa that sequester defensive chemicals could show what physiological changes were involved in the evolution of the ability to sequester those compounds. Such an analysis has been done with regard to the tolerance of swallowtail larvae (Papilionidae) to certain alkaloids and aristolochic acid (Miller and Feeny, 1989), although not, to my knowledge, with sequestration of such compounds.

Table 7.1. Requirements for the evolution of unpalatability due to the sequestration of plant secondary metabolites by phytophagous insects.

Attribute	Consequence
1. Feed on a toxic plant	1. Potential defense compounds in insect's diet
2. Loss of detoxification mechanisms a. inactivation of detoxification enzymes b. protection of defensive compounds	2. Potential defense compounds not broken down
3. Ability of compounds to move through the gut (may require carrier molecules, especially for polar compounds)	3. Potential defense compounds can reach the hemolymph
4. Tissues and organs protected from compounds	4. Tissues and organs are tolerant of compounds or protected from them
5. Concentration of defensive compounds in hemolymph	5. Compounds can accumulate in the hemolymph against an osmotic gradient
6. Retention of compounds through to the pupal and adult stages	6. Tissue and cell tolerance of compounds or protection from them

7.3.1. The Evolution of Sequestration Versus De Novo Synthesis of Defensive Chemicals

Several features are required for the evolution of the ability of phytophagous insects to use plant secondary metabolites as defense compounds (Table 7.1). Although feeding on a plant species that contains potential defensive chemicals is a prerequisite for sequestering those compounds, many phytophagous insects feed on "toxic" plants but do not sequester defensive chemicals (Rothschild, 1972; Isman et al., 1977; Blum, 1983; Bowers and Puttick, 1986; Rowell-Rahier and Pasteels, 1986; Brattsten, 1986). For such insects, some degree of protection may derive from the presence of toxic plant material in the gut (Brower, 1984). However, for those insects, the chemicals do not accumulate during the larval stage, nor are they retained through to the pupal or adult stage.

The evolution of unpalatability via recycling plant defense chemicals requires several other physiological adaptations (Table 7.1). First, the normal detoxification mechanisms must be short-circuited. For example, insect defense compounds that are glucosides must be protected from attack by glucosidases or other metabolic enzymes, or the enzymes must be inactive against those compounds. The same is true of other types of defense compounds. To render them immune to enzymatic attack, these compounds may be stored in tissues or organs where detoxification enzymes do not occur, such as the foregut diverticula used to store defense compounds in some sawfly species (Eisner et al., 1974; Morrow et al., 1976). Alternatively, they may be complexed with carrier molecules, making them resistant to detoxification enzymes, for transport out of the gut (Wilkinson, 1986; Wink, 1988). There may also be changes in the specificity of the detoxification enzymes so that they no longer react with the defense compounds.

Second, the compounds must be able to penetrate the gut to reach the hemolymph (Duffey, 1980). Especially for polar molecules passing through the very non-polar (lipophilic) gut (Duffey, 1980), carrier molecules may be required (Wink, 1988). The hemolymph can store the defense compounds, which may be directed against potential predators by reflex bleeding (Happ and Eisner, 1961; Eisner, 1970; Blum, 1981), or following release upon wounding. For example, juvenile praying mantids, *Tenodera sinensis* (Mantidae) may taste and then reject toxic milkweed bugs, *Oncopeltus fasciatus* (Lygaeidae) (Paradise and Stamp, 1990). Ladybird beetles, *Harmonia axyridis* (Coccinellidae) reject the aphid *Acyrthosiphon nipponicus*, which sequesters iridoid glycosides from its host plant (Nishida and Fukami, 1989). Naive predatory wasps (*Polistes dominulus*, Vespidae) attacked, and rejected after tasting the hemolymph of, the warningly-colored larvae of the catalpa sphinx, *Ceratomia catalpae* (Sphingidae), which store bitter iridoid glycosides in the hemolymph in amounts as high as 50% dry weight (Bowers, Lawrence, and Collinge, unpublished data).

Third, the tissues and organs must be protected from the effects of the defensive compounds, or immune to them. Some insects may store sequestered defensive compounds in specialized organs, such as milkweed bugs which store ingested cardenolides in the dorsolateral space (Duffey et al., 1974, 1978). Defensive compounds synthesized *de novo* by insects are often stored in reservoirs associated with the glands (Noirot and Quennedey, 1974; Tschinkel, 1975c). However, they may also be transported and/or stored in the hemolymph. Larvae of moths in the genus *Zygaena* (Arctiidae) use cyanogenic glycosides as defensive compounds and apparently synthesize them in larval organs such as fat body and gut (Franzl et al., 1986). These compounds are then transported in the hemolymph and stored in specialized cuticular cavities and the hemolymph (Franzl et al., 1986). Tissues of insects adapted to a particular class of defensive chemicals may also be physiologically insensitive to their effects. For example, the tissues of monarch butterflies and milkweed bugs were resistant to the effects of cardenolides on sodium and potassium transport, while those of non-adapted insects were not (Vaughan and Jungreis, 1977; Moore and Scudder, 1986).

Finally, for protection throughout the life cycle, defensive compounds from plants must be continuously stored in hemimetabolous insects and held through to the pupal and adult stages in holometabolous insects. For example, in some lepidopteran species specializing on plants containing iridoid glycosides (buckeye butterflies, *Junonia coenia*, Nymphalidae, and the Catalpa sphinx, *Ceratomia catalpae*, Sphingidae), the larvae sequester these compounds, but they are metabolized or eliminated before the pupal stage, and adults are devoid of these compounds (Bowers and Puttick, 1986; Bowers and Collinge, 1992 and unpublished data).

Thus the ability to accumulate defensive compounds from a host plant is not a passive process, but requires circumvention of the normal detoxification processes found in insects (Terriere, 1984; Ahmad et al., 1986), as well as physiological or morphological mechanisms to prevent autotoxicity. Such adaptations are crucial for the evolution of unpalatability by sequestration of defensive chemicals from plants.

In contrast to the insects described above is the case of the lubber grasshopper, *Romalea guttata* (Acrididae), a generalist species feeding on a broad array of herbaceous and woody plants (Jones et al., 1986, 1987, 1989). Adults produce a frothy defensive secretion from their metathoracic spiracles which may be deterrent to certain potential predators (Eisner et al., 1971; Jones et al., 1989; Blum et al., 1990). When these grasshoppers are raised on a "generalist diet," composed of 26 different herb species that they might naturally encounter, the defensive secretion contains primarily phenolics and quinones (Jones et al., 1989). However, when their diet is restricted to a single species such as onion or catnip, compounds from those plants appear in the defensive secretion (Jones et al., 1989; Blum et al., 1990). Thus, Jones, et al. (1989) and Blum et al. (1990) suggest that "causal" sequestration of defensive compounds from plants, such as occurs in *R. guttata*, resulting in increased protection from potential predators, may initiate the evolution of unpalatability, rather than a high degree of adaptation with a particular host plant from which defensive chemicals may be sequestered.

7.3.2. The Evolutionary Relationship of Sequestration and De Novo Synthesis

At first consideration, it might seem energetically "cheaper" and evolutionarily simpler to recycle defensive chemicals already synthesized by the host plant, rather than manufacture them *de novo*. However, contemplation of the requirements for the evolution of unpalatability described above (see Table 7.1) might indicate otherwise.

Tschinkel's work (1975a, b, c) on defensive secretions of tenebrionids, and Pasteels and co-workers on chrysomelids (Deroe and Pasteels, 1982; Pasteels et al., 1984, 1988a, b, 1990), provide valuable studies of the evolution of *de novo* chemical defenses. In both insect groups, the adults' chemical defense compounds are produced by exocrine glands, with varying degrees of complexity (Tschinkel, 1975c; Deroe and Pasteels, 1982). These defensive glands probably evolved from simple epidermal glands (Noirot and Quennedy, 1974). Most exocrine defense glands develop as invaginations of the epidermis and are lined with cuticle (Noirot and Quennedey, 1974; Chapman, 1980), which protects the gland cells (and other cells) from the secretions. In some species, the glands do not make the final product, but only the precursors. Such systems are called reactor glands (Eisner, 1970), and components initially made by different glands are mixed just prior to discharge (Eisner, 1970; Eisner et al., 1977). Ontogenetically, and probably phylogenetically, such glands originate as simple epidermal invaginations (Noirot and Quennedy, 1974). Because the glandular products are isolated from other tissues, there would be no need to evolve mechanisms to prevent autotoxicity.

From their study of the biology, morphology, and chemistry of defenses in the Chrysomelidae, Pasteels et al. (1990) suggested that in this group, sequestration of chemical defenses was a derived feature and *do novo* biosynthesis was ancestral (plesiomorphic). They suggested that the switch to sequestration of plant-derived defense compounds was the result of enzymatic preadaptation: enzymes and glands used for the *de novo* biosynthesis and storage of defense compounds in the insect could be used to metabolize and store compounds accumulated from the host plant.

7.3.3. Natural Selection and the Evolution of Unpalatability

7.3.3.1. Selective Agents

The selective agents for the evolution of unpalatability and associated characteristics, such as warning coloration and gregariousness, are generally considered to be predators and especially vertebrate predators such as birds and lizards (Fisher, 1930; Cott, 1940; Rothschild, 1972; Boyden, 1976; Brower, 1984). These animals can see color and are visual hunters, thus characteristics such as warning coloration and aggregation, which serve to advertise unpalatability, are effective accompaniments to a noxious taste. Although invertebrate predators and parasitoids may be extremely important sources of mortality for insect populations (Gillaspy, 1979; Evans, 1987; Barbosa, 1988; Stamp and Bowers, 1988; Bernays, 1988), they are generally considered to search using chemical cues and are thus considered less important as selective agents in the evolution of unpalatability and warning coloration. However, predatory wasps can exert strong predation pressure on caterpillar populations (Takagi et al., 1980; Evans, 1987; Stamp and Bowers, 1988; Gould and Jeanne, 1984; Gillaspy, 1979; Damman, 1987; Bernays, 1988), and are visual predators. Mantids are also important insect predators, hunt visually, and may respond to chemical defenses in potential prey (Berenbaum and Milickzy, 1984; Paradise and Stamp, 1990, 1991). Thus, invertebrate as well as vertebrate predators may be important selective agents in the evolution of unpalatability and associated characteristics in larval stages of phytophagous insects.

7.3.3.2. Kin Selection versus Individual Selection and the Evolution of Unpalatability

In 1930, Fisher first suggested that kin selection was a prerequisite for the evolution of unpalatability and warning coloration. Because aposematic animals are often gregarious, the attacked individual could be killed, but in the process, the predator would learn to avoid other similarly-colored individuals, which were likely to be siblings in the same aggregation as the attacked individual. This suggestion became commonly accepted (Benson, 1971; Turner, 1971; Harvey and Greenwood, 1978; Harvey and Paxton, 1981) until a series of experiments suggested that individual selection could also be invoked to explain the evolution of unpalatability and warning coloration (Jarvi et al., 1981; Wiklund and Jarvi, 1982; Sillen-Tullberg, 1985). These experiments showed that several species of aposematic insects, including lepidopteran caterpillars, beetles, and true bugs, could survive attack by birds, and in fact were often released unharmed. Because the attacked individual survived, it was not necessary to invoke kin selection as an explanation for the evolution of unpalatability and warning coloration. Moreover, an aposematic form (red) of the milkweed bug, *Lygaeus equestris* (Lygaeidae) was more likely to survive than a cryptic form (gray), even though there was no difference in the distastefulness of the two forms (Sillen-Tullberg, 1985). This indicated that if an insect were unpalatable, aposematic coloration itself was beneficial to the individual insect and suggested that individual selection could explain the evolution of such coloration.

Data from two of these papers showed that survival of larvae or nymphs was significantly lower than that of the adults when attacked by birds (monarchs, *Danaus plexippus*, Wiklund and Sillen-Tullberg, 1985, and milkweed bugs, *L. equestris*, Sillen-Tullberg et al., 1982). Larval or nymphal stages are often soft-bodied and may be less likely to survive an attack. Thus, kin selection may be more important during such stages or in insects with soft bodies. For example, tests with soft-bodied aposematic aphids (*Aphis nerii*) showed that attacked aphids did not survive and that birds and spiders learned to avoid these unpalatable aphids (Malcolm, 1985). These aphids are gregarious and parthenogenic; thus aphids in an aggregation are likely to be genetically identical (Malcolm, 1985). In such an insect, kin selection rather than individual selection is likely to be of most importance in the evolution of aposematism.

Guilford (1985, 1990) suggests that "green beard selection" as defined by Dawkins (1976) may be important in the evolution of warning coloration. Green beard selection describes a situation where the benefit is directed toward other individuals who must have a copy of that gene themselves (Dawkins, 1976), such as a green beard or a particular warning coloration. In such a scenario, warning coloration will spread because individuals with similar phenotypes will be protected, not only because kin are protected (Guilford, 1990).

7.3.4. Metamorphosis and the Evolution of Unpalatability

Unpalatable insects that are hemimetabolous may sequester defensive compounds throughout their nymphal and adult stages. Many holometabolous, unpalatable insects accumulate defensive compounds only during the larval stage (Lepidoptera and sawflies). Ithomiine and danaine butterflies which ingest pyrrolizidine alkaloids during the adult stage are an exception (Brown, 1984a, b; Vane-Wright and Ackery, 1986). In some phytophagous beetles, larvae and adults may feed on the same host plants and may acquire defensive compounds throughout life, for example, *Dibolia* spp. (Chrysomelidae: Coleoptera) (Parry, 1974). In some chrysomelids, larvae and adults have very different means of chemical defense, and the defense may be sequestered from the host plant in one stage and synthesized *de novo* in the other (Pasteels et al., 1984, 1989).

The very different environments and life-styles of the larval and adult stages of many holometabolous insects result in different selective forces acting on those stages. Insect larvae are relatively immobile compared with the adult stage, and thus cannot exploit the same means of escape from predators used by adults. Thus, it may be advantageous for the larval stage to sequester defensive chemicals and be unpalatable, while the adult is palatable. For example, larvae of some moth species may be gregarious, warningly-colored, and unpalatable, while the adults are solitary, cryptic, and palatable (Rothschild, 1972; Bowers and Puttick, 1986; Bowers and Farley, 1990).

The extensive changes associated with metamorphosis in holometabolous insects may cause differences in the efficacy of physiological mechanisms for coping with plant secondary chemicals among life stages. For example, Maddrell and Gardner (1976) found that while Malpighian tubules (the excretory system of insects) of larval Manduca sexta can transport nicotine at rates averaging 3,290 ($\pm$520) pmol/min, those of the adult (which feed on nectar and hence do not ingest nicotine) transport it only at essentially a passive rate (4.07 pmol/min). Thus, the Malpighian tubules of the adult and larva have very different transport characteristics. In contrast, they found that in both nymphs and adult *Rhodnius prolixus*, a hemipteran with incomplete metamorphosis that would not be exposed to nicotine in

Table 7.2. Iridoid glycoside content of different life stages of *Junonia coenia* (Lepidoptera: Nymphalidae).*

Life Stage	N	Dry Weight (mg) (Mean ± SE)	Iridoid Glycoside Content			
			Amount (mg)		Percent Dry Weight	
			Mean (SE)	Range	Mean (SE)	Range
Newly molted 5th instar	16	13.89 (1.09)	0.87 (0.08)	0.19–1.48	6.83 (0.75)	0.92–11.14
Pupa	14	91.75 (3.13)	0.17 (0.03)	0.0–0.40	0.19 (0.04)	0.0–0.48
Adult	10	37.46 (3.20)	0.0	—	0.0	—

*From Bowers and Collinge (1992).

nature, nicotine was transported at relatively high and similar rates (about 125 pmol/min) (Maddrell and Gardner, 1976).

Data from insects that sequester iridoid glycosides suggest that for holometabolous, unpalatable insects, the pupal stage may present an evolutionary barrier to the storage of defensive chemicals through to the adult stage (Bowers and Collinge, 1992). During the pupal stage, holometabolous insects undergo a massive reorganization of the internal organs and high levels of cell growth and division. This high level of metabolic activity may render the insect particularly susceptible to the autotoxicity of compounds that were used as larval defenses, and previously existing barriers to these toxic effects may be changed or lost. One example is the change in Malpighian tubule activity described above (Maddrell and Gardner, 1976). Another is larvae of the buckeye butterfly, *Junonia coenia* (Nymphalidae) that sequester iridoid glycosides in amounts as high as 10% dry weight (Bowers and Collinge, 1992) (Table 7.2). Pupae contain over an order of magnitude less, while adults are devoid of iridoid glycosides (Table 7.2). In another case, the Catalpa sphinx, *Ceratomia catalpae* (Sphingidae), which feeds only on species of *Catalpa* (Bignoniaceae) which contain iridoid glycosides, has gregarious, warningly-colored larvae and cryptic adults. Larvae may contain as much as 15% dry weight iridoid glycosides, and larval hemolymph may contain 50% dry weight iridoids (Bowers, Collinge, and Lawrence, unpublished data). Pupae are relatively low in iridoid glycosides (< 0.3% dry weight), and adults are devoid of these compounds.

Even species that retain defense chemical through to the adult stage may exhibit this reduction from the larval to pupal to adult stage. The checkerspots, *Euphydryas anicia* (L'Empereur and Stermitz, 1990) and *E. phaeton* (Bowers, Collinge, and Ecker, unpublished data) have significantly higher concentrations of iridoid glycosides in the larvae than in the pupae and adults. In both these species, as well as *C. catalpae*, the pupal meconium (the waste products accumulated during metamorphosis which are eliminated when the adult emerges) contains large amounts of iridoid glycosides (Bowers and Puttick, 1986; L'Empereur and Stermitz, 1990). Thus, the prevention of autotoxicity in the pupal stage may be the critical and difficult step in the evolution of unpalatability in the adult stage.

7.4. Costs of Chemical Defense

7.4.1. Can We Measure Costs?

Consideration of the cost of chemical defense, be it sequestration, *de novo* synthesis, or some combination of these two, can be approached from a physiological as well as an evolutionary perspective. From a

Table 7.3. Insect species investigated for the cost of chemical defense.

Insect Species	Life Stage	Host Plant	Chemical Defense Compounds	Evidence of a Cost?	Reference
Danaus plexippus (Lepidoptera: Nympalidae)	Adult	*Asclepias curassavica* (Asclepiadacease)	cardenolides	yes	Cohen, 1985
Danaus gillippus (Lepidoptera: Nymphalidae)	Adult	*Ascelpias currasavica* (Aslepiadaceae)	cardenolides	no	Cohen, 1985
Euphydryas anicia (Lepidoptera: Nymphalidae)	Adult ♂	*Castilleja integra* or *Besseya alpina* (Scrophulariaceae)	iridoid glycosides	no	Bowers, 1988
(Lepidoptera: Nymphalidae)	Adult ♀		iridoid glycosides	no	Bowers, 1988
Junonia coenia[1] (Lepidoptera: Nymphalidae)	Larva	*Plantago lanceolata* (Plantaginaceae)	iridoid glycosides	yes	Bowers & Collinge, 1992
Neodiprion sertifer (Hymenoptera: Diprionidae)	Pupa		iridoid glycosides	yes	
	Larva	*Pinus sylvestris* (Pinaceae)	pine resin acids	yes	Bjorkman & Larsson, 1991
Phratora tibialis (Coleoptera: Chrysomelidae)	Larva-Adult[2]	*Salix purpurea* (Salicaceae)	methylcyclopentanoid monoterpenes	yes	Rowell-Rahier & Pasteels, 1986
Phratora vitellinae (Coleoptera: Chrysomelidae)	Larva-Adult[2]	*Salix nigricans*	salicyclaldehyde	no[3]	Rowell-Rahier & Pasteels, 1986
Plagiodera versicolora (Coleoptera: Chrysomelidae)	Larva-Adult[2]	*Salix purpurea*	methylcyclopentanoid monoterpenes	yes	Rowell-Rahier & Pasteels, 1986
Chrysomela 20punctata (Coleoptera: Chrysomelidae)	Larva-Adult[2]	*Salix nigricans*	salicylaldehyde and unknown compounds	no	Rowell-Rahier & Pasteels, 1986
		Salix caprea	unknown compounds	yes	Rowell-Rahier & Pasteels, 1986

[1]Adults of *J. coenia* are not included because they do not contain any iridoid glyosides.

[2]Defensive secretions were removed continuously from larvae, and then newly enclosed adults were weighed. These weights were compared to the weights of a control group that did not have the defensive secretions removed.

[3]*Ph. vitellinae* from which defensive secretions were removed had a higher weight than the control group. This was probably due to the glucose obtained from ingested salicin (see Pasteels et al., 1983, and Rowell-Rahier and Pasteels, 1986).

physiological perspective, there is an energetic cost to *de novo* synthesis of chemical defenses due to: (1) enzymes needed to catalyze the reactions to produce the compounds; (2) use of resources that might otherwise be used for growth; and (3) prevention of autotoxicity, such as by the use of carrier molecules. In addition, once used in defense against an attacking predator, a further energetic cost is incurred because these compounds must be replenished. From an evolutionary perspective, the specialized glands for producing these compounds as well as specialized storage organs (if needed) would entail a "cost" in terms of using resources that might otherwise be used for growth or reproductive functions.

Defensive chemicals sequestered from the host plant need not by synthesized by the animal, thus the chemical precursors and catalytic enzymes of *de novo* synthesis are not needed; but there may be other physiological costs associated with sequestration (see above). In addition, these compounds may be converted metabolically in some way (Bernays et al., 1977; Duffey et al., 1978; Brattsten, 1979; Seiber et al., 1980; Gardner and Stermitz, 1988). In some cases, such metabolism may provide a nutritional benefit to the insect. For example, salicin is metabolized by some chrysomelid beetles into salicylaldehyde, which is used as a defense compound, and glucose, which is used as a nutrient (Pasteels et al., 1983).

There have been several attempts to measure the cost of chemical defense in insect species that sequester defense compounds from host plants, as well as some that manufacture their own defenses *de novo* (Table 7.3). Studies listed in Table 7.3 measured the insect's chemical defense content directly, rather than inferring it from the chemicals content of the host plant. Therefore Table 7.3 does not include the work of Smith (1978), who reared larvae of *Danaus chrysippus* (Nymphalidae) on milkweed species with and without cardenolides. He found that larvae grew significantly faster on cardenolide-containing species, but he did not measure the cardenolide content of the larvae. Nor does it include a study showing that *Danaus plexippus* larvae grew equally well on four species of milkweeds in the genus *Asclepias*, ranging from one very high in cardenolide content to two that were quite low (Erickson, 1973). Here again, the cardenolide content of the insects was not measured. The data from experiments that did measure the chemical content of insects are mixed (Table 7.3); in some cases, there was an indication that there was a cost, as indicated by lower weight or lower growth rate; while other showed no indication of cost.

7.4.2. Calculating the Cost of Chemical Defense

What constitutes a "cost of chemical defense?" To a physiologist, a cost might be the relegation of any resources, such as amino acids, glucose or

ATP, to synthesis or storage of chemical defenses. This is analogous to the "direct cost" of Gulmon and Mooney (1986) to a plant producing secondary metabolites. From that perspective, there are certainly physiological costs: precursors for *de novo* synthesis must be obtained and replenished; enzymes are used to catalyze reactions; and sequestered compounds must be moved out of the gut, in some cases stored, and autotoxicity prevented. To an evolutionary biologist, there must be fitness differences (analogous to the "indirect cost" of Gulmon and Mooney, 1986) associated with a high or low investment in chemical defense. Measures of fitness that have been used have included growth rate or weight relative to chemical content (see references in Table 7.3). Of course, the selective milieu may determine what those costs really are. For example, in the presence of effective predators that can taste prey and reject the most unpalatable, the fitness of relatively palatable individuals might be zero, even though those individuals might have been larger and thus potentially able to produce more offspring. Or the availability of a host plant species that results in the most unpalatable insects may be unpredictable, resulting in the use of a more predictable, but less toxic host plant.

Determination of a cost to sequestration of chemical compounds is typically revealed by a negative correlation of chemical concentration (percent dry weight) with body weight, growth rate, or fecundity (Brower and Glazier, 1975; Cohen, 1985) (Figure 7.1). Alternatively, amount rather than concentration can be used to determine whether a cost exists. In Figure 7.1 are plotted three hypothetical data sets in which the amount of sequestered defense compounds (in mg) increased linearly with insect biomass (A); was constant whatever the insect biomass (B); or decreased with increasing biomass (C). The concentration of defensive chemicals resulting from these amounts was calculated from each data set and plotted on each graph as well.

In graph A, the amount of defense compound increases linearly with biomass, resulting in a constant concentration of that compound at all biomasses. Depending on the exact relationship of insect biomass and defense compound content, the concentration could also be positively correlated with biomass, rather than constant with biomass. In either case, there would be no indication of a physiological cost to sequestering these compounds. This assumes that the chemical content of the food is constant and that the insect's efficiency of sequestration does not change as the amount of sequestered chemical increases.

In graph B, the amount of compound sequestered is constant, resulting in a decrease in the concentration with increasing insect biomass. However, whether or not a cost to sequestration is involved would depend on the exact relationship between amount and concentration. If larger insects

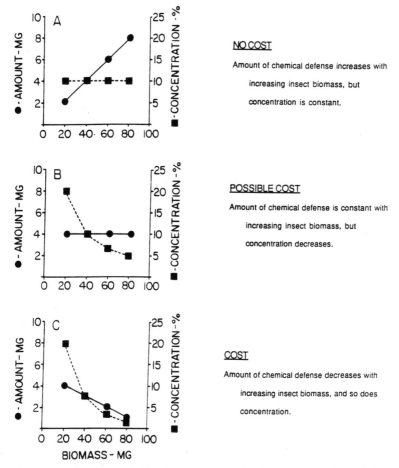

Figure 1. Possible relationships of concentration (percent dry weight) and content (amount) of insect defense chemicals and their interpretation for determining the existence of a possible cost of chemical defense. A detailed description is given in the text.

have a lower concentration of compound, that suggests either that there is an upper limit to the amount that a single insect can sequester, regardless of the insect's size; or that larger insects are larger because they have a lower concentration of compound, and thus have incurred less of a physiological cost. In such a case, it is difficult to determine whether or not there is a physiological cost to sequestration.

In graph C, there is a negative relationship between insect biomass and both content and concentration of defense compound. Thus, the more defense compound an insect has, the smaller it is. This suggests that there is a physiological cost to sequestration so that insects are smaller if they are better chemically-defended.

Both amount and concentration are probably important in determining whether there is a physiological cost to sequestration (Figure 7.1). One milligram of chemical defense synthesized or sequestered in a 10-milligram caterpillar is very different from one-milligram in a 500-milligram caterpillar: although the amount of biochemical machinery needed to synthesize or sequester one milligram may be the same in caterpillars of both sizes and the proportion of the insect's total resources used probably differs greatly. The best demonstration of a cost would thus be a negative correlation of both amount and concentration with growth rate, biomass, fecundity, etc. (Fig. 7.1C).

7.4.3. Automimicry and the Cost of Chemical Defense

If an individual insect does indeed experience a reduction in fitness for being less palatable than other insects in the population; then it is to the advantage of some individuals in the population to be palatable mimics of the unpalatable individuals. Substantial variation in the defensive chemical content of individuals within a population has been documented in a diversity of systems (Brower et al., 1968, 1982, 1984; Bowers, 1980; Gardner and Stermitz, 1988; Jones et al., 1987, 1989). Because the individuals are all the same species, the resemblance is exact. This is the phenomenon of automimicry (Brower et al., 1968, 1970). However, if predators can taste many individuals and eat only the relatively palatable ones, releasing the unpalatable ones relatively unharmed, then automimicry may not be an advantage. The relative abundance of models and automimics in the population, coupled with the predation pressure and the predators' discriminatory ability, will determine the efficacy of the automimcry (Brower et al., 1970).

7.5. Conclusions and Future Directions

The fascination of biologists with chemical defense in insects and other organisms is evidence by the many books and articles in this field in the past decades. The references in this article represent only a small subset of those available. The study of insect chemical defenses represents a unique combination of the fields of evolution, ecology, physiology, behavior, and chemistry. It is those studies that have combined these perspectives that have contributed the most to our understanding of insect defenses. Future research into chemical defenses in insects will be most informative when it continues such multi-faceted approaches. For example, an understanding of the evolution of unpalatability requires knowledge of the physiological mechanisms involved in sequestration and prevention of autotoxicity, a phylogenetic framework in which to interpret these mechanisms, and an understanding of the ecological milieu in which they operate. Future research that combines all these perspectives will enable us to better answer the many question that remain in this area.

I believe that some productive areas for future research include the following.

1. Documentation of chemical variation in defense and assessment of its importance for insect population dynamics.
2. Determination of the biochemical processes involved in insect sequestration of plant secondary metabolites.
3. Study of the evolution of unpalatability through combining phylogenetic and physiological approaches.
4. Comparison of hemimetabolous and holometabolous insects for their physiological adaptations to sequestration of defense compounds.
5. Continuation of efforts to define a cost to chemical defense; and experimentation to determine whether or not such costs are important in the evolution of chemical defense in insects.

Acknowledgements

I thank S. Collinge, E. Fajer, M. Isman, J. Meyers, and N. Stamp for comments on the manuscript. I am especially grateful to J. Pasteels for providing preprints of unpublished material.

References

Abushama, F. T. E. 1972. The repugnatorial gland of the grasshopper *Poecilocerus hieroglyphicus*. *J. Ent. A*. 47:95–100.

Ackery, P. R. and R. I. Vane-Wright. 1984. *Milkweed Butterflies: their Cladistics and Biology*. Cornell University Press. Ithaca, New York.

Ahmad, S., L. B. Brattsten, C. A. Mullin, and S. J. Yu. 1986. Enzymes involved in the metabolism of plant allelochemicals. In *Molecular Aspects of Insect-Plant Associations*, eds. L. B. Brattsten and S. Ahmad, pp. 211–255. Plenum Press, New York.

Aldrich, J. R. and M. S. Blum. 1978. Aposematic aggregation of a bug (Hemiptera: Coreidae): the defensive display and formation of aggregations. *Biotropica* 10:58–61.

Barbosa, P. 1988. Natural enemies and herbivore-plant interactions: influence of plant allelochemicals and host specificity. In *Novel Aspects of Insect-Plant*

Interactions. eds. P. Barbosa and D. Letourneaux, pp. 201–229. Wiley Interscience, New York.

Bates, H. W. 1862. Contributions to an insect fauna of the Amazon valley. *Trans. Linn. Soc. Lond.* 23:495–566.

Belofsky, G., M. D. Bowers, S. Janzen, and F. Stermitz. 1989. Iridoid glycosides of *Aureolaria flava* and their sequestration by *Euphydryas phaeton* butterflies. *Phytochemistry* 28:1601–1604.

Benson, W. W. 1971. Evidence for the evolution of unpalatability through kin selection in the Heliconiinae. *Am. Nat.* 105:213–226.

Berenbaum, M. R. and E. Miliczky. 1984. Mantids and milkweed bugs: efficacy of aposematic coloration against invertebrate predators. *Am. Midl. Natur.* 111:64–68.

Bernays, E. A. 1988. Host specificity in phytophagous insects: selection pressure from generalist predators. *Entomol. Exp. Appl.* 49:131–140.

Bernays, E., J. A. Edgar, M. Rothschild. 1977. Pyrrolizidine alkaloids sequestered and stored by the aposematic grasshopper, *Zonicerus variegatus. J. Zool.* 182:85–87.

Bjorkman, C. and S. Larsson. 1991. Pine sawfly defence and variation in hostplant resin acids: a trade-off with growth. *Ecol. Entomol.* 16:283–289.

Blest, A. D. 1963. Longevity, palatability, and natural selection in five species of new world saturniid moths. *Nature* 197:1183–1186.

Blum, M. S. 1981. *The Chemical Defenses of Arthropods.* Academic Press, New York.

Blum, M. S. 1983. Detoxication, deactivation, and utilization of plant compounds by insects. In *Plant Resistance to Insects*, ed. P. A. Hedin, pp. 265–275. ACS Symposium Series 208. American Chemical Society, Washington, DC.

Blum, M. S., R. F. Severson, R. F. Arrendale, D. W. Whitman, P. Escoubas, O. Adeyeye, and C. G. Jones. 1990. A generalist herbivore in a specialist mode: metabolic, sequestrative, and defensive consequences. *J. Chem. Ecol.* 16:223–244.

Bowers, M. D. 1980. Unpalatability as a defense strategy of *Euphydryas phaeton* (Lepidoptera: Nymphalidae). *Evolution* 34:586–600.

Bowers, M. D. 1988a. Plant allelochemistry and mimicry. In *Novel Aspects of Insect-Plant Interactions*, eds. P. Barbosa and D. Letourneau, pp. 273–311. John Wiley, New York.

Bowers, M. D. 1988b. Chemistry and coevolution: iridoid glycosides, plants and herbivorous insects. In *Chemical Mediation of Coevolution*, ed. K. Spencer, pp. 133–165. Academic Press.

Bowers, M. D. and G. M. Puttick. 1986. The fate of ingested iridoid glycosides in lepidopteran herbivores. *J. Chem. Ecol.* 12:169–178.

Bowers, M. D. and Z. Larin. 1989. Acquired chemical defense in the lycaenid butterfly, *Eumaeus atala* (Lycaenidae). *J. Chem. Ecol.* 15:1133–1146.

Bowers, M. D. and S. D. Farley. 1990. The behaviour of gray jays (*Perisoreus canadensis*) toward palatable and unpalatable Lepidoptera. *Anim. Behav.* 39:699–705.

Bowers, M. D. and S. K. Collinge. 1992. The fate of iridoid glycosides in different life stages of the buckeye (*Junonia coenia*, Nymphalidae) *J. Chem. Ecol.* (in press).

Boyden, T. C. Butterfly palatability and mimicry: experiments with *Ameiva* lizards. *Evolution* 30:73–81. 1976.

Brattsten, L. B. 1979. Ecological significance of mixed-function oxidases. *Drug Metab. Rev.* 10:35–58.

Brattsten, L. B. 1986. Fate of ingested plant allelochemicals in herbivorous insects. In *Molecular Aspects of Insect Plant Associations*, eds. L.B. Brattsten and S. Ahmad, pp. 211–255. Plenum Press, New York.

Brower, L. P. 1984. Chemical defense in butterflies. In *The Biology of Butterflies*, eds. R. I. Vane-Wright and P. R. Ackery, pp. 109–132. Academic Press, New York.

Brower, L. P. and S. C. Gazier. 1975. Localization of heart poisons in the monarch butterfly. *Science* 188:19–25.

Brower, L. P., W. N. Ryerson, L. L. Coppinger, and S. C. Glazier. 1968. Ecological chemistry and the palatability spectrum. *Science* 161:1349–1351.

Brower, L. P., F. H. Pough, and H. R. Meck. 1970. Theoretical investigations of automimicry. I. Single trial learning. *Proc. Natl. Acad. Sci.* 66:1059–1066.

Brower, L. P., J. N. Seiber, C. J. Nelson, S. P. Lynch, and P. M. Tuskes. 1982. Plant determined variation in the cardenolide content, thin-layer chromatography profiles, and emetic potency of Monarch butterflies, *Danaus plexippus*, reared on the milkweed, *Asclepias eriocarpa* in California. *J. Chem. Ecol.* 8:579–633.

Brower, L. P., J. N. Seiber, C. J. Nelson, S. P. Lynch, and M. M. Holland. 1984. Plant determined variation in the cardenolide content, thin-layer chromatography profiles, and emetic potency of monarch butterflies, *Danaus plexippus*, reared on the milkweed *Asclepias speciosa* in California. *J. Chem. Ecol.* 10:601–639.

Brower, L. P. and W. H. Calvert. 1985. Foraging dynamics of bird predators on overwintering Monarch butterflies in Mexico. *Evolution* 39:852–868.

Brown, K. S. 1984a. Adult-obtained pyrrolizidine alkaloids defend ithomiine butterflies against a spider predator. *Nature* 309:707–709.

Brown, K. S. 1984b. Chemical ecology of dehydropyrrolizidine alkaloids in adult ithomiinae (Lepidoptera: Nymphalidae). *Rev. Bras. Biol.* 44:435–460.

Carpenter, G. D. H. 1942. Observations and experiments in Africa by the late C.F.M. Swynnerton on wild birds eating butterflies and the preference shown. *Proc. Linn. Soc. Lond.* 154:10–46.

Chapman, R. 1980. *Insects, Their Structure and Function*. Harvard University Press, Cambridge, MA.

Cohen, J. A. 1985. Differences and similarities in cardenolide contents of queen and monarch butterflies in Florida and their ecological and evolutionary consequences. *J. Chem. Ecol.* 11:85–103.

Coppinger, R. P. 1970. The effect of experience and novelty on avian feeding behavior with reference to the evolution of warning coloration in butterflies. II. Reactions of naive birds to novel insects. *Am. Nat.* 104:323–335.

Cott, H. B. 1940. *Adaptive Coloration in Animals*. Methuen, London, UK.

Damman, A. J. 1987. Leaf quality and enemy avoidance by the larvae of a pyralid moth. *Ecology* 68:88–97.

Dawkins, R. 1976. *The Selfish Gene*. Oxford University Press, Oxford.

Denno, R. F., S. Larsson, and K. L. Olmstead. 1990. Role of enemy-free space and plant quality in host-plant selection by willow beetles. *Ecology* 71:124–137.

Deroe, C. and J. M. Pasteels. 1982. Distribution of adult defense glands in chrysomelids (Coleoptera: Chrysomelidae) and its significance in the evolution of defense mechanisms within the family. *J. Chem. Ecol.* 8:67–82.

Dolinger, P. M., P. R. Ehrlich, W. L. Fitch, and D. E. Breedlove. 1973. Alkaloid and predation patterns in Colorado lupine populations. *Oecologia* 13:191–204.

Duffey, S. S. 1980. Sequestration of plant natural products by insects. *Annu. Rev. Entomol.* 25:447–477.

Duffey, S. S. and G. G. E. Scudder. 1974. Cardiac glycosides in *Oncopeltus fasciatus* (Dallas). I. The uptake and distribution of natural cardenolides in the body. *Can. J. Zool.* 52:283–290.

Duffey, S. S., M. S. Blum, M. Isman, G. G. E. Scudder. 1978. Cardiac glycosides: A physical system for their sequestration by the milkweed bug. *J. Insect Physiol.* 24:639–645.

Edmunds, M. 1974. *Defence in Animals: A Survey of Antipredator Defences*. Longman, Essex, England.

Eisner, T. 1965. Defensive spray of a phasmid insect. *Science* 148:966–968.

Eisner, T. 1970. Chemical defense against predation in arthropods. In *Chemical Ecology*, eds. E. Sondheimer and J. B. Simeone, pp. 157–217. Academic Press, New York.

Eisner, H. E., D. W. Alsop, and T. Eisner. 1967. Defense mechanisms of arthropods. XX. Quantitative assessment of hydrogen cyanide production in two species of millipedes. *Psyche* 74:107–117.

Eisner, T., L. B. Hendry, D. B. Peakall, and J. Meinwald. 1971. 2,5-dichlorophenol (from ingested herbicide?) in defensive secretion of grasshopper. *Science* 172:277–278.

Eisner, T., J. S. Johnessee, J. Carrel, L. B. Hendry, and J. Meinwald. 1974. Defensive use by an insect of a plant resin. *Science* 184:996–999.

Eisner, T., T. H. Jones, D. J. Aneshansley, W. R. Tschinkel, R. E. Silberglied, J. Meinwald. 1977. Chemistry of defensive secretions of bombadier beetles. *J. Insect Physiol.* 23:1383–1386.

Erickson, J. M. 1973. The utilization of various *Asclepias* species by larvae of the monarch butterfly, *Danaus plexippus. Psyche* 80:230–244.

Evans, H. E. 1987. Observations of the prey and nests of *Podalinia occidentalis* Murray (Hymenoptera: Sphecidae). Pan-Pac. *Entomol.* 63:130–134.

Evans, H. E. and M. J. West Eberhard. 1970. *The Wasps.* Univ. Michigan Press, Ann Arbor.

Fisher, R. A. 1930, *The Genetical Theory of Natural Selection.* Clarendon Press, Oxford.

Franzl, S. A. Nahrstedt, and C. M. Naumann. 1986. Evidence for site of biosynthesis and transport of the cyanoglucosides linamarin and lotaustralin in larvae of *Zygaena trifolii* (Insecta: Lepidoptera). *J. Insect Physiol.* 32:705–709.

Gardner, D. R. and F. R. Stermitz. 1988. Host plant utilization and iridoid glycoside sequestration by *Euphydryas anicia* (Lepidoptera: Nymphalidae). *J. Chem. Ecol.* 14:2147–2168.

Gillaspy, J. E. 1979. Management of *Polistes* wasps for caterpillar predation. *Southwest Entomol.* 4:334–350.

Gittleman, J. L. and P. H. Harvey. 1980. Why are distasteful prey not cryptic? *Nature* 286:149–150.

Gittleman, J. L., P. H. Harvey, and P. J. Greenwood. 1980. The evolution of conspicuous coloration: some experiments in bad taste. *Anim. Behav.* 28:897–899.

Gould, W. P. and R. L. Jeanne. 1984. *Polistes* wasps (Hymenoptera: Vespidae) as control agents for lepidopteran cabbage pests. *Environ. Entomol.* 13:150–156.

Guilford, T. 1985. Is kin selection involved in the evolution of warning coloration? *Oikos* 45:31–36.

Guilford, T. 1986. How do "warning colours" work? Conspicuousness may reduce recognition errors in experienced predators. *Anim. Behav.* 34:286–288.

Guilford, T. 1990. The evolution of aposematism. In *Defense*: *Predator-Prey Interactions*, eds. J. O. Schmidt and D. L. Evans, pp. 23–61. SUNY Press, Syracuse, NY.

Gulmon, S. and H. A. Mooney. 1986. Costs of defense and their effects on plant productivity. In *On the Economy of Plant Form and Function*, ed. T. J. Givnish, pp. 681–698. Cambridge University Press. Cambridge.

Happ, G. M. and T. Eisner. 1961. Hemorrhage in a coccinellid beetle and its repellent effect on ants. *Science* 132:329–31.

Harvey, P. H. and P. J. Greenwood. 1978. Anti-predator defence strategies: some evolutionary problems. In *Behavioral Ecology: an Evolutionary Approach*, eds. J. R. Krebs and N.B. Davies, pp. 129–151. Blackwell, Oxford.

Harvey, P. H. and R. J. Paxton. 1981. The evolution of aposematic coloration. *Oikos* 37:391–393.

Isman, M. B. 1977. Dietary influence of cardenolides on larval growth and development of the milkweed bug *Oncopeltus fasciatus. J. Insect Physiol.* 23:1183–1187.

Isman, M. B., S. S. Duffey, and G. G. E. Scudder. 1977. Cardenolide content of some leaf- and stem-feeding insects on temperate North American milkweeds (*Asclepias* spp.) *Can. J. Zool.* 55:1024–1028.

Jarvi, T., B. Sillen-Tullberg, and C. Wiklund. 1981. The cost of being aposematic: an experimental study of predation on larvae of *Papilio machaon* by the great tit *Parus major. Oikos* 36:267–272.

Jones, C. G., T. A. Hess, D. W. Whitman, P. J. Silk, and M. S. Blum. 1986. Idiosyncratic variation in chemical defenses among individual generalist grasshoppers. *J. Chem. Ecol.* 13:749–761.

Jones, C. G., T. S. Hess, D. W. Whitman, P. J. Silk, M. S. Blum. 1987. Effects of diet breadth on autogenous chemical defense of a generalist grasshopper. *J. Chem. Ecol.* 13:283–297.

Jones, C. G., D. W. Whitman, S. J. Compton, P. J. Silk, M. S. Blum. 1989. Reduction in diet breadth results in sequestration of plant chemicals and increases efficacy of chemical defense in a generalist grasshopper. *J. Chem. Ecol.* 15:1811–1822.

Jones, F. M. 1932. Insect coloration and the relative acceptability of insects to birds. *Trans. R. Ent. Soc. Lond.* 80:345–385.

Jones, F. M. 1934. Further experiments on colouration and relative acceptability of insects to birds. *Trans. R. Ent. Soc. Lond.* 82:443–453.

Kidd, N. A. C. 1982. Predator avoidance as a result of aggregation in the grey pine aphid, *S. pineti, J. Anim. Ecol.* 51:397–412.

Larsson, S., C. Bjorkman, and R. Gref. 1986. Responses of *Neodiprion sertifer* (Hym. Diprionidae) larvae to variation in needle resin concentration in Scots pine. *Oecologia* 70:77–84.

L'Empereur, K. M. and F. R. Stermitz. 1990. Iridoid glycoside content of *Euphydryas anicia* (Lepidoptera: Nymphalidae) and its major host plant, *Besseya plantaginea* (Scrophulariaceae), at a high plains Colorado site. *J. Chem. Ecol.* 16:187–197.

Lincoln, D. E. and J. H. Langenheim. 1978. Effect of light and temperature on monoterpenoid yield and composition in *Satureja douglasii. Biochem. Syst. Ecol.* 6:21–32.

Lincoln, D. E. and J. H. Langenheim. 1979. Variation of *Satureja douglasii* monoterpenoids in relation to light intensity and herbivory. *Biochem. Syst. Ecol.* 7:289–298.

Lincoln, D. E. and J. H. Langenheim. 1981. A genetic approach to monoterpenoid variation in *Satureja douglasii. Biochem. Syst. Ecol.* 9:153–161.

Lincoln, D. E. and H. A. Mooney. 1984. Herbivory on *Diplacus aurantiacus* shrubs in sun and shade. *Oecologia* 64:173–177.

Lincoln, D. E., M. J. Murray, and B. M. Lawrence. 1986. Chemical composition and genetic basis for the isopinocamphone chemotype of *Mentha citrata* hybrids. *Phytochemistry* 25:1857–1863.

Louda, S. M. and J. E. Rodman. 1983a. Ecological patterns in the glucosinolate content of a native mustard, *Cardamine cordifolia*, in the Rocky Mountains. *J. Chem. Ecol.* 9:397–421.

Louda, S. M. and J. E. Rodman. 1983b. Concentration of glucosinolates in relation to habitat and insect herbivory for the native crucifer *Cardamine cordifolia*. *Biochem. Syst. Ecol.* 11:199–207.

Maddrell, S. H. P. and B. O. C. Gardner. 1976. Excretion of alkaloids by malpighian tubules of insects. *J. Exp. Biol.* 64:267–281.

Malcolm, S. B. 1986. Aposematism in a soft-bodied insect: a case for kin selection. *Behav. Ecol. Sociobiol.* 18:387–393.

Malcolm, S. B. and L. P. Brower. 1989. Evolutionary and ecological implications of cardenolide sequestration in the monarch butterfly. *Experientia* 45:284–295.

McKey, D. 1979. The distribution of secondary compounds within plants. In *Herbivores: Their Interactions with Secondary Plant Metabolites*, eds. G. Rosenthal and D. H. Janzen, pp. 55–133. Academic Press, New York.

McKey, D., P. G. Waterman, C. N. Mbi, G. N. Gardlan, and T. T. Struhsaker. 1978. Phenolic content of vegetation in two African rainforests: ecological implications. *Science* 202:61–64.

McLain, D. K. and D. J. Shure. 1985. Host plant toxins and unpalatability of *Neacoryphus bicrucis* (Hemiptera: Lygaeidae). *Ecol. Ent.* 10:291–298.

Miller, J. S. and P. P. Feeny. 1989. Interspecific differences among swallowtail larvae (Lepidoptera: Papilionidae) in susceptibility to aristolochic acids and berberine. *Ecol. Entomol.* 14:287–296.

Moore, L. V. and G. G. E. Scudder. 1986. Ouabain-resistant Na,K-ATPases and cardenolide tolerance in the large milkweed bug, *Oncopeltus fasciatus*. *J. Insect Physiol.* 32:27–33.

Morrow, P. A., T. E. Bellas, and T. Eisner. 1976. Eucalyptus oils in the defensive oral discharge of Australian sawfly larvae (Hymenoptera: Pergidae). *Oecologia* 24:193–206.

Nishida, R. and H. Fukami. 1989. Host plant iridoid-based chemical defense of an aphid, *Acyrthosiphon nipponicus*, against ladybird beetles. *J. Chem. Ecol.* 15:1837–1843.

Nishio, S. 1983. The fates and adaptive significance of cardenolides sequestered by larvae of *Danaus plexippus* (L.) and *Cycnia inopinatus* (Hy. Edwards). Ph.D. dissertation, University Microfilms, University of Georgia, Athens, Georgia.

Nishio, S., M. S. Blum, and S. Takahashi. 1983. Intraplant distribution of cardenolides in *Asclepias humistrata* (Asclepiadaceae), with additional notes on their fates in *Tetraopes melanurus* (Coleoptera: Cerambycidae) and *Rhyssomatus lineaticollis* (Coleoptera: Curculionidae). *Mem. Coll. Agric., Kyoto Univ.* 122:43–52.

Noirot, C. and A. Quennedey. 1974. Fine structure of insect epidermal glands. *Annu. Rev. Entomol.* 19:61–80.

Paradise, C. J. and N. E. Stamp. 1990. Variable quantities of toxic diet cause different degrees of compensatory and inhibitory responses by juvenile praying mantids. *Entomol. Exp. Appl.* 55:213–222.

Paradise, C. J. and N. E. Stamp. 1991. Episodes of unpalatable prey reduce consumption and growth of juvenile preying mantids. *Insect Behav.* 4:265–273.

Parry, R. H. 1974. Revision of the genus *Dibolia* Latreille in America north of Mexico (Coleoptera: Chrysomelidae). *Can. J. Zool.* 52:1317–1354.

Pasteels, J. M. and J.-C. Gregoire. 1983. Selective predation on chemically defended chrysomelid larvae. A conditioning process. *J. Chem. Ecol.* 10:1693–1700.

Pasteels, J. M., M. Rowell-Rahier, J. C. Braekman, and A. Dupont. 1983. Salicin from host plant as precursor of salicylaldehyde in defensive secretion of Chrysomeline larvae. *Physiol. Entomol.* 8:307–314.

Pasteels, J. M., M. Rowell-Rahier, J.-C. Braekman, and D. Daloze. 1984. Chemical defences in leaf beetles and their larvae: the ecological, evolutionary and taxonomic significance. *Biochem. Syst. Ecol.* 12:395–406.

Pasteels, J. M., D. Daloze, and M. Rowell-Rahier. 1986. Chemical defense in chrysomelid eggs and neonate larvae. *Physiol. Entomol.* 11:29–37.

Pasteels, J. M., M. Rowell-Rahier and M. J. Raupp. 1988a. Plant derived defense in chrysomelid beetles. In: P. Barbosa and D. Letourneau (eds.). In *Novel Aspects of Insect-Plant Relationships*, eds. P. Barbosa and D. Letourneau, pp. 235–272. John Wiley, New York.

Pasteels, J. M., J. C. Braekman, and D. Daloze. 1988b. Chemical defense in Chrysomelidae. In *The Biology of Chrysomelidae*. eds. P. Jolivet, T.H. Hsiao, and E. Petitpierre, pp. 231–250. Dordrecht.

Pasteels, J. M., M. Rowell-Rahier, J. C. Braekman, D. Daloze, and S. Duffey. 1989. Evolution of exocrine chemical defense in leaf-beetles (Coleoptera: Chrysomelidae). *Experientia* 45:295–299.

Pasteels, J. M., S. Duffey, and M. Rowell-Rahier. 1990. Toxins in chrysomelid beetles: possible evolutionary sequence from *de novo* synthesis to derivation from food plant chemicals. *J. Chem. Ecol.* 16:211–222.

Pasteur, G. 1982. A classificatory review of mimicry systems. *Ann. Rev. Ecol. Syst.* 13:169–199.

Roper, T. J. and R. Wistow. 1986. Aposematic coloration and avoidance learning in chicks. *Quart. J. Exp. Psych.* 38B:141–149.

Rothschild, M. 1972. Secondary plant substances and warning coloration in insects. In *Insect-Plant Relationships*, ed. H.F. van Emden, pp. 59–83. Blackwell, Oxford.

Rothschild, M. 1985. British aposematic Lepidoptera. In *The Moths and Butterflies of Great Britain and Ireland*, eds. J. Heath and A.M. Emmet, pp. 9–62. Harley Books, Essex, England.

Rothschild, M. and N. Marsh. 1978. Some peculiar aspects of danaid/plant relationships. *Entomol. Exp. Appl.* 24:437–450.

Rothschild, M., R. J. Nash, and E. A. Bell. 1986. Cycasin in the endangered butterfly, *Eumaeus atala florida*. *Phytochemistry* 25:1853–1854.

Rowell-Rahier, M. and J. M. Pasteels. 1986. Economics of chemical defense in chrysomelinae. *J. Chem. Ecol.* 12:1189–1203.

Schuler, W. and E. Hesse. 1985. On the function of warning coloration: a black and yellow pattern inhibits prey-attack by naive domestic chicks. *Behav. Ecol. Sociobiol.* 16:249–255.

Scudder, G. G. E. and J. Meredith. 1982. The permeability of the midgut of three insects to cardiac glycosides. *J. Insect Physiol.* 28:689–694.

Scudder, G. G. E., L. V. Moore, and M. B. Isman. 1986. Sequestration of cardenolildes in *Oncopeltus fasciatus*: morphological and physiological adaptations. *J. Chem. Ecol.* 12:1171–1187.

Seiber, J. N., P. M. Tuskes, L. P. Brower, and C. J. Nelson. 1980. Pharmacodynamics of some individual milkweed cardenolides fed to larvae of the monarch butterfly (*Danaus plexippus* L.). *J. Chem. Ecol.* 6:321–339.

Seiber, J. N., L. P. Brower, S. M. Lee, M. M. McChesney, H. T. A. Cheung, C. J. Nelson, and T. R. Watson. 1986. Cardenolide connection between overwintering monarch butterflies from Mexico and their larval food plant. *Asclepias syriaca*. *J. Chem. Ecol.* 12:1157–1170.

Sillen-Tullberg, B. 1985. Higher survival of an aposematic than of a cryptic form of a distasteful bug. *Oecologia* 67:411–415.

Sillen-Tullberg, B. 1988. Evolution of gregariousness in aposematic butterfly larvae: a phylogenetic analysis. *Evolution* 42:293–305.

Sillen-Tullberg, B., C. Wiklund, and T. Jarvi. 1982. Aposematic coloration in adults and larvae of *Lygaeus equestris* and its bearing on Mullerian mimicry: an experimental study on predation on living bugs by the great tit *Parus major*. *Oikos* 39:131–136.

Sillen-Tullberg, B. and O. Leimar. 1988. The evolution of gregariousness in distasteful insects as a defense against predators. *Am. Nat.* 132:723–734.

Smith, D. A. S. 1978. The effect of cardiac glycoside storage on growth rate and adult size in the butterfly *Danaus chrysippus* (L.). *Experientia* 34:845–846.

Smith, S. M. 1975. Innate recognition of coral snake pattern by a possible avian predator. *Science* 187:759–760.

Stamp, N. E. 1980. Egg deposition patterns in butterflies: why do some species cluster their eggs rather than deposit them singly? *Am. Nat.* 115:367–180.

Stamp, N. E. and M. D. Bowers. 1988. Direct and indirect effects of predatory wasps (*Polistes* sp.: Vespidae) on gregarious caterpillars (*Hemileuca lucina*: Saturniidae). *Oecologia* 75:619–624.

Takagi, M., Y. Hirose, and M. Yamasaki. 1980. Prey-location learning in *Polistes jadwigae* Dalla Torre (Hymenoptera, Vespidae), field experiments on orientation. *Kontyu* 48:53–58.

Terriere, L. C. 1984. Induction of detoxication enzymes in insects. *Annu. Rev. Entomol.* 29:71–88.

Tschinkel, W. R. 1975a. A comparative study of the chemical defensive system of tenebrionid beetles: chemistry of the secretions. *J. Insect Physiol.* 21:753–783.

Tschinkel, W. R. 1975b. A comparative study of the chemical defensive system of tenebrionid beetles: defensive behavior and ancillary features. *Ann. Entomol. Soc. Am.* 68:439–453.

Tschinkel, W. R. 1975c. A comparative study of the chemical defensive system of tenebrionid beetles III. Morphology of the glands. *J. Morph.* 145:355–370.

Turner, J. R. G. 1984. Darwin's coffin and Doctor Pangloss—do adaptionist models explain mimicry? In *Evolutionary Ecology*, ed. B. Shorrocks, p. 313–361. Blackwell, Oxford, UK.

Vaughan, G. L. and A. M. Jungreis. 1977. Insensitivity of lepidopteran tissues to ouabain: physiological mechanisms for protection from cardiac glycoisdes. *J. Insect Physiol.* 23:585–589.

Wiklund, C. and T. Jarvi. 1982. Survival of distasteful insects after being attacked by naive birds: a reappraisal of the theory of aposematic coloration evolving through individual selection. *Evolution* 36:998–1002.

Wiklund, C. and B. Sillen-Tullberg. 1985. Why distasteful butterflies have aposematic larvae and adults, but cryptic pupae: evidence from predation experiments on the monarch and the european swallowtail. *Evolution* 39:1155–1158.

Wilkinson, C. F. 1986. Xenobiotic conjugation in insects. *Am. Chem. Soc. Symp.* 299:48–61.

Wink, M. 1988. Carrier-mediated uptake of pyrrolizidine alkaloids in larvae of the aposematic and alkaloid-exploiting moth Creatonotos. *Naturwissenschaften* 75:524–525.

Witz, B. W. 1989. Antipredator mechanisms in arthropods: a twenty-year literature survey. *Florida Entomol.* 73:71–99.

8

Behavioral Plasticity and Patterns of Host Use by Insects

John Jaenike   University of Rochester
and
Daniel R. Papaj   University of Arizona

8.1. Introduction

The study of associations between insects and their hosts has for many years focused on two major questions. The first involves the effects of host chemistry on the diet breadth of herbivorous and frugivorous insects (see Rausher, this volume). While plant chemistry has generally been accorded prominence as a determinant of host range, some workers contend that ecological factors, such as predation or parasitism, are equally or even more important (Bernays and Graham, 1988). Regardless of whether ecological or chemical factors are key to host specificity, though, few deny that a shift to a novel host or an expansion of host range involves at least some adaptations to host chemistry. Chief among such evolutionary changes are those that permit the insect to find and accept the new host for oviposition or feeding (Dethier, 1970; Jermy, 1984).

The second question concerns the possibility of sympatric speciation via the formation of host races in host-specific insects. Although the concept of sympatric speciation in such groups has been around for decades (for reviews, see Thorpe, 1930; Diehl and Bush, 1984), it is only recently that convincing evidence has been published documenting significant genetic divergence between sympatric, conspecific populations of *Rhagoletis pomonella* infesting different host plants (Feder et al., 1988, 1990; McPheron, et al. 1988). Such host races have been postulated to represent the early stages of divergence leading ultimately to the evolution of reproductively-isolated populations, that is, species. This genetic differentiation is believed to result primarily from genetically-based variation in host selection behavior by the ovipositing female (Prokopy et al., 1988). Because mating occurs on the host plant, mate choice and oviposition site-selection are, to some extent, pleiotropic manifestations of the same set of genes, which increases the likelihood of sympatric divergence (Rice, 1987; Rice and Salt, 1988).

Evolutionary change in host-selection behavior is evidently of great importance in affecting both diet breadth and the development of host races. In most population genetic theory pertaining to these questions, insects are characterized as congenitally hard-wired automatons (for perspective on this question, see Dethier, 1964). In fact, individual insects manifest a considerable degree of plasticity in behavior, and this may have a significant impact on evolutionary changes in host use.

There is growing recognition that phenotypic plasticity, defined as the ability of an organism to change phenotypically in response to environmental conditions, can influence evolutionary processes, but there is little agreement on exactly how (Schlichting, 1986; Via and Lande, 1985; Via, 1988; Stearns, 1989; West-Eberhard, 1989; Wcislo, 1989). Waddington (1953), for example, contended that plasticity permitted organisms to adapt readily to novel environments. Wright (1931), by contrast, viewed plasticity as an impediment to adaptive evolution. The lack of consensus notwithstanding, we concur with Stearns (1989, p. 445), who wrote, "Genetics and demography are not sufficient to explain evolution; they must be combined with descriptions of phenotypic plasticity—reaction norms and developmental switches—before the sources of variation can be understood." Despite its potential importance, phenotypic plasticity in general and behavioral plasticity in particular has received scant attention among students of host-specific insects (but see Via, 1988). In this chapter, we focus on two kinds of behavioral plasticity common in host-specific insects—motivation and learning—and address their significance for patterns of host use.

8.1. Motivation and Learning: Definitions

Motivation is manifested as a behavioral change that takes place with the passage of time, even in the absence of an overt experience. Such "time-dependent responsiveness" is commonplace in host-specific insects (Papaj and Rausher, 1983). For example, a butterfly deprived of the opportunity to lay eggs becomes increasingly likely to respond to oviposition stimuli (probably chemical) with time (Singer, 1982). Differences in rate of egg production have also been shown to be associated with differences in host acceptance behavior in Drosophila (Courtney et al., 1989) and beetles (Wasserman and Futuyma, 1981), with more fecund individuals being more likely to lay eggs on poor hosts. Apparently, a greater current egg load, whether resulting from a faster rate of egg production or longer time since the last oviposition, can make an individual prone to accept a host that would otherwise be rejected. While evidence is scant, it is likely that behavioral responses to host chemistry are sensitive to the physiological state of the insect.

In contrast to motivation, learning embraces only behavioral changes contingent upon experience. The change is gradual with continued experience up to an asymptote and wanes in the absence of continued experience of the same kind or as a consequence of a different experience (Papaj and Prokopy, 1989). Learning defined in this way has been demonstrated in a variety of behaviors in insects, including foraging for food by adults and juveniles, mate choice, kin recognition, and host acceptance by ovipositing females (for representative reviews and papers on these respective subjects, see Gould, 1984; Heinrich, 1979; Lewis, 1986; Jermy, 1987; Papaj and Prokopy, 1989; Vet and van Opzeeland, 1984). There is abundant evidence that behavioral responses to host chemistry are learned (see reviews by Papaj and Prokopy 1986, 1989).

8.2. General Approach

With respect to behavioral plasticity (whether due to learning or motivation), most theoretical treatments have addressed the question of which behavior among a number of alternatives an individual should adopt in order to maximize its expected fitness (cf. Levins and MacArthur, 1969; Jaenike, 1978; Iwasa et al., 1984; Charnov and Skinner, 1984; Parker and Courtney, 1984; Mangel, 1987; Mangel and Clark, 1988; Mangel and Roitberg, 1989). In some models, learning and motivational processes are viewed as constraints; given that an animal must learn, for example, what is the best it can do under a specified set of ecological conditions? In others, learning or motivational state is itself the trait to be optimized; given that an animal learns, for example, what form of learning maximizes fitness under specified conditions (Pulliam, 1981)?

Here, we address a different question: regardless of the adaptive significance of learning, how do these processes affect patterns of host use within insect populations? Specifically, we ask if learning alters the magnitude of genetically-conferred differences in host acceptance behavior, including differences attributed to responses to host chemistry. These questions are important ones, bearing directly on repeated claims that learning can facilitate sympatric genetic divergence by augmenting genetically-based differences in host preference (Thorpe, 1945; Bush, 1974). We know of no support for this claim on either theoretical or empirical grounds. Nevertheless, learning has been shown to affect host use in *R. pomonella* in ways that could potentially promote formation of host races thought to exist in that species (Prokopy et al., 1982; Papaj and Prokopy,

1988). It is thus not inconceivable that learning with respect to host selection has a role in host race formation and, hence, in generating species diversity in host-specific insects. Given that such learning involves changes in responses to host chemistry (Papaj and Prokopy, 1986; Papaj et al., 1989), it is also possible that the extent to which host race formation is regulated by chemical factors depends on whether insects learn. In Section 8.3, we present a simple mechanistic model of chemically-based host acceptance behavior incorporating motivation and learning and explore some of the consequences of such behavioral plasticity for patterns of host use.

8.3. The Model

We make the following assumptions about the behavior of an insect seeking oviposition sites:

1. Oviposition can occur on either or both of two host types, and these are encountered in proportion to their abundances (that is, neither learning nor genotype affect whether hosts of different types fall within an insect's "search path").

2. As time spent searching for an oviposition site increases, the probability of acceptance (i.e., oviposition) of an encountered host increases from zero asymptotically to one. This pattern is portrayed by the motivational curve shown in Fig. 8.1. Empirical support for this assumption is provided by the elegant experiments of Singer (1982) on *Euphydryas editha*.

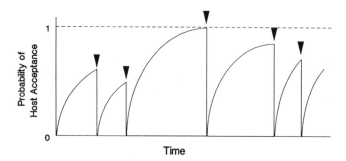

Figure 8.1. As a result of changes in its internal state (motivation), an insect's probability of host acceptance increases with time since last oviposition. Arrows indicate times at which insects oviposit. No learning occurs in this simulation.

3. Eggs are laid singly or in clutches of fixed size. When oviposition occurs, search time (and thus the probability of host acceptance) is reset to zero.

4. The rate at which probability of acceptance of a specific host increases with search time depends both on an individual's genotype and its previous oviposition experience, as summarized in the following equation:

$$P_{it} = 1 - e^{-[tM_i(1+N_iL)]} \qquad (8.1)$$

Here, P_{it} is the probability of accepting host i (if encountered) t time units after the last egg was laid. For now, we assume that acceptance of a host involves a response to host chemistry.

M_i is a genetically-variable parameter affecting the use of host i: the greater M_i, the sooner into the search it is likely to be accepted when encountered. The change in acceptance over time could conceivably reflect an increase in sensitivity to a stimulant, a decrease in sensitivity to a deterrent, or both.

L is a host-independent parameter that reflects the degree to which oviposition on a given host increases the rate at which it subsequently becomes acceptable. If experience has no effect on subsequent behavior, then $L = 0$.

Finally, N_i is a function of the recent history of host acceptance by the insect. In a two-host situation, every oviposition on host i increases N_i by one and decreases N_j by one, with the constraint that neither N_i nor N_j can be less than zero. Thus, the more times an insect oviposits on a given host type, the greater the rate at which that host subsequently becomes acceptable. This pattern is portrayed by the learning curve shown in Fig. 8.2. For now, we assume that N_i and N_j do not change through time other than as a result of oviposition experience (that is, there is no forgetting other than resulting from experience with an alternative host).

Our model thus postulates that an individual learns to accept a host on which it has oviposited and forgets, to some extent, a preference for another host induced previously. An encounter that does not result in oviposition has no effect on P_{it} other than through increasing search time (t). Thus, rejecting a given host does to predispose a female to reject a subsequently encountered host of that type.

Although Equation 8.1 must be inadequate in many respects, it does incorporate the properties of learning and motivation outlined above. We explored the effects of learning on patterns of host use via numerical simulations, as outlined in Fig. 8.3. In these simulations, the relative abundances of hosts 1 and 2, D_1 and D_2, ranged from 0 to 0.5. A random

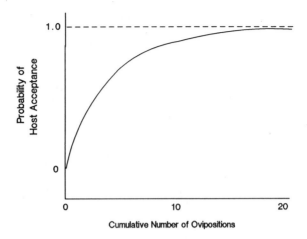

Figure 8.2. Learning changes the time-specific probability of host acceptance. Plotted is the probability of
acceptance of a specific host (one time unit after the
last oviposition) as a function of the cumulative
number of ovipositions on that host type.

number (X) between 0 and 1 was generated to determine which, if either,
host was encountered during a given time interval, as indicated below:

X	Host Encountered
$0 < X < D_1$	A
$D_1 < X < 1 - D_2$	neither
$X > 1 - D_2$	B

If either host was encountered, a second random number (Y) between 0
and 1 was generated to determine if the insect accepted that host for
oviposition. If $Y < P_{it}$, where i represents the host that has been encountered, then oviposition takes place, the values of N_1 and N_2 are adjusted,
and the search time reset to zero. If the host is rejected or if no host is
encountered during a given time interval, then t is increased by one and
the search resumed. The oviposition bout for a given insect is terminated
after 100 time units has elapsed. Except where noted, 1,000 such simulations were carried out for each set of parameters investigated. Time
constraints prevent us from studying the entire parameter space of variation in host abundance, innate preference, and learning ability. Thus,
below we have selected particular sets of parameter values to illustrate the
general manner in which learning and motivation affect patterns of host
use.

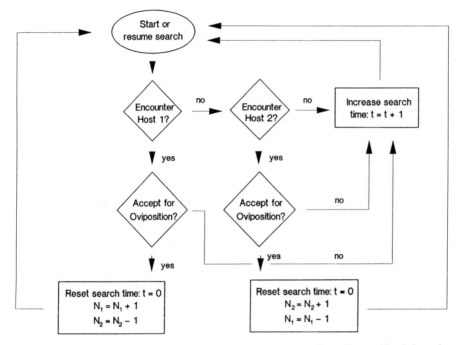

Figure 8.3. Flow diagram for simulation model incorporating effects of both learning and motivational state.

8.4. Learning and Host Use as a Function of Relative Host Abundance

Here, we let the motivation to oviposit for naive individuals be equal for the two hosts ($M_1 = M_2 = 0.1$). The total rate of encounter with both hosts is set at 0.3 per unit time, while their relative abundances are allowed to vary ($D_1 + D_2 = 0.3$). As expected, when previous experience has no effect on probability of host acceptance ($L = 0$), the mean fraction of eggs laid on a given host is simply equal to the relative abundance of that host (Fig. 8.4). When individuals learn, however, there is a sigmoidal relationship between host use and host density. Termed a Type III functional response (Holling, 1965), this sigmoidal pattern has been observed repeatedly in predator-prey and parasite-host systems and is commonly thought to arise via learning (Krebs, 1973). In that respect, our model is consistent with biological reality.

By fostering greater use of a given host (specifically, the host that is currently most abundant), learning should lead to increased selection on the physiological capacities of juvenile insects to develop on that host. As

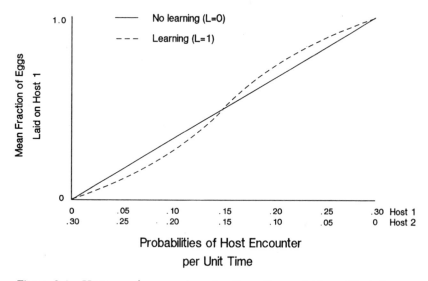

Figure 8.4. Host use (expressed as fraction of eggs laid on Host 1) as a function of learning ability (L) and relative host abundance when hosts are equally acceptable to naive individuals. Total probability of encountering a host plant per unit time ($D^1 + D_2$) was set at 0.3, and $M_1 = M_2 = 0.1$.

juveniles become physiologically adapted, host-seeking individuals (juvenile or adult) that genetically prefer this host will be favored in turn. If a host remains abundant for many generations, learning may thus accelerate both physiological and behavioral adaptation to this host. To the extent that both pysiological and behavioral adaptations involve plant chemistry, learning could influence chemically-mediated evolution of host use.

8.5. Learning, Genetic Variation, and Host Use

Genetic variation in oviposition responses to different hosts has now been found in a number of insects (reviewed in Jaenike and Holt, 1991). Here we illustrate how learning interacts with such variation to influence host use. Our argument is a simple one. If, during a search, the probability of acceptance increases more rapidly for host 1 than for host 2, then when the two hosts are equally abundant, an individual will be more likely to oviposit first on host 1. Learning will make that host even more acceptable subsequently. Our simulations confirmed this intuition (Fig. 8.5). By increasing the use of genetically preferred hosts, learning should amplify genetically-conferred differences among individuals in host preference.

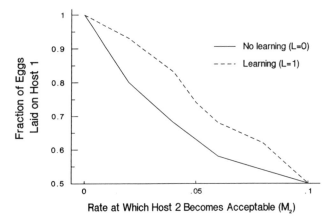

Figure 8.5. Host use as a function of learning ability (L) and variation in the motivational parameter (M) for Host 2. Both hosts were equally likely to be found ($D_1 = D_2 = 0.4$), and $M_1 = 0.1$.

Such augmentation could potentially facilitate the formation of host races within the insect population.

Because, through learning, individuals use a genetically-preferred host more than they would otherwise, competition for food among the offspring of individuals with genetically similar preferences may increase in severity. Such competition can, in theory, facilitate the maintenance of genetic variation for preference within a population (Rausher, 1984). Rather than viewing learning and genetically-programmed preferences as alternative routes to individual specialization, it is conceivable that one actually promotes the other.

We were surprised to discover that use of a genetically-preferred host did not increase monotonically with the magnitude of the learning effect L. This is illustrated in Fig. 8.6 for one set of parameter values: hosts equally abundant ($D_1 = D_2 = 0.4$), with a greater tendency of naive individuals to accept host 1 ($M_1 = 0.1$; $M_2 = 0.05$). Oviposition on the genetically-preferred host is actually maximized at intermediate values of L, about 1.25 in this case. It appears that at larger values of L, individuals tend to become locked onto whatever host they first accept. This host is occasionally the one that is genetically less preferred (Fig. 8.6).

At least within the context of the present model, even rather low values of L can significantly increase use of the genetically more acceptable host. It follows that learning should facilitate the formation of host races in sympatry as long as there is initially some genetic variation among individ-

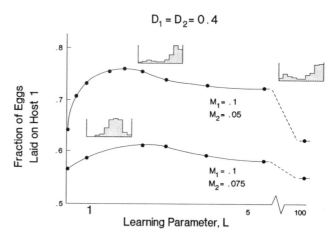

Figure 8.6. Host use as a function of learning ability (L) and difference in the motivational parameter (M) for the two hosts. Hosts were equally abundant ($D_1 = D_2 = 0.4$). $M_1 = 0.1$ and $M_2 = 0.05$ or 0.075. Insets denote frequency distributions of host use within the insect population. Note that proportional use of the genetically preferred host does not increase monotonically with learning ability.

uals in their responses to different hosts. Both learning and genetic variation account for variation in host acceptance in the apple maggot fly, *Rhagoletis pomonella* (Prokopy et al., 1982, 1988), which, as mentioned above, has recently been shown to comprise genetically differentiated apple and hawthorn host races. It is tempting to speculate that learning played a role in this divergence.

8.6. Learning and Genetic Variation: The Effect of Trade-offs

Up until now, we have assumed that an insect that has just learned something about host 1 will, as a consequence, forget part of what it learned about host 2. This phenomenon whereby learning one thing interferes with what is remembered with respect to something else is widespread in animals (Papaj, in press and references within). Referred to simply as "interference," this memory constraint represents a kind of trade-off associated with learning. To examine how such a trade-off might affect genetically-based host preference, we modified our basic model such that the response to a given host was no longer a function of the history of encounters with *both* hosts. Specifically, we supposed that experience with

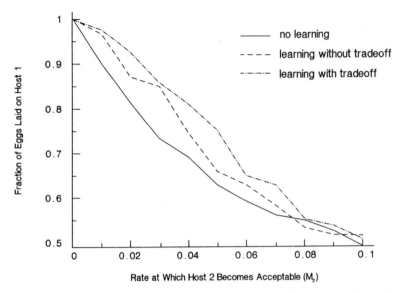

Figure 8.7. Host use when learning involves trade-offs as a function of variation in the motivational parameter (*M*) for Host 2. Parameters as in Fig. 8.5.

one host would increase the probability of accepting that host while experience with the other host would leave that same probability unchanged. In the terms of the model, every oviposition on host i now increases N_i by one but leaves N_j unchanged.

The results of simulations with and without trade-offs are shown in Figure 8.7. When both hosts are equally abundant, learning generally augments a genetically-based preference for a host even in the absence of trade-offs. However, the occurrence of a trade-off in the form of memory constraints generally augments a genetically-based preference further still. By inference, learning which involves interference should facilitate genetic subdivision and host race formation more than learning which does not.

Host race formation aside, this result has significance for our understanding how diet breadth evolves in host-specific insects. Trade-offs have been proposed as one explanation for the extensive degree of host specialization observed in host-specific insects (see Rausher, this volume). Advocates of this proposition have made two implicit assumptions: (1) the trade-off involves a physiological trait, such as processing of toxic secondary chemicals and (2) the trade-off is genetically-based, that is, a genotype that performs well on one host necessarily performs worse on the other. Our results provide an alternative. First, the trade-off discussed here involves not a physiological trait but a behavioral one: an insect that learns to accept one host forgets something about accepting the other host. Second, the trade-off is not genetically-based, but is generated through experience with the host environment and so is environmentally-based. Despite their environmental basis, trade-offs due to learning should nevertheless promote selection for adaptation to a particular host.

8.7. Visual Learning and Its Effects on Chemically-Mediated Host Use

Thus far, we have supposed that the probability of accepting a host, P_{it}, depends only on host chemistry. Even if the parameter M reflects a response to chemical stimuli, however, the effect of experience on host acceptance may conceivably reflect a change in response to some other kind of stimulus. We might imagine, for example, that an inexperienced insect responds only to chemical cues, but after experience, comes to respond to tactile or visual stimuli. Exactly this sort of phenomenon characterizes host selection by ovipositing butterflies. After a butterfly has landed on a host plant, oviposition is elicited by contact with chemical constituents of the plant surface (cf. Feeny et al., 1983 and Feeny, 1990). Experience may or may not affect the response to plant chemistry, but it definitely affects responses to plant color and form. As a consequence of oviposition (or even contact with a host plant), females learn visual properties of their host, notably the color or intensity of the host surface (Traynier, 1984, 1986) or the shape of its leaves (Papaj, 1986). In search of subsequent oviposition sites, experienced females are more likely to land on plants of the appropriate color or leaf shape than on other plants.

To model this phenomenon, we break down the probability of accepting a host given encounter into two components:

$$P_{it}\,(\text{acceptance}|\text{encounter}) = P_{it}(\text{landing}|\text{encounter})$$

$$\times P_{it}\,(\text{acceptance}|\text{landing}) \quad (8.2)$$

Both components on the right-hand side of the equation are assumed susceptible to the effects of learning (see Equation 8.1). We assume that the probability of landing on host i, given encounter at time t, reflects a response to a visual cue (e.g., leaf color or shape) and that the probability of accepting a host for oviposition having landed on it depends on a chemical cue. Finally, we assume that only responses to host chemistry are genetically variable.

Fig. 8.8 shows the results of a simulation in which we assumed that experience affected: (1) the response to the chemical cue only; (2) the

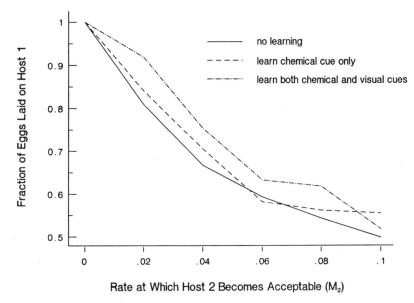

Figure 8.8. Host use as a function of the number of different cues learned and variation in the motivational parameter (*M*). Variation in M_2 is based solely on chemical cues. Parameters as in Fig. 8.5.

response to both chemical and visual cues; or (3) neither component of host selection. As before, learning to respond to the chemically-based host stimuli of a host augments preference for the genetically-preferred host. Note, however, that learning to land in response to a visual cue further augments preference for the genetically-preferred host.

These results demonstrate that differences in host use that are chemically-mediated and genetically-based may be increased by learned responses to non-chemical stimuli. It follows that for insects that learn non-chemical cues, a relatively small genetic change in responses to host chemistry may generate a large change in overall host use (a point made in less formal terms by Rausher [1986] and Papaj [1986]). In this way, learning visual or tactile cues can potentially accelerate natural selection for use of a novel host species, even when the barriers these species pose to the insect are largely chemical in nature.

We now present a possible scenario for a host shift or expansion of host range in a host-specific insect like a butterfly. A female butterfly that has not oviposited recently lands on a non-host plant whose chemistry is similar in certain respects to that of the host. Due partly to her heavy egg load and partly to the chemical resemblance, she "mistakenly" lays eggs on that plant and learns its color or leaf shape. As a consequence, she continues to find and lay eggs on the non-host species. To the extent that her initial propensity for mistakes is genetically-based, any daughters surviving on the non-host might make similar "mistakes" as adults. At the same time, there will be increased selection on the physiological capacities of larvae to develop on that plant. As larvae become physiologically adapted, host-seeking individuals (larval or adult) that genetically are more likely to accept the novel host will be favored in turn.

Mistakes like those supposed above are thought to play a key role in the evolution of host use in butterflies. With respect to swallowtail butterflies (Family Papilionidae), for example, Feeny (1990) postulated that mistakes made by individual females are a critical prerequisite for host shifts and expansions of host range observed at the level of population or species. In our scenario, both learning and motivation can promote mistakes. Without becoming motivated to lay eggs on a usually unacceptable plant, the butterfly would never make a mistake in the first place; without learning the characteristics of the non-host plant, the mistake might not be propagated. In short, the overall frequency of oviposition mistakes may be multiplied greatly through the combined action of motivation and learning. This multiplicative process may account for the rapid expansion of host breadth observed in historical times for insects as diverse as Baltimore checkerspot butterflies and apple maggot flies (Prokopy et al. 1990).

8.8. A Caveat: The Importance of Relative Host Abundance

We have seen that learning enhances use of a genetically preferred host when the two hosts are equally abundant and that learning enhances use of the more abundant of two hosts that are genetically equally acceptable. We now consider how host use is affected when one host is genetically more acceptable but rarer than the other. Fig. 8.9 shows, over a range of relative host abundances, how learning affects host use when one host is genetically-preferred over the other. As expected, learning causes an increase in the use of the genetically-preferred host when it is the more abundant of the two. When this host is substantially rarer than the other, however, individuals that learn actually use less of that host than individuals that do not learn. Here, then, is a major qualification of our earlier predictions: when the genetically-preferred host is relatively rare, learning does not necessarily increase and may even decrease its use. If that host remains rare over many generations, learning may actually accelerate both physiological and behavioral adaptation to the originally less-preferred host. Learning may thereby expedite a shift in overall host preference in a population, but will probably not promote genetic subdivision among members of that population. In fact, it might inhibit subdivision by

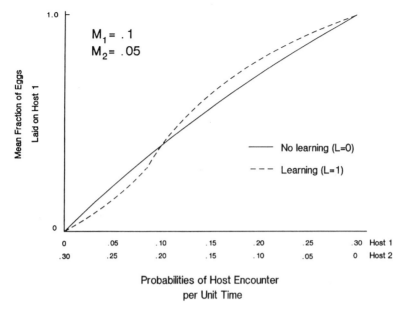

Figure 8.9. Host use as a function of learning ability (L) and relative host abundance when hosts are not equally acceptable to naive individuals. Total probability of host encounter per unit time ($D_1 + D_2$) = 0.3 M_1 = 0.1 and M_2 = 0.05.

ensuring that all individuals, regardless of genetically-based propensities, allocate most of their eggs to one host, that is, the more abundant one.

Consider now what happens when relative host abundance fluctuates through time or varies in a coarse-grained manner spatially. If the degree of fluctuation is moderate, with D_1/D_2 alternating between 1/2 and 2, then individuals that learn ($L = 1$) use slightly more (61%) of the genetically preferred host than do those that do not learn (60%). However, if the fluctuations are more extreme, with D_1/D_2 alternating between 1/14 and 14, then learning actually reduces use of the genetically preferred host from 54% to 51%, even though, over the long run, both hosts are equally abundant. This result should caution against assuming that learning necessarily increases use of a genetically preferred host, even when it is as abundant, on average, as the less preferred host.

8.9. Conclusions

The details of our results depend, of course, on the specific formulation of the probability that an insect will oviposit on a host that it has encountered. The simplicity of our models and their basis on what we believe to be biologically realistic assumptions give us confidence that the following conclusions may be fairly general.

1. Even modest effects of learning can have substantial effects on patterns of host use by insects.

2. By increasing the relative use of abundant hosts, learning will speed up the rate of behavioral and physiological adaptation to such plants.

3. When alternative hosts are equally abundant, learning will usually reinforce genetically-based differences in host preference and thereby facilitate the formation of host races within populations. Host race formation may be facilitated most when the effect of learning is rather moderate.

4. Phenotypic trade-offs generated via memory constraints can potentially augment genetically-based preferences and further facilitate host race formation and adaptation to preferred hosts.

5. Learning visual or other non-chemical cues can promote evolutionary change in responses to host chemistry. This change may affect adaptation to a particular host, genetic subdivision within the insect population, or both.

6. When a genetically preferred host is rare relative to other potential hosts, learning can reduce the use of this host and thus slow the rate at which insects adapt to it.

What lies ahead? In order to understand better the effects of learning and motivation, it is important to get estimates of the various parameters that affect the probability of host acceptance. We believe that experiments carried out along the lines of those pioneered by Singer (1982) may be most effective. Specifically, how do variables such as search time since last oviposition, host type encountered, insect genotype, and history of previous ovipositions affect the likelihood that an insect lays eggs on a host? Answers to these questions are likely to shed considerable light on evolutionary changes in host use in insects.

Acknowledgements

This work was supported in part by NSF grant BSR-8905399 to J. Jaenike. D. R. Papaj was supported by a Center of Insect Sciences postdoctoral traineeship at the University of Arizona. We thank M. Singer for comments on the manuscript.

References

Bernays, E. A. and M. Graham. 1988. On the evolution of host specificity in phytophagous arthropods. *Ecology* 69:886–892.

Buckle, G. R. and L. Greenberg. 1981. Nestmate recognition in sweat bees (*Lasioglossum zephyrum*): does an individual recognize its own odour or only odours of its nestmates? *Anim. Behav.* 29:802–809.

Bush, G. L. 1974. The mechanism of sympatric host race formation in the true fruit flies (Tephritidae). In *Genetic Mechanisms of Speciation in Insects*, ed. M. J. D. White, pp. 3–23. Australian and New Zealand Book Co., Sydney.

Charnov, E. L. and S. W. Skinner. 1984. Evolution of host selection and clutch size in parasitoid wasps. *Fla. Entomol.* 67:5–21.

Courtney, S. P., G. K. Chen, and A. Gardner. 1989. A general model for individual host selection. *Oikos* 55:55–65.

Dethier, V. G. 1964. Microscopic brains. *Science* 143:1138–1145.

Dethier, V. G. 1970. Chemical interactions between plants and insects. In *Chemical Ecology*, eds. E. Sondheimer and J. B. Simeone, pp. 83–102. Academic Press, New York.

Diehl, S. R. and G. L. Bush. 1984. An evolutionary and applied perspective of insect biotypes. *Annu. Rev. Entomol.* 29:471–504.

Feder, J. L., C. A. Chilcote, and G. L. Bush. 1988. Genetic differentiation between sympatric host races of the apple maggot fly *Rhagoletis pomonella*. *Nature* 336:61–64.

Feder, J. L., C. A. Chilcote, and G. L. Bush. 1990. The geographic pattern of genetic differentiation between host associated populations of *Rhagoletis pomonella* (Diptera: Tephritidae) in the eastern United States and Canada. *Evolution* 44:570–594.

Feeny, P. P. 1990. Chemical constraints on the evolution of swallowtail butterflies. In *Herbivory: Tropical and Temperate Perspectives*, eds. P. W. Price, T. M. Lewinsohn, W. W. Benson, and G. W. Fernandes. John Wiley, New York (in press).

Feeny, P., L. Rosenberry, and M. Carter. 1983. Chemical aspects of oviposition behavior in butterflies. In *Herbivorous Insects*, eds. S. Ahmad, pp. 27–76. Academic Press, New York.

Gould, J. L. 1984. Natural history of honey bee learning. In *The Biology of Learning*, eds. P. Marler and H. Terrace, pp. 149–180. Springer-Verlag, Berlin.

Heinrich, B. 1979. 'Majoring' and 'minoring' by foraging bumblebees, *Bombus vagans*: an experimental analysis. *Ecology* 60:245–255.

Holling, C. S. 1965. The functional response of predators to prey density and its role in mimicry and population regulation. *Mem. Entomol. Soc. Can.*, No. 45.

Iwasa, Y., Y. Suzuki, and H. Matsuda. 1984. Theory of oviposition strategy of parasitoids. I. Effect of mortality and limited egg number. *Theor. Pop. Biol.* 26:205–227.

Jaenike, J. 1978. On optimal oviposition behavior in phytophagous insects. *Theor. Pop. Biol.* 14:350–356.

Jaenike, J. and R. D. Holt. 1991. Genetic variation for habitat preference: evidence and explanations. *Am. Nat.* 137:567–590.

Jermy, T. 1984. Evolution of insect/host plant relationships. *Am. Nat.* 124:609–630.

Jermy, T. 1987. The role of experience in the host selection of phytophagous insects. In *Perspectives in Chemoreception and Behavior*, eds. R. F. Chapman, E. A. Bernays, and J. G. Stoffolano, pp. 143–157. Springer-Verlag, New York.

Krebs, J. R. 1973. Behavioral aspects of predation. In *Perspectives in Ethology*, Vol. 1, eds. P. P. G. Bateson and P. H. Klopfer pp. 73–109. Plenum Press, New York.

Levins, R. and R. H. MacArthur. 1969. An hypothesis to explain the incidence of monophagy. *Ecology* 50:910–911.

Lewis, A. C. 1986. Memory constraints and flower choice in *Pieris rapae*. *Science* 232:863–865.

Mangel, M. 1987. Oviposition site selection and clutch size in insects. *J. Math. Biol.* 25:1–22.

Mangel, M. and C. W. Clark. 1988. *Dynamic Modeling in Behavioral Ecology*. Princeton University Press, Princeton, NJ.

Mangel, M. and B. D. Roitberg. 1989. Dynamic information and host acceptance by a tephritid fruit fly. *Ecol. Entomol.* 14:181–189.

McPheron, B. A., D. C. Smith, and S. H. Berlocher. 1988. Genetic differences between *Rhagoletis pomonella* host races. *Nature* 336:64–66.

Papaj, D. R. 1986. Conditioning of leaf-shape discrimination by chemical cues in the butterfly, *Battus philenor*. *Anim. Behav.* 34:1281–1288.

Papaj, D. R. 1990. Interference with learning in pipevine swallowtail butterflies: behavioural constraint or possible adaptation? *Symp. Biol. Hung.* 39:89–101.

Papaj, D. R and M. D. Rausher. 1983. Individual variation in host location by phytophagous insects. In *Herbivorous Insects: Host-seeking Behavior and Mechanisms*, ed. S. Ahmad, pp. 77–124. Academic Press, New York.

Papaj, D. R. and R. J. Prokopy. 1986. Phytochemical basis of learning in *Rhagoletis pomonella* and other herbivorous insects. *J. Chem. Ecol.* 12:1125–1143.

Papaj, D. R. and R. J. Prokopy. 1988. The effect of prior adult experience on components of habitat preference in the apple maggot fly (*Rhagoletis pomonella*). *Oecologia* 76:538–543.

Papaj, D. R. and R. J. Prokopy. 1989. Ecological and evolutionary aspects of learning in phytophagous insects. *Annu. Rev. Entomol.* 34:315–350.

Papaj, D. R., S. B. Opp, R. J. Prokopy, and T. T. Y. Wong. 1989. Cross-induction of fruit acceptance by the medfly *Ceratitis capitata*: the role of fruit size and chemistry. *J. Insect Behav.* 2:241–254.

Parker, G. A. and S. P. Courtney. 1984. Models of clutch size in insect oviposition. *Theor. Pop. Biol.* 26:27–48.

Pfennig, D. W., G. J. Gamboa, H. K. Reeve, and J. Shellman Reeve. 1983. The mechanism of nestmate discrimination in social wasps (Polistes, Hymenoptera: Vespidae). *Behav. Ecol. Sociobiol.* 13:299–305.

Prokopy, R. J., A. L. Averill, S. S. Cooley, and C. A. Roitberg. 1982. Associative learning in egglaying site selection by apple maggot flies. *Science* 218:76–77.

Prokopy, R. J., S. R. Diehl, and S. S. Cooley. 1988. Behavioral evidence for host races in *Rhagoletis pomonella* flies. *Oecologia* 76:138–147.

Pulliam, H. R. 1981. Learning to forage optimally. Pp. 379–388 in A. C. In *Foraging Behavior*. eds. A. C. Kamil and T. D. Sargent, pp. 379–388. Garland, New York.

Rausher, M. D. 1984. The evolution of habitat selection in subdivided populations. *Evolution* 38:596–608.

Rausher, M. D. 1986. Variability for host preference in insect populations: mechanistic and evolutionary models. *J. Insect Physiol.* 31:873–889.

Rice, W. R. 1987. Speciation via habitat specialization. *Evol. Ecol.* 1:301–314.

Rice, W. R. and G. W. Salt. 1988. Speciation via disruptive selection on habitat preference: experimental evidence. *Am. Nat.* 131:911–917.

Schlichting, C. D. 1986. The evolution of phenotypic plasticity in pants. *Ann. Rev. Ecol. Syst.* 17:667–693.

Singer, M. C. 1982. Quantification of host preference by manipulation of oviposition behavior in the butterfly *Euphydryas editha*. *Oecologia* 52:230–235.

Stearns, S. C. 1989. The evolutionary significance of phenotypic plasticity. *BioScience* 39:436–445.

Thorpe, W. H. 1930. Biological races in insects and allied groups. *Biol. Rev.* 5:177–212.

Thorpe, W. H. 1945. The evolutionary significance of habitat selection. *J. Anim. Ecol.* 14:67–70.

Tompkins, L., R. W. Siegel, D. A. Gailey, and J. C. Hall. 1983. Conditioned courtship in *Drosophila* and its mediation by association of chemical cues. *Behav. Genet.* 13:565–578.

Traynier, R. M. M. 1984. Associative learning in the ovipositional behavior of the cabbage butterfly, *Pieris rapae*. *Entomol. Exp. Appl.* 9:465–472.

Traynier, R. M. M. 1986. Visual learning in assays of sinigrin solution as an oviposition releaser for the cabbage butterfly, *Pieris rapae*. *Entomol. Exp. Appl.* 40:25–33.

Vet, L. E. M. and K. van Opzeeland. 1984. The influence of conditioning on olfactory microhabitat and host location in *Asobara tabida* (Nees) and *A. rufescens* (Foerster) (Braconidae:Alysilinae) larval parasitoids of Drosophilidae. *Oecologia* 63:171–177.

Via, S. 1988. Genetic constraints on the evolution of phenotypic plasticity. In *Genetic Constraints on Adaptive Evolution*, ed. V. Loeschke, pp. 47–71. Springer-Verlag, Berlin.

Via, S. and R. Lande. 1985. Genotype-environment interaction and the evolution of phenotypic plasticity. *Evolution* 39:505–522.

Waddington, C. H. 1953. Genetic assimilation of an acquired character. *Evolution* 7:118–126.

Wasserman, S. S. and D. J. Futuyma. 1981. Evolution of host plant utilization in laboratory populations of the southern cowpea weevil, *Callosobruchus maculatus* Fabricius (Coleoptera:Bruchidae). *Evolution* 35:605–617.

Wcislo, W. T. 1989. Behavioral environments and evolutionary change. *Ann. Rev. Ecol. Syst.* 20:137–169.

West-Eberhard, M. J. 1989. Phenotypic plasticity and the origins of diversity. *Ann. Rev. Ecol. Syst.* 20:249–278.

Wright, S. 1931. Evolution in Mendelian populations. *Genetics* 16:97–159.

9

Evolution of Sex Pheromones and The Role of Asymmetric Tracking

P. L. Phelan

The Ohio State University

> On the morning of the 6th of May a female emerged from her cocoon in my presence on my laboratory table...I had no particular intentions regarding her; I imprisoned her from mere habit; the habit of an observer always on the alert for what may happen. I was richly rewarded. About nine o'clock that evening when the household was going to bed, there was a sudden hubbub in the room next to mine. Little Paul, half undressed, was rushing to and fro, running, jumping, stamping, and overturning chairs as if possessed. I heard him call me. "Come quick!" he shrieked; "come and see these butterflies! Big as birds! The room's full of them!"...It was a memorable sight—the Night of the Great Peacock! Come from all points of the compass, warned I know not how, here were forty lovers eager to do homage to the maiden princess that morning born in the sacred precincts of my study.
>
> J. H. Fabré (1911)
> *Social Life in the Insect World*

9.1. Introduction

Since the earliest observations of insect mate-finding by Fabré and others, the power of female insects to lure males has astounded biologists. From his observations and simple experiments, Fabré was so impressed by the distances over which males must be attracted that he rejected the idea that this response could be mediated by a chemical odor which he himself could not detect. The absurdity of such an idea would be like suggesting that one could "redden a lake with a grain of carmine; to fill a space with a mere nothing." Rather than molecular emanations, he concluded that moth olfaction must take place in a domain different from ours, which he termed "etheric waves," whose relation to odor particles he likened to that between X-rays and visible light. Although his conclusion was wrong, Fabré's analogy of a grain of dye in a lake, as much as it strains credulity, was correct, albeit a small lake. Even though today we understand the mechanisms of chemically-mediated sexual communication to a much greater degree than ever imagined by Fabré and his contemporaries, wonderment has not diminished. In fact, the ever-growing body of information concerning sex pheromones has only heightened our appreciation of the sensitivity and selectivity of pheromone-receptor systems and of the complexity of integration among various sensory and motor systems necessary for the deceptively simple behavior of odor-modulated mate-finding.

I recognize the challenges of discussing a subject that has been so extensively addressed, and the reader is referred to Cardé and Baker (1984), Baker (1985), Cardé (1986), and Löfstedt (1990) for some excellent discussions of the evolutionary aspects of insect sex pheromones. Many of the issues addressed here have been discussed by those authors, as well. However, work during recent years, particularly with regard to inter-individual variance and the genetic inheritance of sex-pheromone systems, allows us to better test and extend the hypotheses previously proposed concerning the evolution of sex pheromones in insects. Although a desire to better control pest insects has been the primary driving force behind the research of insect pheromones, we have amassed a considerable volume of information that will allow us to address important fundamental issues as well, such as the evolution of mating systems, communication, and speciation. Particularly with the Lepidoptera, there probably is no better-studied mate-communication system in any other animal group, with exceptional opportunities to test ideas concerning these evolutionary questions. Yet this sexual-communication database has remained largely untapped by evolutionary biologists outside the field of insect pheromones. In this chapter, I shall use the insect-pheromone literature to: (1) challenge some widely-held theoretical frameworks concerning the evolution of mate signalling in general, and (2) discuss in detail male lepidopteran courtship pheromones, which have received only limited attention and which provide an informative contrast to their female counterparts for understanding the evolution of sex pheromones.

9.2. Definitions

9.2.1. Types of Evolution: Adaptive versus Nonadaptive

Due to a prevalent underlying assumption that natural selection acts as an optimizing agent, most work has emphasized the adaptive significance of traits when attempting to understand the means by which they have evolved. As a result, nonadaptive forces, such as developmental and phyletic constraints, pleiotropy, and genetic drift, are frequently overlooked as potentially important evolutionary factors (see Rausher, this volume). Also significant are cases of changes due to selection on some aspect of a mate-signalling system, but not for the purpose of reproductive efficiency. Similarly, models of speciation based on founder effect emphasize that differences in mating characters arise not as a means to effect reproductive isolation, but rather result from changes in gene frequency due to stochastic factors (Carson and Templeton, 1984).

On the other hand, the value of the adaptationist approach, as argued by Williams (1966) and Alcock (1989), is in generating testable hypotheses. One can then determine through comparative study and experimentation the relative significance of the roles played by adaptive and nonadaptive processes. From an adaptationist perspective, one would predict that a communication system generally would evolve to encode *only* as much information as is necessary; however, stochastic processes may counter this expectation. There may be selection on some other trait with which aspects of the communication system are genetically linked, or a reduction in the genetic variability of the communication system in an isolated portion of the species due to founder effect. Measuring information as the uncertainty in a system (Shannon and Weaver, 1949), there would be an overall increase in the information content in both examples because a response may be elicited by a signal in one portion of the population, but not in the other.

9.2.2. Types of Selection: Natural versus Sexual

More than a little confusion has resulted from the lack of a clear delineation of sexual selection. In spite of Darwin's own apparent distinction between natural and sexual selection (Darwin, 1871), most writers today consider the latter to be only a subset of natural selection. One exception to this is Arnold (1983), who views Darwin's concept of sexual selection "as arising from variance in mating success and natural selection as arising from variance in all other components of total fitness." The primary motivation for distinguishing the two forms is "that structures that confer mating success may hinder the male in the struggle for survival" (Arnold, 1983). However, the direction of different forms of natural selection are also frequently in conflict, leading to some intermediate condition for a character that is not optimal for either parameter. In this sense, sexual selection is not *qualitatively* distinct from natural selection. With regard to the evolution of mate-signalling systems, the predicted outcomes of natural and sexual selection will at times be in conflict and at times be confounded. For example, a possible explanation for the extraordinary sensitivity of the males of many moths to female sex pheromone is based on intersexual selection, as females producing less pheromone would select for more sensitive males which, assuming a genetic basis, would be more likely to produce pheromone-sensitive sons (Greenfield, 1981; Cardé and Baker, 1984). Natural selection *sensu stricto* also would favor a female producing small quantities of pheromone if it allowed her to evade detection by predators (Greenfield, 1981) and would favor those males more efficient in finding females, that is, males that expend less time and energy in searching, and that reduce the probability of predation.

In such cases, it may be very difficult to partition the relative contribution of these two forces. On the other hand, it *is* important to draw attention to the dynamics of character spread and elaboration possible through sexual selection that make it unique from natural selection. Fisher (1930) was the first to point out that when there is variance in the mating success among members of one sex that is due to a preference for some character by members of the other sex (intersexual selection), there is the potential for rapid evolution of that character due to a positive-feedback selection between the character and the degree of preference for that character. A multitude of quantitative genetic models attest to the feasibility of such a process of runaway sexual selection under a variety of conditions (reviewed by Arnold, 1987). Thus, in terms of population genetics, this subset of sexual selection has no equal under natural selection. As I shall argue below, subsuming sexual selection under natural selection has led many evolutionary biologists to overlook the potential for adaptive divergence of mate signalling within the context of reproductive isolation.

9.2.3. Level of Selection: Group / Species versus Individual

Although the framework of group- or species-level selection largely has been repudiated as a major evolutionary force (Williams, 1966), it may still be encountered in discussions of sexual behavior and reproductive isolation. Selection at the population level, except under certain restricted conditions (Wilson, 1975), is inconsistent with the basic tenets of Darwinian evolution and with our present understanding of population genetics. The Achilles' heel of any evolutionary argument based on group selection lies in the fact that evolution will normally proceed more rapidly at the individual level and that "selfish" traits will spread more rapidly than those that benefit the group at the expense of the individual, unless special circumstances counteract this tendency. The issue has been specifically addressed for the evolution of bark-beetle pheromones (Alcock, 1982), but the point is worth reiterating since the problem persists. For example, Borden's (1985, pp. 258–259) argument concerning the evolution of the aggregation pheromones found in many nonsocial insects is based largely on species-selectionist thinking: "In many coleopterans, aggregation pheromones have apparently evolved as indicators of a potentially suitable food source or habitat.... Stored-grain insects, as well as bark and timber beetles, often form aggregations in scarce or isolated habitats, *e.g.*, collections of stored food or moribund trees. It is of selective advantage to utilize these rare habitats maximally." It is hard to deduce the selective advantage for a stored-product insect who has already located the resource in producing a pheromone that aids other individuals in locating the rare resource.

Similarly, the debate concerning the evolution of reproductive isolation has at times been muddled by arguments (however unintentional) presented from the perspective of species advantage. For example, Spiess (1987, p. 105) states, "Positive assortative mating must then be established for efficiency of species reproduction and for lessening or preventing the mixing of populational adapted genomes." The question of whether reproductive isolation is brought about as a result of selection for assortative mating or as an incidental outcome of other processes is presently an open one. However, to make this debate meaningful, discussions of the adaptive selection for reproductive isolation must be based on the benefit that may be accrued to individuals. Individual-selectionist thinking opens us to the likelihood for within-species competitive interactions, leading to a variety of different strategies for maximizing reproductive success. As we shall see, this difference in strategies particularly between the sexes likely has been a major force in the evolution of sexual communication and does much to explain the variety of mate-communication systems that we observe.

9.3. Selective Forces in The Evolution of Sexual Communication

9.3.1. Differential Parental Investment

It is widely believed that when two organisms engage in sexual reproduction, there will naturally develop an asymmetry in the relative size of gametes (anisogamy), which is the initial stage in the potential evolution of greater asymmetry in the contribution toward offspring (Stearns, 1987). Trivers (1972) significantly advanced the discussion of sexual selection by extending the notion of sexual asymmetry due to anisogamy with the concept of parental investment, which includes "any investment by the parent in an individual offspring that increases the offspring's chance of surviving at the cost of the parent's ability to invest in other offspring." From the perspective of individual selection, it is clear that differences in parental investment between the sexes will give rise to different strategies for maximizing reproductive fitness. If one sex invests significantly more than the other sex in an individual offspring, the reproductive fitness of that parent will be limited by the resources available for provisioning offspring. This sex, typically the female, then will maximize its reproductive fitness by a strategy that stresses the quality of mate and/or the resources made available through mating. In contrast, the sex that invests little in its offspring maximizes its success by engaging in as many matings as possible, with little regard to the quality of the mate.

It is easy to see how this difference in mating strategies would extend to mate communication. In fact, I would argue that differential parental investment represents the most significant factor in the evolution of mate-signalling systems, acting as the evolutionary arena within which all other adaptive and nonadaptive forces will have their effect. For example, one would predict that the limiting sex that invests more in offspring will invest less in mate finding. The ubiquitous nature of female-emitted sex pheromones in moths to which males respond from a distance is a forceful demonstration of this argument (Greenfield, 1981). The production of nanogram quantities of pheromone by females is obviously less expensive than searching for a mate, the latter comprising costs related both to locomotion and to increased predation. In contrast to female pheromone production in moths, in species using long-distance auditory communication, the male is typically the signal producer. This role reversal again is consistent with the high energy expenditure of sound production relative to chemical communication (Alcock, 1989) and the considerable susceptibility of the auditory modality to "eavesdropping" by predators and parasites (Dicke and Sabelis, this volume).

9.3.2. Coordination of Pheromone Production and Sexual Response

9.3.2.1. Signal-Response Coevolution

Another force generally cited in the evolution of mate-signalling systems is the requirement for a coordinated signal and response, leading to reciprocal stabilizing selection between the signaller and responder (see McNeil, this volume). For moths, this stabilizing selection is seen to favor "those males most sensitive to the blend emitted by the majority of females, and those females emitting the blend to which males are most sensitive" (Cardé and Baker, 1984). "Strong stabilizing selection" frequently is invoked to explain the highly canalized ratio of pheromonal components (Glover et al., 1990). Understanding the evolution of a coordinated mate-signalling system is particularly germane to understanding the process of speciation. If the mating signal and response are coevolved by mutual stabilizing selection, then the establishment of a new signal-response system is made difficult since both would have to evolve concurrently to establish a new population norm. Paterson (1985) points to this expectation as evidence against the feasibility of mate signals diverging during speciation by reinforcement as envisaged by Dobzhansky (1940). According to Paterson (1980), "In sexual species the achievement of syngamy is of fundamental significance. It is therefore not surprising that all sexual organisms possess adaptations which have evidently evolved to facilitate the achieving of fertilization." Paterson concludes that as a subset of these adaptations, the specific-mate recognition system (SMRS) is under strong stabilizing selection, and thus "adaptive change can only occur in small steps with coadaptation between male and female being re-established at each step."

In this framework, changes in the mate-signalling system due to hybrid matings will be resisted by two factors. First, any directional change in the system will be buffered by stabilizing selection for the population norm resulting from coadaptation of signal and response. Thus, the mate-signalling system of a species is inherently homeostatic (Lambert et al., 1987). Secondly, the hypothesized strong selection will reduce the inherent variability in the mate-signalling system, which is necessary for directional selection to act. Thus, a coevolved signal-response system should show low variance both within and among populations of a species. Lambert et al. (1987) used the literature of insect sex pheromones to test the concept that the SMRS is under strong mutual stabilizing selection and found substantial support in the observation that "pheromones are generally highly specific and show low compositional and proportional variability." They also point to examples of heterospecific cross-attraction in the field as favoring the concept, since "the SMRS is not viewed as a mechanism moulded by natural selection for the efficient prevention of interbreeding between species." More convincing demonstration of the potential stability of sex pheromones was provided by Haynes and Baker (1988), who measured ratios and amounts of ZZ-/ZE-7,11-hexadecadienyl acetate (ZZ/ZE-7,11-16:Ac) produced by *Pectinophora gossypiella* from seven locations around the world. Emissions from females of the seven populations only ranged from 57.5% $\pm$ 4.0(SD) to 61.5% $\pm$ 2.3 ZZ (Table 9.1).

Table 9.1. Variation in ratio of $(Z,Z){:}(Z,E)$-7,11-hexadecadienyl acetate and emission rates of *Pectinophora gossypiella* females from populations around the world. Means within a column followed by the same letter are not significantly different.[*]

Population	%Z,Z $\pm$ S.D.	Z,Z(ng/min) $\pm$ S.D.
United States	61.5 $\pm$ 2.3a	0.119 $\pm$ 0.043a
Mexico	61.9 $\pm$ 3.5a	0.085 $\pm$ 0.046bc
Brazil	62.9 $\pm$ 3.0a	0.082 $\pm$ 0.043bc
Argentina	63.0 $\pm$ 2.4a	0.091 $\pm$ 0.038ab
Egypt	61.5 $\pm$ 2.4a	0.103 $\pm$ 0.041ab
Pakistan	63.1 $\pm$ 2.1a	0.092 $\pm$ 0.036ab
China	57.5 $\pm$ 4.0b	0.057 $\pm$ 0.035c

[*]From Haynes and Baker (1988).

In contrast, traps baited with ratios ranging from 40:60 to 70:30 and deployed in the Coachella Valley of California all caught significant numbers of males, so that the worldwide level of variation in pheromone production appears very low.

9.3.2.2. Genetic Linkage

An alternative to the coevolution model explains the congruence of signal and response as a result of genetic coupling (Alexander, 1962; O'Donald, 1962; Hoy et al., 1977). Genetic linkage can result from two phenomena: pleiotropy and/or linkage disequilibrium (Kirkpatrick, 1987; Rausher, this volume). The genetic-coupling model bypasses the problem cited by Paterson, since if signal and response are genetically linked, then a change in one would automatically be accompanied by a corresponding change in the other, allowing a rapid shift in the mate-signalling system. This would be true whether the change came about as a response to selective forces or by stochastic events and could theoretically lead to rapid isolation of a family of individuals from the original population. Linkage due to pleiotropy may be illustrated by examples of acoustic communication in which the carrier frequency of male calls and the tuning of the female auditory receptor may each be inversely correlated with body size (Ewing, 1989). Such circumstances could potentially lead to size-related assortative mating, with possible divergence in the SMRS in populations where environmental conditions favor larger individuals, in the manner suggested by Paterson (1978). Similarly, production and perception of an acoustic signal might be coupled by some common feature, such as central pattern generators (Doherty and Hoy, 1985).

I am not aware of any similar examples of signal-response linkage caused by pleiotropy in the pheromone literature. This may be because pheromone biosynthesis and the mechanisms of chemoreception are less likely both to be correlated by chance with some third trait. In a possible case of linkage disequilibrium, Grula and Taylor (1979) determined that in the sulphur butterfly, *Colias eurytheme*, two male-produced sexual signals of different modalities are inherited on the X chromosome: an ultraviolet (UV) reflectance pattern specific to wings of male *C. eurytheme*, and the predominant pheromone component, 13-methylheptacosane. Subsequently, genetic control of female mate selection also was found to be located on the X chromosome (Grula and Taylor, 1980), and they argued that the courtship communication systems of *C. eurytheme* and the closely related *C. philodice* are under the control of a coadapted gene complex or "supergene" residing on the X chromosome, although some aspects are also controlled autosomally.

However, a review of the studies of inheritance in acoustic communication suggests that genetic coupling of signal and response mechanisms is certainly not a universal phenomenon (Doherty and Hoy, 1985; Ewing, 1989), and parallel studies in pheromonal systems have also produced results inconsistent with this model. The pheromone system whose genetic architecture is best understood to date is that of the European corn borer,

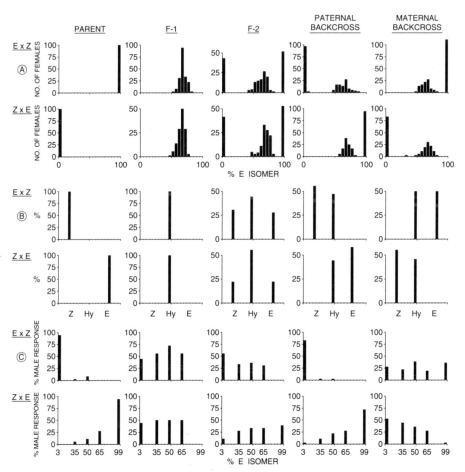

Figure 9.1. Inheritance of the sex-pheromone system in two strains of *Ostrinia nubilalis* based on parental stocks, reciprocal F_1 and F_2 progenies, and backcross progenies for: (A) female pheromone isomer production; (B) proportion of male antennal sensilla responsive to each isomer; and (C) male flights to pheromone source in the wind tunnel. Reproduced from Roelofs et al., 1987.

Ostrinia nubilalis. Two pheromonal races of this species have been found both in the United States and in Europe, from which the two U.S. populations apparently originated independently (Roelofs et al., 1985). In the so-called E-race, females produce and males respond optimally to a 99:1 blend of (*E*)- and (*Z*)-11-tetradecenyl acetate (11-14:Ac), while members of the Z-race utilize a 3:97 blend of these same components (Kochansky et al., 1975). Where these populations overlap geographically, there is evidence of strong assortative mating, but some interbreeding does occur and viable hybrids are produced (Klun and Huettel, 1988). Klun and Maini (1979) determined that female production of the pheromone-blend ratio was controlled by a single autosomal factor with two alleles. This was later confirmed and extended by Roelofs et al. (1987), who found that male behavioral response was inherited on the Z sex chromosome (males are usually the homogametic sex in Lepidoptera) (Fig. 9.1). However, the most extraordinary finding of this study was that while male behavior was determined by a sex-linked gene, the inheritance of antennal receptor types sensitive to *E* and Z11-14:Ac was controlled by a single autosomal gene with two alleles. Subsequently, Löfstedt et al. (1989) determined that the autosomal factors controlling female pheromone production and male olfactory-cell response assorted independently.

Only incomplete information is available concerning the genetics of pheromone communication in other species, with either signal or response investigated but not both, for example, *Drosophila melanogaster* (Scott and Richmond, 1988), *Ctenopseustis obliquana* (Hansson et al., 1989), and *Trichoplusia ni* (Hunt et al., 1990). Although these investigations do not allow conclusions on linkage between signal and response, all studies suggested polygenic control of the character studied, usually involving both autosomes and sex chromosomes. Thus, these systems apparently are not under the control of a coadapted gene complex as suggested for *Colias* butterflies (Grula and Taylor, 1980).

9.3.2.3. Evolution of Sexual Communication via Asymmetric Tracking

Another alternative to the coevolution model for signal-response coordination is what I shall term "asymmetric tracking." The term is meant to emphasize the view that mate-signalling systems rarely will be under strong mutually stabilizing selection, as is widely argued. Rather, selection on signal and response will be asymmetric, the direction and intensity of which will be positively correlated with that of mating competition in the population. The basic presuppositions of this framework are as follows:

1. In species characterized by differential parental investment, the sexes will differ in their "strategies" for maximizing reproductive fitness.

2. Sexual selection, as driven by differential parental investment, represents the principal arena within which all other adaptive and nonadaptive forces will act in the evolution of mate-signalling systems.

Thus, the asymmetric-tracking perspective is a logical extension of the concepts of parental investment and sexual selection to mate communication. Although these concepts are often cited as potential selective forces in the evolution of mate signalling, I do not believe their impact has been sufficiently appreciated. The widespread assumption of mutual stabilizing selection between signaller and responder, which is inconsistent with sexual selection, is evidence of this fact. Comparative studies demonstrate a strong correlation between the sexual differential in parental investment and the type of mating system employed (Kirkendall, 1983; Thornhill and Alcock, 1983). Should we not expect a similar asymmetry in the level of selection between signal and response, as well? Rather than expecting a condition of mutual stabilizing selection, a different set of predictions for the evolution of mate signalling emerges from the belief that the sexes generally differ in the reproductive strategies that maximize their fitness. From this perspective, even when males and females, or emitters and receivers, are subjected to the same adaptive forces (for example, the need to find a mate or the avoidance of hybrid matings), we would expect that the evolutionary response frequently would differ between the sexes, resulting in different patterns of gene spread and genetic inheritance.

More specifically, I would propose the following predictions in applying the asymmetric-tracking model to the evolution of sex pheromones:

1. Sexual assignment of the roles of responder or emitter will be determined by the relative cost of those roles, with the limiting sex generally assuming the less costly of the two (Greenfield, 1981; Hammerstein and Parker, 1987).

2. A pheromone signal of the limiting sex will be under only weak selection due to the response of the nonlimiting sex, and its evolution will be determined primarily by stochastic factors and by the avoidance of mating mistakes, that is, inter- or intraspecific matings that result in offspring of significantly reduced fitness.

3. The corresponding response of the nonlimiting sex will be driven by the pheromone signal of the limiting sex and by intrasexual competition, and thus will "track" evolutionarily the limiting sex; however, hybrid matings will rarely represent a significant force on the nonlimiting sex. For the latter, the actual intensity of selection will be determined by the levels of parental investment and/or the costs of mate finding relative to the fitness of hybrid offspring (Hammerstein and Parker, 1987).

4. When the limiting sex is the pheromone emitter and the nonlimiting sex the receiver, evolution of the signalling system will correspond with changes in population gene frequency expected via natural selection.

When these roles are reversed, a pheromone signal of the nonlimiting sex, somewhat counterintuitively, will track the response of the limiting sex. This signalling system may likely be subject to runaway evolution as envisioned by Fisher (1930), due to positive feedback between the preference of the limiting sex and the signal of the nonlimiting sex. As such, the system would evolve much more rapidly and may become established at a point of equilibrium different from that expected through natural selection.

Although the degree of correspondence between a species' mating system and these predictions will be modified by a number of other factors, such as costs of searching and operational sex ratio (Emlen and Oring, 1977), virtually all of these additional factors exert their influence by altering the intensity of sexual selection. The shift in emphasis that characterizes the asymmetric-tracking perspective makes it easier to reconcile the conflict between a coordinated mating signal-response system and divergence of this system during speciation, as raised by Paterson (1980). Secondly, it better defines the relative roles of adaptive and stochastic forces in shaping sex-pheromone communication, as well as the relative roles of natural and sexual selection.

At least one aspect of the predicted asymmetry in selection within the sexual communication system has been identified previously for pheromones. Löfstedt (1990), citing DeJong (1988), argues that in moth pheromones, sexual selection should be realized as a low inter-individual variance in male response to sex pheromone and a relatively large among-female variation in aspects of pheromone production. Although limited in number, studies of population variance appear to support this contention. Cardé et al. (1976) conducted an attraction-marking-reattraction study of the Oriental fruit moth, *Grapholita molesta*, to measure the natural variation of male response to isomeric ratios of female sex pheromone. Males attracted to three-component lures containing 3%, 8%, or 11% (*E*)-8-dodecenyl acetate relative to the (*Z*) isomer contacted different fluorescent dyes before flying off. Males were then captured on the subsequent night by traps baited with each of the same three ratios. There was no difference among marked males in their response to different blends on the second night, suggesting that the variation in male response was due primarily to intra-individual variation rather than a genetically-based preference for different isomeric blends. In a field experiment of similar design, Haynes and Baker (1988) also found no significant among-male variation in *P. gossypiella* response to different ratios of ZZ/ZE-7,11-16:Ac.

Studies of female variation in sex-pheromone production suggest the opposite relationship for component ratios. Measurements of intra-individual and inter-individual variation in pheromone-component ratios by

Du et al. (1987) for *Yponomeuta padellus*, Barrer et al. (1987) for *Ephestia cautella*, and Witzgall and Frérot (1989) for *Cacoecimorpha pronubana* all demonstrated high levels of repeatability for ratios produced by an individual female measured during different calling periods (usually > 80%). In contrast to low intra-individual variability, variance among females for ratios of components other than geometric isomers was high, with coefficients of variability usually greater than 50% (see Section 9.3.4.2 for greater discussion of variability in component ratio). As predicted by asymmetric tracking, this high among-female variability suggests that most component ratios are under only weak selection, since traits that are closely associated with fitness generally exhibit low heritabilities (Mousseau and Roff, 1987). The pattern of variability in male response among Lepidoptera also conforms to the asymmetric-tracking model, with stronger selection on the male response suggested by low among-male variability and thus low heritability.

Lambert et al. (1987) cite low variability in insect sex pheromones as evidence of the homeostatic nature of the SMRS. It is certainly true that one of the recurrent themes in most discussions of sex pheromones has been the specificity of pheromone response and production, affording a narrowly-tuned and species-specific mate-signalling system. However, with a heightened appreciation of the evolutionary significance of trait variance (Löfstedt, 1990), more recent studies have demonstrated high levels of variation and, in some cases, polymorphism, for pheromone-blend production and response. Such variation is predicted by the asymmetric-tracking model, as nonadaptive shifts through founder effect or other means occur in the female pheromone system and provide a new directional selection on the male response to match the new pheromone signal. The polymorphic pheromone system of *O. nubilalis* is one example that has already been discussed (Section 9.3.2.2). In addition, three pheromone types of the tortricid, *Amorbia cuneana*, have been detected based on the ratio of (E,E)- and (E,Z)-10,12-tetradecadienyl acetate (10,12-14:Ac) produced by females, with means of 37.2% *EZ* (CV = 10%), 58.3% *EZ* (CV = 5%), and 88.5% *EZ* (CV = 4%) (Bailey et al., 1986). Pheromone polymorphism also has been demonstrated in the turnip moth, *Agrotis segetum* (Löfstedt et al., 1986b; Hansson et al., 1990). We cannot say with certainty whether these examples of pheromone variation have evolved in the absence of selection; however, for *A. segetum*, apparently some of the divergence has occurred between geographically-isolated populations (Hansson et al., 1990). And Bengtsson and Löfstedt (1990) provide evidence that assortative mating between *O. nubilalis* pheromone races occurs in the absence of selection.

Although the level of competition for mates has rarely been measured directly for sex pheromones, live females are used routinely in field studies to measure the relative activity of synthetic lepidopteran pheromones, and these studies provide some indication of male competition. Although absolute numbers would be expected to vary with population density, these studies show that large numbers of males commonly are lured each night by traps containing live females, suggesting that the probability of a female attracting a male is high, and thus selection deriving from conspecific-male response is low. In the absence of selection, the probability of nonadaptive shifts in the female pheromone is increased. This is not to say that stochastic change is assured, as pheromone production may be maintained by biochemical canalization or adaptive selection in some other context. While male response will not represent a selective pressure, it will itself be under directional selection if a shift in the female pheromone occurs. The male response that matches the new female pheromone signal will be favored, and once the mean male response "catches up" with the female shift, it will be under stabilizing selection around the new signal mean.

In discussing evolutionary determinants of the genetic architecture underlying a sexual-communication system, Grula and Taylor (1980) argue "that linkage among genes underlying ethological isolation tightens as speciation proceeds," so that the variation in genetic control among species is a reflection of the time since divergence between the two populations. It is interesting, and I would say noteworthy, that in examples cited for genetic linkage, both from acoustic systems (Ewing, 1989) and in pheromone communication (Grula and Taylor, 1980), the male provides little or no parental care, and the female (the limiting sex) is the receiver. In contrast, in the European corn borer, where there is a demonstrated lack of genetic linkage, males are the responding sex. This is a pattern consistent with the predictions of asymmetric tracking. When females are the responders and there is strong competition among males for mates, female response acts as a selective force on the male signal and generates the potential for runaway sexual selection between signal and response, leading to linkage disequilibrium between these two traits (Kirkpatrick, 1982; Lande, 1982). Thus, the fact that in *Colias eurytheme*, genes controlling male pheromone and UV patterns, as well as female preference for those male traits, are found on the same sex chromosome may be due to runaway selection (Grula and Taylor, 1980). Under similar conditions of mate competition, but where males respond to a female-produced signal, male response does not constitute significant selection on the female, and linkage disequilibrium is less likely to evolve. Thus, in contrast to the hypothesis of Grula and Taylor (1980), asymmetric tracking would predict that the level of genetic linkage is more determined by which sex is the signaller and by the degree of sexual asymmetry in parental investment. The limited evidence from studies of the genetic control cited above support this argument. Further comparative studies will be necessary to determine if genetic linkage and coadaptive gene complexes for sexual signal production and response are more likely to evolve in those systems disposed to runaway sexual selection.

9.3.3. Species-Specific Sex Pheromones and the Evolution of Reproductive Isolation

To some, the suggestion that insect sex pheromones have evolved for species recognition is so obvious that a discussion of the subject seems trivial. Most comparative studies discussing the sex-pheromone systems of related species assume that interspecific differences *function* to prevent cross-matings. However, although there is a clear pattern of species specificity in sex pheromones, it is difficult to discern whether that specificity is a result of selection against interspecific mating mistakes or if differences in the sexual communication system that have arisen in some other context have caused the fission of a species. Thus, in a very real sense, the relationship between speciation and divergence in important mating characters is a "chicken-and-egg" debate, that is, which came first? The adaptive evolution of reproductive isolation, a position once widely held as the primary mechanism underlying species specificity in sexual communication, more recently has been largely dismissed by many authors as a significant force for mating systems. Thus, the current debate over the evolution of species specificity in mate communication can be summarized by three questions: Did specificity by narrowing or diverging of signal and/or response) evolve (1) prior to contact between two genetically-compatible populations, thus preadapting them to a reproductively isolated coexistence in sympatry, (2) after contact between potentially interbreeding populations, thus representing an adaptive response to selection against hybrid matings, or (3) after post-mating barriers to gene flow are in place, thus representing a response to "competition" in the sexual-communication channel? Coordination of mating signal and response has special relevance to the evolution of ethological isolation (Section 9.3.2), and as we shall see, the asymmetric-tracking perspective may allow us to better frame this debate.

9.3.3.1. Evolution of Reproductive Isolation: Theory

The concept of reproductive isolation has played a prominent role in discussions of speciation, and it forms the cornerstone of the Biological Species Concept (Mayr 1970). During the past half-century, discussions of reproductive isolation have been dominated by the framework provided by Dobzhansky (1940). Although not originating with him, Dobzhansky most effectively argued the position that "if races are to become species, isolating mechanisms must arise when the distinct adaptive complexes are exposed to the risk of disintegration due to interbreeding" (Dobzhansky, 1940). Two clear predictions for the evolution of ethological isolation that follow from this position are: (1) those species that overlap in geographic distribution with related species should be more distinct in their sexual signals than species that do not, and (2) sympatric populations of two related species will show greater differences in the sexual signal than will allopatric populations of these species. This correlation between interspecific contact and divergence in mating signals has been variously termed reinforcement (Blair, 1955), reproductive character displacement (Brown and Wilson, 1956), and the Wallace Effect (Grant, 1966). Recently, however, there has been a widespread challenge among evolutionary biologists to the concept of an adaptive evolution of reproductive isolation. For example, Templeton (1981) summarizes a discussion of reinforcement by stating: "Experimental and theoretical population-genetic studies indicate that effective pre-mating isolation is far more likely to evolve before secondary contact than after." Similarly, Littlejohn (1981) concludes that: "It is evident... that homogamic mechanisms will evolve, or be maintained, through the direct action of natural selection only under a rather restricted range of conditions." And finally, Paterson (1985) states, "I find it doubtful, to put it very mildly, that any 'premating isolating mechanism' was ever evolved to serve the function of reproductively isolating the members of one species from those of another." In place of an adaptive evolution of reproductive isolation, many authors have argued for an *incidental* divergence of sexual characters prior to contact that preadapts two populations for a sympatric existence without interbreeding between the populations.

Criticism of the Dobzhansky model has been based on both theoretical considerations and on field observations. The number of field studies demonstrating patterns of intraspecific signal divergence consistent with reinforcement has been small; considering the bias against publishing negative data, one has to wonder if even these few cases do not represent a case of biased sampling (Mayr, 1970). In a systematic study of geographic variation in songs of katydids and crickets, Walker (1974) found "no convincing examples" of mating calls diverging in the direction consistent with reproductive character displacement out of a possible 164 pairs of related species with overlapping geographic ranges. Furthermore, studies showing that those species having contact with related species differ more in their sexual calls than species not subject to interspecific sexual challenge also fail to impress the critics of the Dobzhansky model in that such a pattern would be predicted even if mate signalling evolved in the absence of interspecific contact. According to these *incidental* models, if populations of a species become geographically isolated and then at some later time reestablish contact as envisioned by allopatric speciation, then one of two outcomes can be expected. If these populations have diverged in their mating calls while isolated, then members of the two populations will not recognize each other as potential mates, and thus, the populations will be preadapted for sexual isolation and will coexist geographically. If call divergence has not occurred, then either: (1) the two populations will form a single interbreeding population again (when interbreeding produces viable offspring), or (2) one of the populations will be locally driven to extinction (when hybrids are less fit, sterile, etc.) (Paterson, 1978). The cumulative effect of these outcomes is to produce a pattern of greater call differences in related sympatric species than in allopatric ones.

In addition to the paucity of field support for reinforcement, a number of theoretical criticisms have been raised from a population-genetics perspective. The reader is referred to the reviews of Paterson (1985), Templeton (1989), and Butlin (1989) for discussion of the theoretical inconsistencies of the Dobzhansky model. Briefly stated, these objections relate primarily to the difficulties of establishing assortative mating patterns based on diverging signals in the face of even low levels of gene flow between two populations.

What Dobzhansky did not recognize in the question of the possible spread of a new character was the potential for sexual selection to drive this evolution. The role of sexual selection in the evolution of reproductive isolation was first considered by Fisher (1930), although the significance of this argument to the evolution of the reproductive isolation question has not been fully realized, as most writers do not differentiate it from the scenario of Dobzhansky. The significance of Fisher's intersexual-selection mechanism is illustrated by the pattern of signal divergence predicted between allopatric and sympatric populations of a species. In the Dobzhansky model, natural selection is seen to cause the mating signals of sympatric populations of related species to diverge because of the disadvantage of producing less-fit hybrids. In allopatric populations free from interspecific sexual challenge, this selective force does not exist and thus the signals remain unchanged. However, if the selective force is intersexual selection rather than natural selection, then the new sexual character would not be restricted to the zone of interspecific overlap, but would spread rapidly via the positive feedback system of runaway sexual selection, triggered by an initial advantage of avoiding interspecific mating errors. Mathematical models by Lande (1982) suggest the potential power of intersexual selection to initiate and/or amplify differences in a male sexual character across a large geographic range, leading to widespread reproductive isolation between two populations.

Does the asymmetric-tracking framework allow us to better predict the relative significance of the adaptive and incidental models? Where mate-finding and/or selection is mediated by a pheromone emitted by the nonlimiting sex, asymmetric tracking would predict a much higher potential for runaway evolution throughout the range of the species than does a system mediated by a pheromone of the limiting sex. Evolution of reproductive isolation via intersexual selection would appear more likely when the nonlimiting sex is the pheromone emitter. In contrast, the weak selection on a pheromone signal from the limiting sex by the receiving sex allows stochastic factors to come into play, giving rise to the geographic variation in female lepidopteran pheromones discussed in Section 9.3.2.3. If such nonadaptive shifts in the female pheromone are sufficiently great between geographically-isolated populations and the male-response mechanism has sufficient time to parallel this shift before the two populations reestablish contact, incidental reproductive isolation could then result. Incidental reproductive isolation is expected to be less likely for pheromones of the nonlimiting sex, since these should be under stronger selection by the receiver, resulting in lower genetic variability and reducing the potential for stochastic events. Thus, the asymmetric-tracking framework embraces both the adaptive and incidental evolution of reproductive isolation and makes specific predictions as to the circumstances under which each would play the greater role.

9.3.3.2. Evolution of Reproductive Isolation: Evidence From Pheromonal Systems

As discussed above, Paterson and others have argued that those behavioral and/or morphological factors that generally are referred to as reproductive isolation mechanisms are more likely to have evolved prior to contact between two populations, thus preadapting them to coexistence, rather than resulting from selection against hybrids. This incidental-evolution view has led Paterson (1985) to call for a new concept of species, the Recognition Concept, in which species are delimited by the sharing of a common SMRS rather than by their reproductive isolation from other species, as defined in the Biological Species Concept (Mayr, 1970) or, as Paterson terms it, the Isolation Concept. The geographic variation in the pheromone system of *Agrotis segetum*, *Amorbia cuneana* and others suggests that divergence in the mate signal *can* occur outside of selection for reproductive isolation and is consistent with incidental models. On the other hand, a prevalent trait of sex-pheromone communication systems that argues against the divergence of the SMRS as a preadaptation is heterospecific pheromone antagonism. There is a great deal of heterospecific overlap in the chemical signals used by insects for long-range sexual communication, with considerable potential for interspecific cross-attraction. Lambert et al. (1987) not only state that the demonstration of pheromonal heterospecific cross-attraction is consistent with the SMRS and Recognition Concept while problematic for the Isolation Concept, but further suggest that under the Recognition Concept, "there is no reason why cross-attraction should not persist indefinitely." However, in a great many insects, the responding sex has evolved an avoidance of specific compounds found in the pheromone blend of other species. One example is found in the interspecific interactions of the cabbage looper, *T. ni*, and the soybean looper, *Pseudoplusia includens*. These species, which are synchronic, broadly sympatric, and overlap in host range, also share three components of their respective pheromones: (*Z*)-7-dodecenyl acetate (*Z*7-12:Ac) (major component for both species), dodecyl acetate, and 11-dodecenyl acetate (Bjostad et al., 1984; Linn et al., 1987). Additional components that are produced by females but unique to each species are: (*Z*)-5-dodecenyl acetate, (*Z*)-7-tetradecenyl acetate, and (*Z*)-9-tetradecenyl acetate for *T. ni* (Bjostad et al., 1984), and (*Z*)-7-dodecenyl proprionate and (*Z*)-7-dodecenyl butanoate for *P. includens* (Linn et al., 1987). Wind-tunnel investigations measured no cross-attraction of *P. includens* males to a synthetic *T. ni* pheromone blend and only low levels of attraction in the reverse direction (Linn et al., 1988). The basis for discrimination in *T. ni* was concluded to be due to a lower threshold for response to the full blend of *T. ni* female-produced components, while *P. includens* discrimination was due to interruption of response by minor components of *T. ni* pheromone. In the latter case, heterospecific interruption is mediated by a dedicated antennal receptor type in *P. includens*, which is specifically sensitive to (*Z*)-5-dodecenyl acetate (Grant et al., 1987).

Discrimination between the species *Heliothis virescens* and *H. zea* also appears to be mediated in part by pheromonal antagonism. These species share four pheromone components, but *H. virescens* females produce three additional compounds: tetradecanal, (*Z*)-9-tetradecenal (*Z*9-14:Ald), and (*Z*)-11-hexadecenol (Klun et al., 1979). While cross-attraction of *H. zea* males to *H. virescens* females normally does not occur in the field, when areas were permeated with any of the above three compounds, Stadelbacher et al. (1983) measured a significant increase in attraction of *H. zea* males to traps baited with *H. virescens* females and in matings with tethered *H. virescens* females. This effect was particularly pronounced in fields permeated with *Z*9-14:Ald, and in subsequent wind-tunnel studies, attempts by *H. zea* males to "copulate" with a synthetic *H. virescens* pheromone source increased from 0% to 60.4% when the wind tunnel was permeated with this compound. These studies suggest that interruption of cross-attraction is due primarily to the antagonistic effect of *Z*9-14:Ald, which disappears when *H. zea* males are adapted to this compound in permeated air. Pheromonal antagonism is particularly adaptive in this case, as heterospecific matings are fatal due to incompatible genitalia (Stadelbacher et al., 1983).

Species specificity in pheromone communication can be effected in two ways: response synergism due to conspecific pheromone components and response antagonism due to heterospecific components; examples of each are available from a number of insect groups. A synergistic male response to components emitted by the female is completely consistent with Paterson's Recognition Concept, as evolution selects a more efficient means of mate location, even in the absence of related species. In the case of pheromone antagonists, however, it is difficult to conceive how such behavioral responses could have evolved incidentally. Evidence that these antagonistic responses represent adaptations to heterospecific matings is particularly compelling in those species for which electrophysiological recordings reveal specific receptors tuned to the pheromone compounds of heterospecific females.

9.3.3.3. Reinforcement versus Reproductive Character Displacement

The terms "reinforcement" and "reproductive character displacement" generally have been used interchangeably to designate the pattern of adaptive divergence in mate signalling due to interactions between populations. However, Butlin (1987, 1989) has suggested that these terms be used to differentiate between divergence that occurs prior to the evolution of post-mating barriers and after these barriers are established, respectively. In so doing, he has further refined the body of evidence that may be used to support an adaptive evolution of reproductive isolation. Reinforcement is seen to be qualitatively different from reproductive character displacement since in the latter case, gene flow is interrupted and speciation, as defined by the Biological Species Concept, has already occurred.

Although it would appear unlikely that antagonistic responses to heterospecific pheromone components evolved prior to contact between two populations, there still remains the question of whether pheromonal antagonism represents true reinforcement of reproductive isolation or if it is better characterized as reproductive character displacement (Butlin, 1989). In most of the examples of heterospecific avoidance given in the preceding section, it is probable that antagonism has not arisen as a mechanism to avoid hybrid matings, since they represent interactions between species from different genera. On the other hand, examples of pheromonal antagonism exist that are suggestive of reinforcement as narrowly defined by Butlin (1989). *Pissodes strobi* and *P. approximatus* are sibling species of pine weevil that are ecologically divergent and sympatric in the northeastern United States. Males of both species produce grandisol and grandisal which, when combined with host odors, act as an aggregation pheromone for (at least) *P. approximatus* (Booth et al., 1983). Despite this pheromonal overlap, these species appear to be reproductively isolated by differences in behavioral response. White pine leaders (breeding site of this species) that are baited with *P. approximatus* are no more attractive to *P. strobi* than unbaited pine leaders, and *P. approximatus* response to conspecific pheromone is interrupted by components produced by *P. strobi* (Phillips and Lanier, 1986). Hybrid matings in the laboratory produce fertile offspring that are less fit since they cannot colonize host materials used by either parent species (Phillips and Lanier, 1983). Similar evidence is found in the ermine moth *Y. rorellus*, which is sympatric with a number of close relatives in the *Y. padellus* complex. The female pheromone system of *Y. rorellus* is based on a single compound, tetradecyl acetate (14:Ac) (Löfstedt et al., 1986a); volatile emissions from virgin females contain none of the unsaturated compounds that dominate the pheromone blends of its close relatives. Pheromone emissions of these relatives do contain 14:Ac (Löfstedt and van der Pers, 1985), allowing the possibility of cross-attraction of *Y. rorellus* males. Attraction to 14:Ac, however, is prevented by the addition of only 1% of Z11-14:Ac, which is the major pheromone component of other *padellus*-complex members (Löfstedt et al., 1990). This interruption of response is adaptive in that heterospecific matings produce progeny that are less fit than those from conspecific matings (Menken, 1980).

Heterospecific pheromonal antagonism appears to be widespread in insects and, for most of the species involved, the necessary studies of interspecific crosses have not been conducted to determine viability of hybrid offspring. Thus, it is difficult to determine the relative likelihood of antagonism evolving prior to or after post-mating barriers are in place. However, if it is true that males who provide little or no parental investment are likely to persist in interspecific matings, as suggested by Hammerstein and Parker (1987), then we would expect that male avoidance of heterospecific pheromone components is most likely to evolve between populations where hybrid progeny are sterile or of severely reduced fitness, or where no offspring are produced.

9.3.4. Ecological Considerations

9.3.4.1. Optimizing Signal-Noise Ratio

Another factor that frequently is cited as an important consideration for the evolution of sexual communication is the abiotic aspect of the environment (Paterson, 1978; Cardé and Baker, 1984). Environmental considerations dictate, to some extent, the sexual-communication modality, that is, chemical, visual, or auditory. Alcock (1989) has discussed the relative advantages and disadvantages of each modality with respect to environmental considerations and the costs of signal production. Baker (1985) has taken a decidedly mechanistic approach to chemical communication, using precepts from information theory (Shannon and Weaver, 1949) to describe the flow of information across a "noisy" channel. This is a particularly useful framework for developing adaptive evolutionary hypotheses concerning the evolution of communication systems. As such, the adaptive strategy of mate-signalling systems is one of maximizing the signal-noise ratio for conspecifics, but minimizing it for potential "eavesdroppers," such as predators and parasites (Dicke and Sabelis, this volume). The means by which such an adaptive strategy may be achieved will be dependent again on the asymmetry in parental investment by receiver and sender. For the receiver, this may be accomplished by a chemoreceptor system that is highly sensitive and that is also highly specific to the information-bearing chemicals, while filtering out environmental chemicals that represent background noise (Isman, this volume). The antennae of male moths that are loaded with thousands of receptors that are narrowly tuned to a single or a few chemicals emitted by the female is certainly a good example of this expectation.

In addition to maximizing signal:noise, another strategy for improving information transfer in a noisy channel that is derived from information theory is through "redundancy." Nearly all systems of communication use redundant signals, and redundancy is the predominant means by which error-free messages are transmitted in a noisy channel (Bharath, 1987). Redundancy of components in a pheromone blend in theory could improve sexual communication. For example, in species that share one or more pheromone components, interference of the signal perception due to interspecific competition in the pheromone channel might be reduced by redundancy of components emitted and perceived. Redundancy has been invoked to describe the pheromone system of *T. ni*, in which any five-component combination of the six female-produced compounds (except one lacking $Z7$-12:Ac) and several four-component blends elicited attraction in a wind tunnel comparable to the entire six-component combination (Linn et al., 1984). However, redundancy as defined by information theory has not yet been demonstrated for *T. ni*. To do so, we must distinguish between interchangeability of compounds due to broad tuning of pheromone receptors and that due to perception by separate receptors. If the interchangeable compounds are perceived at the same receptor site, then redundancy would not be achieved, since chemical noise in the environment would interfere with the detection of all compounds interacting with that site. To my knowledge, there are no published accounts clearly demonstrating redundancy in the information-theoretical sense in sex pheromones, but there are no theoretical reasons for ruling out its possibility.

9.3.4.2. Physical Aspects of the Communication Channel

In addition to chemical noise in the environment, abiotic aspects of the pheromone channel, such as wind turbulence and variable temperatures, will also impact transmission of the pheromone signal (Cardé and Baker, 1984). If information is to be encoded by the ratio of components in a pheromone blend, as is frequently the case, then the specificity of this information will be dependent on the reliability of the signal under different environmental conditions, such as fluctuating temperatures. Because the release of pheromone components will increase with increasing temperature, but those with different functionalities and/or chain-length may do so at different rates, a particular ratio of components present in a pheromone gland would produce an emission with a ratio that varies with temperature. This is the general explanation given for the widespread observation that ratios between geometric isomers of an unsaturated pheromone compound (whose emission ratio would not change with temperature) are usually relatively invariant in Lepidoptera, whereas ratios between components with different functionalities are usually much more variable. For example, females of the moth *Y. padellus* emit a blend of seven acetates: tetradecyl acetate (14:Ac), (E)-11-tetradecenyl acetate ($E$11-14:Ac), (Z)-11-tetradecenyl acetate ($Z$11-14:Ac), hexadecyl acetate (16:Ac), (Z)-9-hexadecenyl acetate ($Z$9-16:Ac), and (Z)-11-hexadecenyl acetate ($Z$11-16:Ac). Comparing the release of pheromone components from individual females, Du et al. (1987) found that relative to $Z$11-14:Ac, the release of $E$11-14:Ac had a coefficient of variation (CV) of only 15%, while the CVs for other components ranged from 46% to 61%. Recent physiochemical studies of pheromone compounds by McDonough et al. (1989), however, suggest that the effect of temperature on emitted ratios may not be as important as previously supposed. The amounts of $Z$9-14:Ac and $Z$11-16:Ac applied to a rubber septum that would produce an emitted 50:50 ratio at 30°C would only shift to 54:46 (in favor of $Z$9-14:Ac) at 20°C, and a 10:90 ratio would only shift to 9:91 over the same temperature range, even though the absolute rate of release would change by about three- to four-fold. Further evidence against the physiochemical constraint argument is revealed by looking at positional isomers. Calculations from the data of McDonough et al. (1989) show that a 50:50 ratio of $Z$7-12:Ac to $Z$9-12:Ac, $Z$9-14:Ac to $Z$11-14:Ac, or $Z$11-16:Ac to $Z$13-16:Ac would change by less than 0.5% from 30°C to 20°C, yet examples of low variance in the ratio of positional isomers in pheromone blends do not appear common. Compared to the low variance in the E/Z11-14:Ac ratio in *Y. padellus* discussed above, using the data of Du et al. (1987), one can calculate that the CV for the $Z$9/$Z$11-16:Ac ratio was 59%. Measurements by Witzgall and Frérot (1989) of variance in pheromone production among female *Cacoecimorpha pronubana* showed tight regulation of the ratio of emitted Z- and $E11$-14:Ac, as has been reported previously for a number of other tortricids. However, the among-female variance of the $Z11$-14:Ac/$Z11$-14:OH ratio also was very low, ranging from only 0.06 to 0.11, in spite of the expectation of different chemical functionalities possessing different release dynamics. Furthermore, pheromonal components with similar emission characteristics varied widely; the release of $Z9$-14:Ac varied from 4% to 117% relative to $E11$-14:Ac, and the ratio of $Z9$/$Z11$-14:Ac also differed significantly among females.

Thus, these recent measurements of population variance in female pheromone emission, as well as studies of temperature-vapor pressure relationships, suggest that earlier arguments that the regulation of component ratios is constrained by physiochemical properties of those compounds may only represent part of the picture. It is noteworthy that in comparing the pheromone production of a lab strain of *Argyrotaenia velutinana* with field populations, Miller and Roelofs (1980) measured a statistically significant shift in the E/$Z11$-14:Ac ratio, with the lab population having a mean of 7% E compared to 9.1% E in the wild population; however, the CVs of these populations were identical for the E/Z ratio, at 9.7%. These results suggest two relevant points. First, despite low variance in ratio production, shifts in this ratio can occur. Secondly,

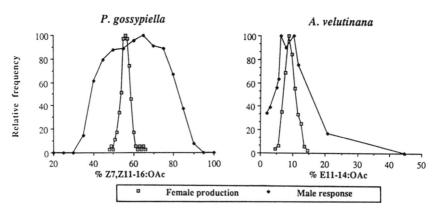

Figure 9.2. Variation in male response to and female production of pheromomal isomer ratios in *Pectinophora gossypiella* and *Argyrotaenia velutinana*. Female pheromone production and male response in *P. gossypiella* is based on data from Collins and Cardé (1985) and Linn and Roelofs (1985), respectively, and that in *Argyrotaenia velutinana* is based on Miller and Roelofs (1980) and Roelofs et al. (1975), respectively. Reproduced from Löfstedt (1990).

whatever the adaptive or nonadaptive basis for this shift, it appears to have occurred independent of a change in variance for the ratio. It is also interesting that in two examples comparing male-response windows and variance in female production of geometric isomeric components (Fig. 9.2), female variance was narrow irrespective of the width of the male-response window. Thus, variance in the ratio of pheromonal geometric isomers also may be independent of variance in male response. Alternatively, variation in different pheromone-component ratios may be more a reflection of physiological canalization than physiochemical constraints or stabilizing selection on the pheromone signal, with compounds sharing more biosynthetic pathways showing lower ratio variance (Bjostad et al, 1987).

9.4. Male-Produced Courtship Pheromones in the Lepidoptera

Mate-finding in moths is brought about almost universally by a long-range male response to female-produced pheromones and in butterflies generally by male response to female color patterns; however, in a large number of species, males also produce pheromones that are employed once the male has located the female. In addition, for a smaller number of species scattered across the Lepidoptera, male-emitted pheromones mediate long-distance mate-finding by females. Structures for disseminating pheromones in lepidopteran males are variously termed androconia, hair-pencils, coremata, and scent brushes and have long fascinated tax-onomists. Reviews of male pheromones and scent structures are provided by Boppré (1984) for the butterflies and by Fitzpatrick and McNeil (1988) and Birch et al. (1990) for the moths. Debate surrounding these organs began with their initial recognition and focused on proximal function, for example, Bailiff (1825) considered them deformities without functional significance. Although we now recognize the primary function to be in odor dissemination, the debate concerning their evolutionary significance has yet to be resolved. This chemical-communication system is represented by an extraordinary degree of diversity in both pheromone chemistry and morphology of the androconia (Fig. 9.3). Furthermore, appearance of these structures across most taxa correlates poorly with presumed taxonomic relatedness. For this reason, most taxonomists, although keenly interested in their morphology, consider androconia as poor taxonomic characters above the level of the species. There probably is no other character in insects that shows as much heterogeneity in form or that is as ephemeral in appearance as male androconia. Male pheromones in the Lepidoptera, however, represent more than a taxonomic curiosity; rather, they provide a useful contrast to their female counterparts and allow a

Figure 9.3. Male androconia in the Lepidoptera: A) forewing costal-fold scent scales of *Cadra cautella* (Pyralidae), B) foretibial hairpencils of *Zanclognatha laevigata* (Noctuidae), C) abdominal hairpencils of *Atteva punctella* (Yponomeutidae), D) abdominal coremata of *Estigmene acrea* (Arctiidae, photo by M. A. Willis), E) mesotibial scent brush of *Catacola concumbens* (Noctuidae), and F) hindwing costal hairpencils of *Ithomyia* sp. (Ithomiidae).

more robust analysis of the predictions of asymmetric tracking when sexual roles of sender and receiver are reversed.

9.4.1. Male Pheromones in the Phycitinae (Pyralidae)

The lepidopteran group for which we have the most extensive information concerning close-range courtship and the role of male pheromones is the pyralid subfamily, Phycitinae. Particularly well-studied are those species belonging to the so-called stored-product phycitine complex. This is due to both their economic standing and the fact that they employ a rather complex courtship sequence prior to copulation. In addition, pheromone-emitting organs have been identified from males of many phycitines (Heinrich, 1956; Roesler, 1973). Presumed scent structures, which are located primarily on the costal margin of the forewing and/or on the eighth abdominal segment, are found in about 80% of species. The importance of male-emitted pheromones to mating success has been characterized for several Phycitinae. Grant (1974) found only 37% of *Plodia interpunctella* males who had their wing pheromone glands removed were successful in courtship, compared with 93% of nonoperated males. Similarly, when females were antennectomized, normal males were successful in only 27% of courtships. Parallel studies in *Vitula edmandsae* yielded similar results; courtship success declined from 79% between normal males and females to 5% when either males were without the wing glands or females had antennae removed (Grant, 1976). In *Ephestia elutella*, courtship success was reduced from 83% to 32% when males had their pheromone wing glands removed, and the glandless males who were successful in courtship took more than twice as long to achieve copulation than successful intact males (Phelan and Baker, 1986b). In a comparative study of the courtship of twelve species of Phycitinae, Phelan and Baker (1990a) found courtship sequences clustered into two types: simple and interactive. In the simple courtship, males copulated by simply lining up parallel to the female and reaching under the female's wing with the abdomen. In the interactive courtship, females took a much more active role in the mating sequence and were to a large extent able to control the outcome of the courtship attempt. In this courtship, the male typically lined up facing the female. He then raised his abdomen over his head and struck the female on the head with it, after which he attempted intromission by a dorsolateral thrust of the abdomen. If receptive, the female would facilitate genital contact by elevating her abdomen; otherwise, copulation was made unlikely.

Two general conclusions were drawn in this study with regard to the evolution of courtship and male pheromones in the Phycitinae. First, courtship in this group appears to have evolved for the sole function of making delivery of the male pheromones more efficient. The interactive courtship was found only in those species where males possessed pheromone-emitting organs. Furthermore, differences in male behaviors among species with interactive courtship corresponded with the relative development of the two pheromone organs. For example, *P. interpunctella* males possess costal hairpencils, but have only vestigial abdominal hairpencils. In this species, the male did not strike the female with his abdomen, a behavior that in other species has the effect of bringing abdominal hairpencils into close proximity to the female's antennae. The second general conclusion was that the relative complexity of courtship showed no correlation with taxonomic groupings, that is, closely related species were sometimes found to have different courtship types, while species from distantly related tribes demonstrated very similar courtship sequences. Thus, it would appear that the interactive courtship and associated male pheromones arose early in the evolution of the subfamily, but they probably have been independently lost many times.

9.4.2. Ultimate Functions

Despite their widespread distribution and their presence being recognized for well over a century, our understanding of these male-produced lepidopteran pheromones lags well behind that of female pheromones. On a practical level, their potential for the development of economic control strategies (the primary force driving the study of female lepidopteran pheromones) seems negligible. Secondly, putative male pheromones many times do not elicit an overt response in females. Coming into play late in the mating sequence, their proximal function is frequently hypothesized to be the induction of quiescence in the female (Brower et al., 1965; Birch, 1970). Such an effect obviously does not lend itself well to a behavioral bioassay. Even when discrete behaviors seen in courtship can be elicited by synthetic pheromones (Phelan et al., 1986; McLaughlin 1982), we might expect bioassays based on these behaviors to be considerably less discriminating than those requiring the long-distance attraction of males by female pheromones. Thirdly, even in species with well-developed androconia employed during courtship, removal of these structures may not reduce mating success (Birch et al., 1989). Males of some species expose their androconia at times other than courtship, for example, when being handled, as if their function may be more defensive (Grant and Eaton, 1973). This impression is reinforced by the fact that some of the constituents from scent brushes are related to defensive compounds. Finally, some researchers have referred to any chemicals identified from androconial secretions as pheromones without any evidence for their behavioral role. As a result of these factors, our understanding of male pheromones is very murky. At present, we have no concensus as to the ultimate role that these pheromones play nor to the forces driving their evolution.

A number of functions have been postulated for the evolution of male lepidopteran pheromones. Those receiving the greatest consideration are: (1) predator repellent (2) mating deterrent against conspecific males, or (3) species-specific signal to prevent interspecific matings (Birch, 1974). Some authors have suggested that, given the diversity in androconial morphology and in pheromone composition, it is foolhardy to suggest that all male lepidopteran pheromones have evolved to serve a single function (Boppré, 1984). While this is undoubtedly true, there are a couple of factors that should be considered. First, we must attempt to distinguish initial functions from derived functions, that is, the selective pressures or the functions that maintain such a system may not have been the pressures that led to their original evolution. Secondly, while invoking a number of functions may be necessary for explaining the evolution of *all* male pheromones, we are primarily interested in the relative contribution of these mechanisms and whether a limited number of functions can explain the evolution of most of these pheromones. Despite the pessimism expressed concerning a unifying view of male pheromones, I would argue that the majority of these pheromones and their associated structures for dissemination can be explained by two models: (1) an adaptive response to mating errors (Phelan and Baker, 1987; Baker, 1989; Phelan, 1991), and/or (2) a male manipulation of nonsexual female behaviors.

That the function of these pheromones lies more in the interaction among conspecific males rather than between male and female was supported by the work of Hirai et al. (1978). In this study, it was found that the sexual response of *Psuedaletia unipuncta* males placed downwind of calling females was significantly diminished if conspecific males were placed between the males and females. Similar studies on the effect of male pheromone on male-male interactions in *Adoxophyes orana* provided similar conclusions concerning the behavioral function of this species as well (Bijpost et al., 1985). The "adaptive significance" of this effect is seen to be "related to the increased reproductive efficiency that results if multiple males are prevented from competing for a single female" (Hirai et al., 1978).

Although a male-inhibitory effect is convincingly demonstrated in both of these studies, there remain two difficulties with the argument that male pheromones have evolved to repel reproductively competing males. First, it is difficult to understand the selective forces that would have given rise to such a communication system. Why should the receiving male respond to the male pheromone of the upwind male by aborting his orientation to the female? From an individual-selection perspective, this could only evolve if the probability of getting a mating with that female is markedly lower than cutting off this mating attempt and then locating a new female. However, if the probability of a successful mating when competing with another male is so low, then what is the selective force driving the production of the male pheromone in the first place? In other words, the higher the probability of a late-coming male displacing the original male, the stronger the selection for production of a male-inhibiting pheromone, but the lower the selection for a late-coming male that "cooperates" by aborting his mate-finding response. The second difficulty is an experimental one, in that in both studies, the reduction in male response was only accomplished by placing multiple males in the upwind position, 30-45 for Hirai et al. (1978). If odor from several males is necessary for the inhibitory response, it is unlikely that this represents their original function.

Fitzpatrick et al. (1988) later demonstrated that when tested at more biologically relevant levels, *P. unipuncta* hairpencil secretions do not act to repel males in the female pheromone plume. At best, it would seem that pheromone-mediated male-male inhibition represents an effect that may have evolved secondarily, that is, once the pheromone had evolved through some other means, males might be selected that avoided a female that already had several males associated with her (as determined by high levels of male pheromone), since the probability of a successful mating under these conditions probably would be low. This, however, would not represent the true "function" of the pheromone, but only a secondary effect.

9.4.3. Adaptive Response Through Intersexual Selection

Whatever the function(s), if any, driving the evolution of male lepidopteran pheromones, the asymmetric-tracking perspective would predict that this communication system would be more prone to runaway evolution by intersexual selection than that of female pheromones. Müller (1878) appears to have been the first to invoke sexual selection as responsible for the evolution of androconia in the Lepidoptera, Darwin (1871) having just recently proposed this concept. On the other hand, Scudder (1877) rejected sexual selection as an explanation, since these structures were so small, making them different from the types of outlandish male-limited structures that Darwin was addressing. In more recent times, Baker and Cardé (1979) were the first to apply detailed behavioral studies to support the role of intersexual selection in the evolution of male pheromomes. They argued for an origin of androconia through female preference for male odor that: (1) indicated higher male quality, (2) signalled gender or species identity, or (3) deterred predators.

There is a considerable body of circumstantial evidence favoring an intersexual selection model for male pheromones. The first of these concerns the extraordinary morphological diversity represented in male androconia discussed previously. This is best illustrated by comparing male lepidopteran pheromone systems to that of their female counterparts. Lepidopteran females almost universally produce pheromone from glandular tissue located on the intersegmental membranes between the eighth and ninth abdominal sternites, with no specialized structures increasing the efficiency of their release. In contrast, male pheromone glands can be found on virtually every part of the body, from legs to wings to abdomen. In addition, these glands almost always have associated with them scales specialized for the transport of chemicals that increase the rates of pheromone release. Not only do we see this diversity when looking across the Lepidoptera as a whole, but we find non-homologous pheromone organs within a family (see review of Birch et al., 1990). A similar contrast can be made when looking at the chemistry of the pheromones. Evolution of female pheromone chemistry has been very conservative in the Lepidoptera, with most being composed of chemicals that are 12–18 carbons in length and possessing one of three basic functionalities: aldehydes, acetates, or alcohols (Tamaki, 1985). Furthermore, within a family or subfamily, one frequently finds many species sharing at least the major component of the pheromone blend, and female pheromones have been suggested as a good character for aiding in taxonomic classification of higher groups (Roelofs and Brown, 1982; Renou et al., 1988). Although considerably fewer male pheromones have been characterized, we nonetheless see a much greater chemical diversity than in female-produced pheromones (Table 9.2). Finally, contrasts can be made in the behavior leading to delivery of the pheromone signal. Whereas females typically signal by sitting motionless with their pheromone gland exposed, complex courtship rituals may evolve in association with male pheromone, such as in the Phycitinae (Section 9.4.1). These general observations, along with the capricious nature of their appearance and disappearance, combine to strongly suggest a much more radical evolution of male pheromones than female pheromones.

In their sexual-selection model, Baker and Cardé (1979) hypothesized the evolution of hairpencil display and pheromone emission in *G. molesta* as a ritualization of clasper extension and incidental genital odors. They particularly emphasized the possibility that the process was initiated by female preference for a species-specific male odor that allowed her to avoid matings with males of the related *G. prunivora*, which are cross-attracted to *G. molesta*. Although the concept of reproductive isolation evolving as an adaptive mechanism has come under broad theoretical

Table 9.2. Examples of male-produced pheromones identified in the Lepidoptera (female response demonstrated).

Species	Components	Reference
Arctiidae		
Creatonotus transiens	R-(-)-Hydroxydanaidal	Wunderer et al. (1986)
Estigmene acrea	Hydroxydanaidal	Krasnoff & Roelofs (1989)
Noctuidae		
Psuedaletia unipuncta	Benzyl alcohol, benzaldehyde, acetic acid	Fitzpatrick & McNeil (1988)
Trichoplusia ni	*d*-Linalool, *m*-cresol, *p*-cresol	Landolt & Heath (1990)
Phlogophora meticulose	6-Methyl-5-hepten-2-one, 6-methyl-5-hepten-2-ol, 2-methylbutyric acid	Aplin & Birch (1970); Birch (1970)
Pyralidae		
Ephestia elutella	(*E*)-Phytol, γ-decalactone γ-undecalactone	Phelan et al. (1986)
Plodia interpunctella	Palmitic, oleic, and linoleic acids and ethyl esters	Phelan (in prep.)
Eldana saccharina	Vanillin, *p*-hydroxyben- zaldehyde, (*E*)-3-methyl- 4-dimethylallyl-γ-lactone	Kunesch et al. (1981); Zagatti et al. (1981)
Corcyra cephalonica	(*E,E*)- & (*Z, E*)-Farnesal	Zagatti et al. (1987)
Tortricidae		
Grapholita molesta	*R*(-)-Mellein, ethyl-*trans*- cinnamate, methyl jasmonate, 2-epijasmonate	Baker et al. (1981)

attack (Section 9.3.3.1), it still remains a compelling possibility for male pheromones, based on observations from a number of different lepidopteran groups. For example, in the stored-product Phycitinae with their widespread male androconia, the need for reproductive isolation seems particularly acute as they overlap in seasonal distribution and may be found co-habiting the same grain storehouse (Levinson and Buchelos, 1981). Furthermore, some species share female pheromone components and show high levels of cross-attraction in the wind tunnel (Phelan and Baker, 1986a) and in the field (Ganyard and Brady, 1972). Although there are some interspecific differences in the time of night at which females emit pheromone (Krasnoff et al., 1983), the temporal window of male response is much broader (Phelan and Baker, 1986a), suggesting a high potential for interspecific sexual encounters. A reproductive-isolation role also is supported by studies that suggest species specificity in the composition of the phycitine male odors, as well as an ability on the part of females to act on these chemical differences. In courtships between *Plodia interpunctella* and *E. cautella*, female rejection of heterospecific males appeared to be mediated by male odor (Grant et al., 1975). *E. elutella* females exhibited a behavioral response to extracts of conspecific males, but not to extracts of *E. cautella*, *E. figulilella*, *E. kuehniella*, or *P. interpunctella* males (Krasnoff and Vick 1984). *E. elutella* and *E. figulilella* readily court interspecifically, but females of both species reject heterospecific males by covering their genital opening after males present their abdominal hairpencils, which occurs in a fraction of a second before the male's copulatory attempt (Phelan and Baker, 1990b). Ecological and pheromonal overlap also characterize many species of tortricid moths, and parallel arguments can be made for the correspondence between high potential for interspecific sexual challenge and greater prominence of male androconia in these groups (Phelan and Baker, 1987). In contrast, partitioning of the long-range female-pheromone channel among species of sesiid moths was determined to be complete, with cross-attraction unlikely in the field among members of this group (Greenfield and Karandinos, 1979). Male pheromones or androconia have not been observed in this family.

The hypothesis that male lepidopteran pheromones have evolved within the domain of reproductive isolation was tested directly by Phelan and Baker (1987), who examined the relationship between interspecific contact and the presence of male androconia in five families of moths. A simple prediction of any adaptive hypothesis for reproductive isolation is that those species that encounter closely related species are more likely to possess the character thought to advance sexual isolation. Using the sharing of a larval host between congeneric species of moths as an indicator of interspecific contact, we determined the degree to which male androconia correlated with this factor in five independent groups: the Phycitinae of North America and Europe, the Tortricidae and Noctuidae of Great Britain, the Yponomeutidae of Japan, and the Ethmiidae of the western United States. In each case, there was a significant positive correlation between the presence of androconia in a species and that species showing host-plant overlap with a congener (Fig. 9.4). In two of three groups reexamined using geographical range as a predictor of interspecific contact, the North American Phycitinae and the British Tortricidae, but not the Japanese Yponomeutidae, again showed a significant positive correlation. Thus, these results are consistent with that predicted by a reproductive-isolation function.

The choice of host-plant associations as an approximate measure of interspecific contact rather than the more-conventional overlap in geo-

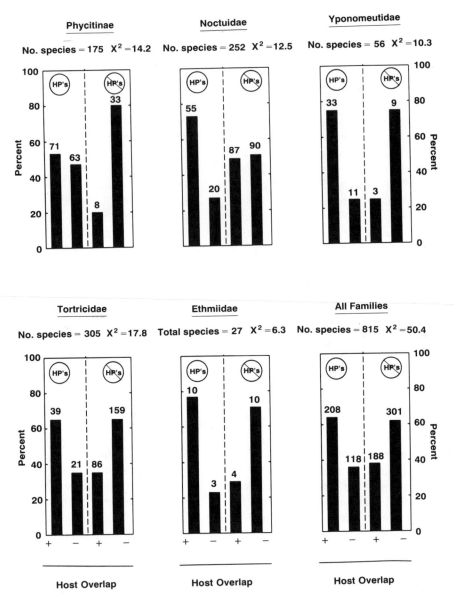

Figure 9.4. Relationship between host-plant sharing by congeners (as a measure of interspecific contact) and the presence of male androconia in five families or subfamilies of moths (HPs = hairpencils, number on bar = number of species). Based on data from Phelan and Baker (1987).

graphical range allowed us to extend the interpretation of this data set one step further. As indicated earlier (Section 9.3.3.1), one of the difficulties in using a positive relationship between geographic overlap and greater differences in mating signals as proof of an adaptive evolution of reproductive isolation is that a similar relationship could also be explained by an incidental evolution of mating characters before secondary contact. Arguing from an incidental viewpoint, one could suggest that sympatry may only be established between allopatric populations that already show considerable premating isolation. However, this argument becomes untenable when host overlap is used as the measure of interspecific contact, as one would have to assume that the major factor determining whether species overlap on a host is whether male pheromones are in place. Host-plant utilization is obviously a much more complex matter than this. Thus, if these structures have evolved for reproductive isolation, then the data suggest that they represent an adaptive response to mating errors, since this pattern is not predicted by an incidental evolution of isolation.

One objection that has been raised against the possible role of male pheromones in species recognition is that they come into play so late in the mating sequence, after the male has already located the female. Thus, as a mechanism for reproductive isolation, male courtship pheromones represent a very inefficient system. However, inefficiency is a problem only if one assumes that the evolution of male pheromones is driven by natural selection. On the other hand, what appears as inefficient when viewed at the level of the species is completely consistent with evolution through intersexual selection and the concept of asymmetric tracking. There has been a somewhat artificial distinction made between avoidance of hybrid matings and intraspecific mate choice. Mating with a heterospecific male is not *qualitatively* different from mating with a conspecific male of low fitness. Thus, we expect the same selective regimes underlying female choice between conspecific males to be operative when females are challenged by males with an incompatibly-adapted genome. In fact, we would expect female choice to be more prominent in this milieu, as the differences between males of different populations are likely to be greater than among males within a population. Seen in this light, the evolution of a mechanism by which females can distinguish males of her kind from other males, even if it requires considerable wastage of time on the part of the male, is readily understandable. The asymmetric-tracking model, and especially the mathematical treatments of Hammerstein and Parker (1987), suggest that viable hybrid matings typically represent a conflict in the mating strategies of males and females. This contrasts with the traditional view that hybridization should be avoided by both sexes, but that selection for this is greater on the female. Thus, if males who persist in such matings are actually favored, assortative mating will only result if females evolve an ability to distinguish between male types. Selection against male persistence in cross-mating attempts will not occur until the probability of success in such attempts becomes very low or fitness of the resulting offspring is very small.

Given the variable distribution of scent structures and courtship in the Lepidoptera, it would appear that if they evolved adaptively, then the function for which they were selected is a rather transitory one, otherwise one would expect the system to have been more extensively conserved. In addition to the evidence for parallel evolution, it appears that these structures have been lost independently many times, as well. The best evidence for loss of male pheromones comes from Birch's (1972a, b) systematic study of male abdominal brush organs in the Noctuidae. This scent-dissemination system consists of five components: a scent-producing gland, a brush for dispersing the odor, a pocket for housing the brush when not in use, and a lever and apodeme for everting the brush during courtship. In a number of species, Birch (1972b) found the system to be lacking in one or more of these elements; 95% of these incomplete systems lacked the gland, next most commonly missing was the brush, and then the pocket. Given this pattern, it is a reasonable assumption that these incomplete systems represent stages in the loss of the organ, since it is difficult to imagine how such an organ could evolve with a scent gland as the final element. Behavioral and morphological evidence also suggest a vestigial state for male-pheromone systems in some species of Phycitinae (Grant et al., 1975; Phelan and Baker, 1990a) and Arctiidae (Krasnoff and Roelofs, 1990). Although sexual isolation would not be expected to be transitory, male pheromones may very well be temporary in this role, made superfluous by subsequent evolution of the female-pheromone system. To illustrate how this may have happened, I offer the following hypothetical scenario:

Step 1: The pressure of hybrid disadvantage or interspecific mating mistakes causes females to become more selective, favoring males of her own population based on incidental body odors.

Step 2: Female preference for odor gives rise to runaway sexual selection in which males are selected for elaboration of odor dissemination. Male odor and female preference for this odor evolve jointly and spread rapidly throughout the range of the population.

Step 3: Once female preference is established, low success of cross-mating attempts causes male response to long-range female pheromones to change through natural selection. Disruptive selection on the female pheromone signal may result if this reduces "harassment" by heterospecific males.

Step 4: Female mating prerequisite for male odor wanes as interspecific sexual challenge is withdrawn.

Step 5: Male pheromone apparatus is gradually lost as mating advantage leading to its initial appearance declines.

Since the olfactory accoutrements of moths generally are well-developed and mating takes place during the night for most species, it seems likely that if female discrimination of males did develop, the chemical modality would be most widely adopted for the purpose. The pheromone may represent an elaboration of the natural body odor of the male (Baker and Cardé, 1979), which may be largely determined by larval host constituents or by odors of plants that males frequent as adults (Baker, 1989; Phelan, 1989). Although this is not a requirement of the model, there are theoretical reasons why this might be the case. If the female uses odors for host finding or for food finding, the receptor system and neural wiring would already be in place for behavioral response. It would seem a relatively short evolutionary step for this response to be employed within another context. Also, there is observational evidence of a connection between host chemistry and male pheromones. For example, the male wing pheromone of *P. interpunctella* is a mixture of 16- and 18-carbon fatty acids and their ethyl esters (Phelan, in prep.). When exposed to a synthetic blend of these components, virgin females respond (at levels comparable to response to male extract) by turning in a circle, a behavior essential to successful copulation (Grant et al., 1975). The long-range host-finding of ovipositing female *P. interpunctella* is mediated by the same long-chain fatty acids found in the male pheromone (Phelan and Roelofs, in prep.). Similarly, Barrer (1977) found that extracts of male *E. cautella*, but not of females, elicited increased oviposition by mated conspecific females. This effect was more pronounced when extracts were made from the male genitalia, the site of male pheromone production, than extracts from either wings or remaining body parts.

Finally, one of the components of Oriental fruit moth male pheromone, ethyl-*trans*-cinnamate (Table 9.2; Baker et al., 1981), is a volatile component of rotting fruit to which adult females and males of this species are attracted (Nishida et al., 1985).

9.4.4. Evolution Through Female Manipulation

Although I would maintain that the bulk of the observational evidence supports a reproductive-isolation role for male pheromones, there are some lepidopteran species for which such an explanation is unsatisfying because the chemical composition of androconia does not appear to be species specific. For example, a number of moths and butterflies

(Ctenuchidae, Arctiidae, Ithomiinae, and Danainae) have been found to be attracted to plants containing pyrrolizidine alkaloids (PAs), upon which they feed as adults and/or oviposit (Pliske, 1975). Males of many of these species possess androconia that contain compounds derived from PAs, which males obtain from the plant either as larvae or adults. These compounds may be derived from either half of the PA molecule (Schneider, 1987). In the case of danaines, androconial components synthesized from the alkaloid nucleus, primarily danaidone, danaidal, and hydroxydanaidal, are widespread (Boppré, 1984). These same compounds have been found in a number of arctiid males, as well (Birch et al., 1990). In the Ithomiinae, male secretions of many species appear to be based on the esterifying acid moiety of PAs (Edgar et al., 1976). With the exception of one primitive tribe, hairpencils on the male hindwing costal margin are almost universal among the Ithomiinae; Edgar et al. (1976) identified a dihydroxy-γ-lactone in microgram quantities from 7 of the 10 Venezuelan species that they surveyed. Likewise, male abdominal scent brushes in many noctuid species contain one or more of the compounds, benzaldehyde, benzyl alcohol, and 2-phenylethanol (Birch et al., 1990). These compounds, which again are present in microgram levels, may be found in members of several different subfamilies. Such sharing of pheromone components seems inconsistent with a possible role in species recognition. This information, however, may be misleading as the chemical composition of most of these systems has not been fully determined, and the actual male pheromone may consist of a more complex species-specific blend. Boppré (1984) cites the presence of as many as 33 unidentified compounds in danaine hairpencils as evidence of the incomplete chemical picture. Equally significant is the paucity of comparative behavioral studies in these groups, which are necessary for defining which androconial components constitute the actual pheromone. Such behavioral studies are also required for determining to what degree response is species specific. Thus, once these male-scent systems are more completely characterized both behaviorally and chemically, they may be found to be more species specific than presently perceived.

If, however, further behavioral investigations support the lack of species specificity in these male pheromones, an alternative route of evolution can be conceived, one that is not entirely exclusive of the reproductive isolation model. This is based on the ideas of Krebs and Dawkins (1984) who see animal signalling as having evolved through manipulation of a responder and involves male exploitation of the female's host-finding system. If females use odors to locate their host, either for feeding or for oviposition, then males with an odor similar to the host blend may have greater access to females. In the extreme case, males might have attracted from a distance females that were either searching for ovipositional sites or for food. Alternatively, after the male locates the female, the odor may act short-range by eliciting host-selection or feeding behaviors in the female.

In any case, the male odor may arrest the movement of the female by "mimicking" plant odors. In the case of the Ithomiinae butterflies discussed above, attraction of adults to PA-containing plants is mediated by long-distance attraction to volatiles derived from esterifying acids of the PAs (Pliske et al., 1976). The chemical similarity between male secretion and odors for plant location is striking. Although the attraction is strongly male-biased in most species, a smaller percentage of females is attracted, and in the primitive genus *Tithorea*, which probably represents the ancestral condition, both sexes are attracted to baits in equal numbers (Pliske et al., 1976). Although Eisner and Meinwald (1987) have argued that PA-based male pheromones have evolved as an indicator of the male's ability to provide the female with chemical defense, this may represent a secondary function (Baker, 1989). If males emitted volatiles to which females responded by attraction or arrestment, then elaboration of this odor could have evolved initially without any direct advantage being realized by the female. An association between male pheromone and the chemical nuptial gift may have then evolved subsequently and might underlie the maintenance of this communication system. A similar connection can be made with the aromatic constituents of noctuid male brushes, as these can be common components of floral odor blends. Since adult noctuids are attracted to flowers from which they forage for nectar, this may be mediated by such chemicals. This possibility is strengthened by an early observation that several noctuids (both male and female) were attracted to traps baited with phenylacetaldehyde, benzyl ether, benzyl acetate, and other structurally-related chemicals (Smith et al., 1943). Such a signal-as-manipulation framework might also explain the rather perplexing mating system of *Estigmene acrea*, an arctiid moth in which males form leks (at least in the loose sense of the word) for calling and to which females are attracted (Willis and Birch, 1982). This species maintains a dual mating system in that at another period of the night, females call and attract males.

9.5. Conclusions and Future Directions

Anyone attempting to survey the systems of insect mate-signalling based on sex pheromones has to marvel at the great variety of "strategies" employed to accomplish the biologically profound act of mating. There is also a profusion of potential forces underlying this diversity. However, I have attempted to argue here that understanding the evolutionary basis for sex pheromones and ferreting out the relative role of these forces is not as futile a task as it might first seem. Taking fundamental Darwinian concepts of individual selection and maximizing reproductive success to their logical end, we can start to frame a model for unifying this great array of sexual-communication systems. Recognizing the conflict between sexes in maximizing fitness, one is compelled to expect not a communication system that is under strong reciprocal selection, but one of asymmetric selection, the intensity of which reflects the degree of mate competition inherent in the species. This shift in paradigm allows us to explain at once how signal and response are coordinated and how shifts may occur readily during the process of speciation.

This emphasis on the adaptive nature of mate-signalling systems nevertheless allows for a nontrivial role by nonadaptive processes. In fact, due to weak selection on the signal when produced by the limiting sex, asymmetric tracking predicts that in the absence of other selective forces, such as avoidance of hybrid matings or physiological canalization, signal shifts might result from stochastic processes. This explains the geographic variation in lepidopteran sex pheromones reported in recent studies, which is inconsistent with predictions of models of reciprocal stabilizing selection. Also explained in large degree is the evolution of male courtship pheromones in the Lepidoptera. The taxonomically untidy distribution of these pheromones, the diversity of their chemistry, and the specialized structures for their emission, form a sharp contrast with similar parameters of the female-pheromone system, and all are consistent with that predicted by the asymmetric-tracking model.

Although the study of insect sex pheromones has suggested some of the relationships between ecological parameters and the evolution of chemical communication, we obviously need more information in order to develop a more robust understanding of the evolutionary processes involved. As stated in the Introduction of this chapter, such information not only aids us in understanding the evolution of insect pheromones, but more importantly allows us to address larger questions of the evolution or reproductive isolation, speciation, and the relative contribution of sexual selection. Insect chemical ecology has the potential for major contributions to these fundamental areas of biology, although this potential has not been fully utilized. Evolutionary biologists could make better use of the database of insect sex pheromones in addressing fundamental questions of the evolution of mating systems and speciation processes. Recent studies emphasizing the nature of variance in pheromone production and response, as well as investigations of the patterns of inheritance, have been most instructive in this regard. Greater emphasis in these areas will allow a better appreciation of the feasibility of postulated evolutionary pathways. Specifically, asymmetric tracking would predict that future studies on the genetic control of pheromone systems will find that in limiting-sex-based pheromones, signal and response are more likely to be independently inherited, while systems such as male lepidopteran pheromones, for example, are more likely to be genetically linked. Comparing the genetic architecture of male- and female-pheromone systems, particularly within a single species, will provide an informative test of the role of intersexual selection in pheromone systems.

Finally, as with many evolutionary questions, the power of our conclusions is constrained by limitations in our phylogenetic understanding of the groups with which we are working. Thus, the biggest advances in the area of pheromone evolution that may be made in the future may be through collaborations between chemical ecologists and taxonomists, particularly in showing direction of evolution for characters or behavior (Roitberg, this volume). Those working with groups for which there is good phylogenetic information have a decided advantage in addressing these important fundamental questions.

Acknowledgements

I would like to thank R. B. Mellon and an anonymous reviewer for critical revisions to the manuscript. Also, I am greatly appreciative of the editors' extradordinary patience in its preparation.

References

Alcock, J. 1982. Natural selection and communication in bark beetles. *Fla. Entomol.* 65:17–32.

Alcock, J. 1989. *Animal Behavior: An Evolutionary Approach*. Sinauer, Sunderland, MA.

Alexander, R. D. 1962. Evolutionary change in cricket acoustical communication. *Evolution*, 16:443–467.

Aplin, R. T. and M. C. Birch. 1970. Identification of odorous compounds from male Lepidoptera. *Experentia* 26:1193.

Arnold, S. J. 1983. Sexual selection: the interface of theory and empiricism. In *Mate Choice*, ed. P. Bateson, pp. 67–107. Cambridge University Press, Cambridge, England.

Arnold, S. J. 1987. Quantitative genetic models of sexual selection: a review. In *The Evolution of Sex and Its Consequences*, ed. S. C. Stearns, pp. 283–315. Birkhäuser Verlag, Basel.

Bailey, J. B., L. M. McDonough, and M. P. Hoffmann. 1986. Western avocado leaf roller, *Amorbia cuneana* (Walsingham), (Lepidoptera: Tortricidae): discovery of populations utilizing different ratios of sex pheromone components. *J. Chem. Ecol.*, 12:1239–1245.

Bailliff, L. 1825. Examen microscopique de la Lupuline (matiere active de Houblon). *J. Chem. Med. Pharm. Tox., Paris*, 2:501–502.

Baker, T. C. 1985. Chemical control of behavior. In *Comprehensive Insect Physiology, Biochemistry, and Pharmacology*, eds. G. A. Kerkut and L. I. Gilbert, pp. 621–672. Pergamon Press, Oxford, England.

Baker. T. C. 1989. Origin of courtship and sex pheromones of the oriental fruit moth and a discussion of the role of phytochemicals in the evolution of lepidopteran male scents. In *Proc. Symp. Phytochemical Ecology: Allelochemicals, Mycotoxins, and Insect Pheromones and Allomones*, eds. C. H. Chou and G. R. Waller, pp. 401–418. Academia Sinica Monograph Series No. 9, Taipei, R.O.C.

Baker, T. C. and R. T. Cardé. 1979. Courtship behavior of the oriental fruit moth (*Grapholitha molesta*): experimental analysis and consideration of the role of sexual selection in the evolution of courtship pheromones in the Lepidoptera. *Ann. Entomol. Soc. Am.*, 72:173–188.

Baker, T. C., Nishida, R. and Roelofs, W. L. (1981). Close-range attraction of female oriental fruit moths to herbal scent of male hairpencils. *Science* 214:1359–1361.

Barrer, P. M. 1977. The influence of airborne stimuli from conspecific adults on the site of oviposition of *Ephestia cautella* (Lepidoptera: Phycitidae). *Ent. Exp. Appl.* 22:13–22.

Barrer, P. M., M. J. Lacey, and A. Shani. 1987. Variation in relative quantities of airborne sex pheromone components from individual female *Ephestia cautella* (Lepidoptera: Pyralidae). *J. Chem. Ecol.* 13:639–653.

Bengtsson, B. O. and C. Löfstedt. 1990. No evidence for selection in a pheromonally polymorphic moth population. *Am. Nat.* 136:722–726.

Bharath, R. 1987. Information theory. *Byte* December, 291–298.

Bijpost, S. C. A., G. Thomas, and J. P. Kruijt. 1985. Olfactory interactions between sexually active males in *Adoxophyes orana* (F.v.R.) (Lepidoptera: Tortricidae). *Behaviour* 45:121–137.

Birch, M. C. 1970. Pre-courtship use of abdominal brushes by the nocturnal moth, *Phlogophora meticulosa* (L.) (Lepidoptera: Noctuidae). *Anim. Behav.* 18: 310–316.

Birch, M. C. 1972a. Male abdominal brush-organs in British noctuid moths and their value as a taxonomic character. Part I. *The Entomol.* 105:185–205.

Birch, M. C. 1972b. Male abdominal brush-organs in British noctuid moths and their value as a taxonomic character. Part II. *The Entomol.* 105:233–244.

Birch, M. C. 1974. Aphrodisiac pheromones in insects. In *Pheromones*, ed. M. C. Birch, pp. 115–134. North Holland Publishing, Amsterdam.

Birch, M. C., D. Lucas, and P. R. White. 1989. The courtship behavior of the cabbage moth, *Mamestra brassicae* (Lepidoptera: Noctuidae), and the role of male hairpencils, *J. Insect Behav.* 2:227–239.

Birch, M. C., G. M. Poppy, and T. C. Baker. 1990. Scents and eversible scent structures of male moths. *Annu. Rev. Entomol.* 35:25–58.

Bjostad, L. B., C. E. Linn, J.-W. Du, and W. L. Roelofs. 1984. Identification of new sex pheromone components in *Trichoplusia ni*, predicted from biosynthetic precursors. *J. Chem. Ecol.* 10:1309–1323.

Bjostad, L. B., W. A. Wolf, and W. L. Roelofs. 1987. Pheromone biosynthesis in lepidopterans: desaturase and chain shortening. In. *Pheromone Biochemistry*, eds. G. D. Prestwich and G. L. Blomquist, pp. 77–120. Academic Press, New York.

Blair, W. F. 1955. Mating call and stage of speciation in the *Microhyla olivacea-M. carolinensis* complex. *Evolution* 9:469–480.

Booth, D. C., T. W. Phillips, A. Claesson, R. M. Silverstein, G. N. Lanier, and J. R. West. 1983. Aggregation pheromone components of two species of *Pissodes* weevils (Coleoptera: Curculionidae): isolation, identification, and field activity. *J. Chem. Ecol.* 9:1–12.

Boppré, M. 1984. Chemical mediated interactions between butterflies. In *The Biology of Butterflies*, eds. R. I. Vane-Wright and P. R. Ackery, pp. 259–275. Academic Press, New York.

Borden, J. H. 1985. Aggregation pheromone. In *Comprehensive Insect Physiology, Biochemistry, and Pharmacology*, Vol. 9, eds. G. A. Kerkut and L. I. Gilbert, pp. 257–285. Pergamon Press, Oxford, England.

Brower, L. P., J. Van Zant Brower, and F. P. Cranston. 1965. Courtship of the queen butterfly, *Danaus gilippus berenice* (Cramer). *Zoologica* 50:1–39.

Brown, W. L., Jr. and E. O. Wilson. 1956. Character displacement. *Syst. Zool.* 5:49–64.

Butlin, R. 1987. Speciation by reinforcement. *Trends Ecol. & Evol.* 2:8–13.

Butlin, R. 1989. Reinforcement of premating isolation. In *Speciation and its Consequences* eds. D. Otte and J. A. Endler, pp. 158–179. Sinauer, Sunderland, MA.

Cardé, R. T. 1986. The role of pheromones in reproductive isolation and speciation of insects. In *Evolutionary Genetics of Invertebrate Behavior*, ed. M. D. Huettel, pp. 303–317. Plenum Press, New York.

Cardé, R. T., T. C. Baker, and W. L. Roelofs. 1976. Sex attractant responses of male oriental fruit moths to a range of component ratios: pheromone polymorphism? *Experientia*, 32:1406–1407.

Cardé, R. T. and T. C. Baker. 1984. Sexual communication with pheromones. In *Chemical Ecology of Insects* eds. W. J. Bell and R. T. Cardé, pp. 355–383. Chapman and Hall, London.

Carson, H. L. and A. R. Templeton. 1984. Genetic revolutions in relation to speciation phenomena: the founding of new populations. *Annu. Rev. Ecol. Syst.* 15:97–131.

Collins, R. D. and Cardé. 1985. Variation in and heritability of aspects of pheromone production in the pink bollworm moth, *Pectinophora gossypiella* (Lepidoptera: Gelechiidae). *Ann. Entomol. Soc. Am.* 78:229–234.

Darwin, C. 1871. *The Descent of Man and Selection in Relation to Sex*. John Murray, London.

DeJong, M. C. M. 1988. *Evolutionary Approaches to Insect Communication Systems*. Ph.D. dissertation, University of Leiden, Leiden, The Netherlands.

Dobzhansky, T. 1940. Speciation as a stage in evolutionary divergence. *Am. Nat.* 74:312–321.

Doherty, J. A. and R. R. Hoy. 1985. The auditory behavior of crickets: some views of genetic coupling, song recognition, and predator detection, *Q. Rev. Biol.* 60:457–472.

Du, J.-W., C. Löfstedt, and J. Löfqvist. 1987. Repeatability of pheromone emissions from individual female ermine moths *Yponomeuta padellus* and *Yponomeuta rorellus. J. Chem. Ecol.* 13:1431–1441.

Edgar, J. A., C. C. J. Culvenor, and T. E. Pliske. 1976. Isolation of a lactone, structurally related to the esterifying acids of pyrrolizidine alkaloids, from the costal fringes of male Ithomiinae. *J. Chem. Ecol.* 2:263–270.

Eisner, T. and J. Meinwald. 1987. Alkaloid-derived pheromones and sexual selection in Lepidoptera. In *Pheromone Biochemistry.* eds. G. D. Prestwich and G. J. Blomquist, pp. 251–269. Academic Press, New York.

Emlen, S. T. and L. W. Oring. 1977. Ecology, sexual selection, and the evolution of mating systems. *Science* 197:215–223.

Ewing, A. M. 1989. *Arthropod Bioacoustics: Neurobiology and Behavior.* Cornell University Press, Ithaca, New York.

Fabré, J. H. 1911. *Social Life in the Insect World.* Ernest Benn Ltd., London.

Fisher, R. A. 1930. *The Genetical Theory of Natural Selection.* Oxford University Press, Oxford, England.

Fitzpatrick, S. M. and J. N. McNeil. 1988. Male scent in lepidopteran communication: the role of male pheromone in mating behaviour of *Pseudoletia unipuncta* (Haw.) (Lepidoptera: Noctuidae). *Mem. Ent. Soc. Can.* 146:131–151.

Fitzpatrick, S. M., J. N. McNeil, and S. Dumont. 1988. Does male pheromone effectively inhibit competition among courting true armyworm males (Lepidoptera: Noctuidae)? *Anim. Behav.* 36:1831–1835.

Ganyard, M. C. and U. E. Brady. 1972. Interspecific attraction in Lepidoptera in the field. *Ann. Entomol. Soc. Am.* 65:1279–1282.

Glover, T., M. Campbell, P. Robbins, and W. Roelofs. 1990. Sex-linked control of sex pheromone behavioral responses in European corn-borer moths (*Ostrinia nubilalis*) confirmed with TPI marker gene. *Arch. Insect Biochem. and Physiol.* 15:67–77.

Grant, A. J., R. J. O'Connell, and A. M. Hammond. (1987). A comparative study of the neurophysiological response characteristics of olfactory receptor neurons in two species of noctuid moths. In: *Olfaction and Taste IX* (S. D. Roper, and J. Atema, eds.) pp. 311–314. New York Academy of Sciences, New York.

Grant, G. G. 1974. Male sex pheromone from the wing glands of the Indian meal moth, *Plodia interpunctella* (Hbn.) (Lepidoptera: Phycitidae). *Experientia* 30:917–918.

Grant, G. G. 1976. Courtship behavior of a phycitid moth, *Vitula edmandsae. Ann. Entomol. Soc. Am.* 69:445–449.

Grant, G. G. and J. L. Eaton. 1973. Scent brushes of the male tobacco hornworm *Manduca sexta* (Lepidoptera: Sphingidae). *Ann. Entomol. Soc. Am.* 66:901–904.

Grant, G. G., E. B. Smithwick, and U. E. Brady. 1975. Courtship behavior of phycitid moths. II. Behavioral and pheromonal isolation of *Plodia interpunctella* and *Cadra cautella* in the laboratory. *Can. J. Zool.* 53:827–832.

Grant, V. 1966. The selective origin of incompatibility barriers in the plant genus *Gilia*. *Am. Nat.* 100:99–118.

Greenfield, M. D. 1981. Moth sex pheromones: an evolutionary perspective. *Fla. Entomol.* 64:4–17.

Greenfield, M. D. and M. G. Karandinos. 1979. Resource partitioning of the sex communication channel in clearwing moths (Lepidoptera: Sesiidae) of Wisconsin. *Ecol. Monogr.* 49:403–426.

Grula, J. W. and O. R. Taylor. 1979. The inheritance of pheromone production in the sulfur butterflies, *Colias eurytheme* and *C. philodice*. *Heredity* 42:359–371.

Grula, J. W. and O. R. Taylor. 1980. The effect of X-chromosome inheritance on mate-selection behavior in the sulfur butterflies, *Colias eurytheme* and *C. philodice*. *Evolution* 34:688–695.

Hammerstein, P. and G. A. Parker. 1987. Sexual selection: games between the sexes. In *Sexual Selection: Testing the Alternatives*, eds. J. W. Bradbury and M. B. Andersson, pp. 119–142. John Wiley, Chichester, England.

Hansson, B. S., C. Löfstedt, S. P. Foster. 1989. Z-linked inheritance of male olfactory response to sex pheromone components in two species of tortricid moths, *Ctenopseustis obliquana* and *Ctenopseustis sp. Entomol. Exp. Appl.* 53:137–145.

Hansson, B. S., M. Tóth, C. Löfstedt, G. Szöcs, M. Subchev, and J. Löfqvist. 1990. Pheromone variation among eastern European and a western Asian population of the turnip moth, *Agrotis segetum. J. Chem. Ecol.* 16:1611–1622.

Haynes, K. F. and T. C. Baker. 1988. Potential for evolution of resistance to pheromones: worldwide and local variation in chemical communication system of pink bollworm moth, *Pectinophora gossypiella. J. Chem. Ecol.* 14:1547–1560.

Heinrich, C. 1956. American moths of the subfamily Phycitinae. *U.S. Natl. Mus. Bull.* 207:1–581. Smithsonian Institute, Washington, DC.

Henderson, N. R. and D. M. Lambert. 1982. No significant deviation from random mating of worldwide populations of *Drosophila melanogaster*. *Nature* 300:437–440.

Hirai, K., H. H. Shorey, and L. K. Gaston. 1978. Competition among courting male moths: male-to-male inhibitory pheromone. *Science* 202:644–645.

Hoy, R. R., J. Hahn, and R. C. Paul. 1977. Hybrid cricket auditory behavior: evidence for genetic coupling in animal communication. *Science* 195:82–84.

Hunt, R. E., B.-G. Zhao, and K. F. Haynes. 1990. Genetic aspects of interpopulational differences in pheromone blend of cabbage looper moth, *Trichoplusia ni. J. Chem. Ecol.* 16:2935–2946.

Kirkendall, L. R. 1983. The evolution of mating systems in bark and ambrosia beetles (Coleoptera: Scolytidae and Platypodidae) *Zool. J. Linn. Soc.* 77:293–352.

Kirkpatrick, M. 1982. Sexual selection and the evolution of female choice. *Evolution* 36:1–12.

Kirkpatrick, M. 1987. The evolutionary forces acting on female mating preferences in polygynous animals. In *Sexual Selection: Testing the Alternatives* eds. J. W. Bradbury and M. B. Andersson, pp. 67–82. John Wiley, Chichester, England.

Klun, J. A. and S. Maini. 1979. Genetic basis of an insect chemical communication system: the European corn borer. *Environ. Entomol.* 8:423–426.

Klun, J. A., J. R. Plimmer, B. A. Bierl-Leonhardt, A. N. Sparks, and O. L. Chapman. 1979. Trace chemicals: the essence of sexual communication. *Science* 204:1328–1330.

Klun, J. A. and M. D. Huettel. 1988. Genetic regulation of sex pheromone production and response: interaction of sympatric pheromonal types of European corn borer, *Ostrinia nubilalis* (Lepidoptera: Pyralidae). *J. Chem. Ecol.* 14:2047–2061.

Kochansky, J., R. T. Cardé, J. Liebherr, and W. L. Roelofs. 1975. Sex pheromone of the European corn borer. *J. Chem. Ecol.* 1:225–231.

Krasnoff, S. B., K. W. Vick, and R. W. Mankin. 1983. Female calling behavior in *Ephestia elutella* and *E. figulilella* (Lepidoptera: Pyralidae). *Fla. Entomol.* 66:249–254.

Krasnoff, S. B. and K. W. Vick. 1984. Male wing-gland pheromone of *Ephestia elutella*. *J. Chem. Ecol.* 10:667–679.

Krasnoff, S. B. and W. L. Roelofs. 1989. Quantitative and qualitative effects of larval diet on male scent secretions of *Estigmene acrea, Phragmatobia fulginosa* and *Pyrrharctia isabella* (Lepidoptera: Arctiidae). *J. Chem. Ecol.* 15:1077–1093.

Krasnoff, S. B. and W. L. Roelofs. 1990. Evolutionary trends in the male pheromone systems of arctiid moths: evidence from studies of courtship in *Phragmatobia fuliginosa* and *Pyrrharctia isabella* (Lepidoptera: Arctiidae). *Zool. J. Linn. Soc.* 99:319–338.

Krebs, J. R. and R. Dawkins. 1984. Animal signals: mind-reading and manipulation. In *Behavioral Ecology: An Evolutionary Approach*, 2nd ed., eds. J. R. Krebs and N. B. Davies, pp. 380–402. Sinauer Associates, Sunderland, MA.

Kunesch, G., P. Zagatti, J. Y. Lallemand, A. Debal, and J. P. Vigneron. 1981. Structure and synthesis of the wing gland pheromone of the male African sugar-cane borer: *Eldana saccharina* (Wlk.) (Lepidoptera: Pyralidae). *Tetrahedron Lett.* 22:5271–5274.

Lambert, D. M., B. Michaux, and C. S. White. 1987. Are species self-defining? *Syst. Zool.* 36:196–205.

Lande, R. 1982. Rapid origin of sexual isolation and character divergence in a cline. *Evolution* 36:213–223.

Levinson, H. Z. and C. Th. Buchelos. 1981. Surveillance of storage moth species (Pyralidae, Gelechiidae) in a flour mill by adhesive traps with notes on the pheromone-mediated flight behaviour of male moths, *Z. ang. Entomol.* 92:233–251.

Linn, C. E. Jr., L. B. Bjostad, J.-W. Du, and W. L. Roelofs. 1984. Redundancy in a chemical signal: behavioral responses of male *Trichoplusia ni* to a six-component sex pheromone blend. *J. Chem. Ecol.* 10:1635–1658.

Linn, C. E. and W. L. Roelofs. 1985. Response specificity of male pink bollworm moths to different blends and dosages of sex pheromone. *J. Chem. Ecol.* 11:1583–1590.

Linn, C. E. Jr., J.-W. Du, A. Hammond, and W. L. Roelofs. 1987. Identification of unique pheromone components for the soybean looper moth *Psuedoplusia includens.* *J. Chem. Ecol.* 13:1351–1360.

Linn, C. E. Jr., A. Hammond, J.-W. Du, and W. L. Roelofs. 1988. Specificity of male response to multicomponent pheromones in noctuid moths *Trichoplusia ni* and *Pseudoplusia includens.* *J. Chem. Ecol.* 14:47–57.

Littlejohn, M. J. 1981. Reproductive isolation: a critical review. In *Evolution and Speciation*, eds. W. R. Atchley and D. S. Woodruff, pp. 298–334. Cambridge Univ. Press, Cambridge, England.

Löfstedt, C. 1990. Population variation and genetic control of pheromone communication systems in moths. *Entomol. Exp. Appl.* 54:199–218.

Löfstedt, C. and J. N. C. van der Pers. 1985. Sex pheromones and reproductive isolation in four European small ermine moths. *J. Chem. Ecol.* 11:649–666.

Löfstedt, C., W. M. Herrebout, and J. -W. Du. 1986a. Evolution of the ermine moth pheromone tetradecyl acetate. *Nature* 323:621–623.

Löfstedt, C., J. Löfqvist, B. S. Lanne, J. N. C. van der Pers, and B. S. Hansson. 1986b. Pheromone dialects in European turnip moths *Agrotis segetum*. *Oikos* 46:250–257.

Löfstedt, C., B. S. Hansson, W. L. Roelofs, and B. O. Bengtsson. 1989. No linkage between genes controlling female pheromone production and male pheromone response in the European corn borer, *Ostrinia nubilalis* Hübner (Lepidoptera: Pyralidae). *Genetics* 123:553–556.

Löfstedt, C., B. S. Hansson, H. J. Dijkerman, and W. M. Herrebout. 1990. Behavioral and electrophysiological activity of unsaturated analogues of the pheromone tetradecyl acetate in the small ermine moth *Yponomeutidae rorellus*. *Physiol. Entomol.* 15:47–54.

Mayr, E. 1970. *Populations, Species, and Evolution*. Belknap Press, Cambridge, MA.

McDonough, L. M., D. F. Brown, and W. C. Aller. 1989. Insect sex pheromones: effect on temperature on evaporation rates of acetates from rubber septa. *J. Chem. Ecol.* 15:779–790.

McLaughlin, J. R. 1982. Behavioral effect of a sex pheromone extracted from forewings of male *Plodia interpunctella*. *Environ. Entomol.* 11:378–380.

Menken, S. B. J. 1980. Inheritance of allozymes in *Yponomeuta* II. Interspecific crosses within the padellus complex and reproductive isolation. *Proc. Kon. Ned. Acad. Wet.* (C) 83:425–431.

Miller, J. R. and W. L. Roelofs. 1980. Individual variation in sex pheromone component ratios in two populations of the redbanded leafroller moth, *Argyrotaenia velutinana*. *Environ. Entomol.* 9:359–363.

Mousseau, T. A. and D. A. Roff. 1987. Natural selection and the heritability of fitness components. *Heredity* 59:181–197.

Müller, F. 1878. Scent organs in butterflies. *Trans. Entomol. Soc.* (*London*) 1878:211–221.

O'Donald, P. 1962. The theory of sexual selection. *Heredity*. 17:541–552.

Paterson, H. E. H. 1978. More evidence against speciation by reinforcement. *South African J. Sci.* 74:369–371.

Paterson, H. E. 1980. A comment on 'mate recognition systems.' *Evolution* 34:330–331.

Paterson, H. E. H. 1985. The recognition concept of species. In *Species and Speciation*, ed. E. S. Vrba, pp. 21–29. Transvaal Museum Monograph No. 4. Transvaal Museum, Pretoria.

Paterson, H. 1989. A view of species. In *Dynamic Structures in Biology*, eds. B. Goodwin, A. Sibatani, and G. Webster, pp. 77–88. Edinburgh Univ. Press, Edinburgh.

Phelan, P. L. 1989. The evolution of male courtship pheromones in the Lepidoptera. In *Proc. 6th Annu. Internat. Soc. Chem. Ecol.*, Göteborg, Sweden.

Phelan, P. L. and T. C. Baker. 1986a. Cross-attraction of five species of stored-product Phycitinae (Lepidoptera: Pyralidae) in a wind tunnel. *Environ. Entomol.* 15:369–372.

Phelan, P. L. and T. C. Baker. 1986b. Male-size-related courtship success and intersexual selection in the tobacco moth. *Ephestia elutella. Experientia* 42:1291–1293.

Phelan, P. L., P. J. Silk, C. J. Northcott, S. H. Tan, and T. C. Baker. 1986. Chemical identification and behavioral characterization of male wing pheromone of *Ephestia elutella* (Pyralidae). *J. Chem. Ecol.* 12:135–146.

Phelan, P. L. and T. C. Baker. 1987. Evolution of male pheromones in moths: reproductive isolation through sexual selection. *Science* 235:205–207.

Phelan, P. L. and T. C. Baker. 1990a. Comparative study of courtship in twelve phycitine moths (Lepidoptera: Pyralidae). *J. Insect. Behav.* 3:303–326.

Phelan, P. L. and T. C. Baker. 1990b. Information transmission during intra- and interspecific courtship in *Ephestia elutella* and *Cadra figulilella*. *J. Insect Behav.* 3:589–602.

Phillips, T. W. and G. N. Lanier. 1983. Biosystematics of *Pissodes* Germar (Coleoptera: Curculionidae): feeding preference and breeding site specificity in *P. strobi* and *P. approximatus*. *Can. Entomol.* 115:1627–1636.

Phillips, T. W. and G. N. Lanier. 1986. Interspecific activity of semiochemicals among sibling species of *Pissodes* (Coleoptera: Curculionidae). *J. Chem. Ecol.* 12:1587–1601.

Pliske, T. E. 1975. Attraction of Lepidoptera to plants containing pyrrolizidine alkaloids. *Environ. Entomol.* 4:455–473.

Pliske, T. E., J. A. Edgar, and C. C. J. Culvenor. 1976. The chemical basis of attraction of ithomiine butterflies to plants containing pyrrolizidine alkaloids. *J. Chem. Ecol.* 2:255–262.

Renou, M., B. Lalanne-Cassou, D. Michelot, G. Gordon, and J.-C. Doré. 1988. Multivariate analysis of the correlation between Noctuidae subfamilies and the chemical structure of their sex pheromones or male attractants. *J. Chem. Ecol.* 14:1187–1215.

Roelofs, W. L., A. Hill, and R. T. Cardé. 1975. Sex pheromone components of the redbanded leaf roller, *Argyrotaenia velutinana* (Lepidoptera: Tortricidae). *J. Chem. Ecol.* 1:83–89.

Roelofs, W. L. and R. L. Brown. 1982. Pheromones and evolutionary relationships of Tortricidae. *Annu. Rev. Ecol. Syst.* 13:395–422.

Roelofs, W. L., J.-W. Du, X.-H. Tang, P. S. Robbins, and C. J. Eckenrode. 1985. Three European corn borer populations in New York based on sex pheromone and voltinism. *J. Chem. Ecol.* 11:829–836.

Roelofs, W. L., T. Glover, X.-H. Tang, I. Sreng, P. Robbins, C. Eckenrode, C. Löfstedt, B. S. Hansson, and B. O. Bengtsson. 1987. Sex pheromone production and perception in European corn borer moths is determined by both autosomal and sex-linked genes. *Proc. Natl. Acad. Sci. USA* 84:7585–7589.

Roesler, R. V. 1973. *Die trifinen Acrobasiina der Phycitinae* (*Lepidoptera*, *Pyralidae*). G. Fromme, Vienna.

Schneider, D. 1987. The strange fate of pyrrolizidine alkaloids. In *Perspectives in Chemoreception and Behavior* eds. R. F. Chapman, E. A. Bernays, and J. G. Stoffolano, Jr. pp. 123–142. Springer-Verlag, New York.

Scott, D. and R. C. Richmond. 1988. A genetic analysis of male-predominant pheromones in *Drosophila melanogaster*. *Genetics* 119:639–646.

Scudder, S. H. 1877. Antigeny or sexual dimorphism in butterflies. *Proc. Am. Acad. Arts Sci.* 12:150–158.

Shannon, C. E. and W. Weaver. 1949. *The Mathematical Theory of Communication.* Univ. of Illinois Press, Urbana.

Smith, C. E., N. Allen, and O. A. Nelson. 1943. Some chemotropic studies with *Autographa sp. J. Econ. Entomol.* 36:619–621.

Spiess, E. B. 1987. Discrimination among perspective mates in *Drosophila* In *Kin Recognition in Animals*, eds. D. J. C. Fletcher and C. D. Michener, pp. 75–119. John Wiley, New York.

Stadelbacher, E. A., M. W. Barry, A. K. Raina, and J. R. Plimmer. 1983. Fatal interspecific mating of two *Heliothis* species induced by synthetic sex pheromone. *Experientia* 39:1174–1176.

Stearns, S. C. 1987. Why sex evolved and the difference it makes. In *The Evolution of Sex and Its Consequences*, ed. S. C. Stearns pp. 15–31. Birkhäuser Verlag, Basel.

Tamaki, Y. 1985. Sex pheromones. In *Comprehensive Insect Physiology, Biochemistry, and Pharmacology*, Vol. 9, eds. G. A. Kerkut and L. I. Gilbert, pp. 145–192. Pergamon Press, Oxford.

Templeton, A. R. 1981. Mechanisms of speciation: a population genetic approach. *Annu. Rev. Ecol. Syst.* 12:23–48.

Templeton, A. R. 1989. The meaning of species and speciation: a genetic perspective. In *Speciation and Its Consequences*, eds. D. Otte and J. A. Endler, pp. 3–27. Sinauer, Sunderland, MA.

Thornhill, R. and J. Alcock. 1983. *The Evolution of Insect Mating Systems*. Harvard Univ. Press, Cambridge.

Trivers, R. L. 1972. Parental investment and sexual selection. In *Sexual Selection and the Descent of Man, 1871–1971*, ed. B. Campbell, pp. 136–179. Aldine Publishing, Chicago.

Walker, T. J. 1974. Character displacement and acoustic insects. *Amer. Zool.* 14:1137–1150.

White, M. J. 1978. *Modes of Speciation*. W.H. Freeman, San Francisco.

Willis, M. A. and M. C. Birch. 1982. Male lek formation and female calling in a population of the arctiid moth, *Estigmene acrea*. *Science* 218:168–170.

Williams, G. C. 1966. *Adaptation and Natural Selection*. Princeton Univ. Press, Princeton.

Wilson, E. O. 1975. *Sociobiology: The New Synthesis*. Harvard Univ. Press, Cambridge, MA.

Witzgall, P. and B. Frérot. 1989. Pheromone emission by individual females of the carnation tortrix, *Cacoecimorpha pronubana*. *J. Chem. Ecol.* 15:707–717.

Wunderer, H., K. Hansen, T. W. Bell, D. Schneider, and J. Meinwald. 1986. Sex pheromones of two Asian moths (*Creatonotus transiens*, *C. gangis*; Lepidoptera: Arctiidae): behavior, morphology, chemistry and electrophysiology. *Exp. Biol.* 46:11–27.

Zagatti, P., G. Kunesch, and N. Morin. 1981. La vanilline, constituant majoritaire de la sécrétion aphrodisiaque émise par les androconies du mâle de la Pyrale de la Canne à sucre: *Eldana saccharina* (Wlk.) (Lépidoptère, Pyralidae, Gallerinae). *C. R. Acad. Sc. Paris* 292: Série III, 633–635.

Zagatti, P., G. Kunesch, F. Ramiandrasoa, C. Mallose, D. R. Hall, R. Lester, and B. F. Nesbitt. 1987. Sex pheromones of rice moth. *Corcyra cephalonica* Stainton. I. Identification of male pheromone. *J. Chem. Ecol.* 13:1561–1573.

10

Semiochemicals and Insect Sociality

Mark L. Winston

Simon Fraser University

Semiochemicals (message-bearing chemicals) are the glue that holds social insect colonies together. In the most profound sense, individual insects in their societies are bound to each other by the odors they produce and acquire, allowing the colony to develop as much more than a simple sum of individual behaviors. While social insects are rich in communicative mechanisms besides semiochemicals, it is chemical communication which is most important in mediating the complex interactions which characterize the social insects.

It is not surprising that this should be so, since even the nonsocial insects exhibit great diversity and ability in chemical communication. Insects are rich in glands that can produce message-carrying chemicals and have extraordinarily acute receptive and integrative systems to utilize the information carried by semiochemicals. Even completely solitary insects produce semiochemicals, particularly for mate attraction and nest- and forage-marking. In the social insects, semiochemicals are used in alarm behavior, recruitment, the construction of odor trails, kin and colony recognition, the control of reproduction, caste determination, trophallaxis, nest entrance-marking, and the regulation and integration of colony activities (reviewed by Bradshaw and Howse, 1984; Duffield et al., 1984; Free, 1987; Howse, 1984; Nault and Phelan, 1984).

While much is known about many of these more specific functions, the role of semiochemicals in colony organization is still poorly understood, with few chemicals identified and the relationship between specific odors and behaviors often obscure. It is not my intention in this chapter to review the vast literature on social insect pheromones. Rather, I will concentrate on discussing the function and evolution of social insect semiochemicals in: (1) kin and colony recognition, (2) the control of reproduction and queen number, and (3) the distribution of semiochemicals through the nest. Also, I will only discuss the truly social (eusocial) insects, in which colonies consist of one or more reproductive queens and workers that are not reproductive, or at least reproduce notably less than the queen. These include all of the termites (Isoptera) and the ants (Hymenoptera), and many of the wasps and bees (Hymenoptera).

10.1. Semiochemicals and Recognition

Recognition of kin, nest, and colony is central to social behavior, since it is only through recognition that behaviors can be performed that favor relatives and nestmates over strange, unrelated individuals. The theoretical need for recognition mechanisms has been understood for some time, particularly since the development of kin recognition theory emphasized the importance of individuals performing behaviors that favored close relatives (Hamilton, 1964). According to kin selection theory, individuals should act the most altruistically toward other individuals with whom they have the largest proportion of alleles in common. These arguments are particularly compelling for the hymenopterous social insects since, because of their haplo-diploid mechanism of sex determination, asymmetries of relatedness are found in nests such that sister workers are more closely related to each other than they would be to their own daughters.

The altruistic acts that are predicted by this theory depend on the ability to discriminate nestmates from non-nestmates, and sisters from less-related relatives. While the basis of this recognition is largely olfactory, not a single compound or blend has been identified to date that can conclusively be labelled as a recognition odor. In the last ten years, though, much has been learned about the functioning of chemically-based recognition, and the importance of these systems in social insect colony organization is clear.

Some level of kin and/or colony recognition has been found in virtually all social species that have been examined, including ants, bees, wasps and termites, although very little is known about termites (Thorne, 1982). The classic study on recognition was performed by Greenberg (1979) on *Lasioglossum zephyrum*, but was preceded by numerous studies, especially at the University of Kansas (particularly Barrows et al., 1975; for a detailed review, see Michener and Smith, 1987). *L. zephyrum* is a primitively eusocial sweat bee in the family Halictidae, whose nests are in earthen burrows. Colonies are generally founded by a single queen, and after adults are produced, colonies consist of that queen and 2 to 20 females. A few of these non-reproductive females act as guards at the nest entrance, while the remaining females are workers, performing tasks such as foraging and nest construction, or remaining largely inactive in the nest. Nests are aggregated, so that the possibility of individuals drifting between nests is high. Further, since many females establish their nests near their parental nest (Kukuk and Decelles, 1986), potential drifters could include both related and unrelated individuals. Thus, the ability to discriminate between nestmates and to discern the degree of relatedness of non-nestmates would be beneficial in this species.

Greenberg's (1979) experimental design, which has been utilized in many variations by subsequent researchers, involved determining the relationship between relatedness and various inter-individual behaviors. In *L. zephyrum*, Greenberg examined the interactions at nest entrances between guard bees and incoming bees, reasoning that closely-related individuals should be admitted more readily than distantly-related or unrelated individuals. Using known matings that produced sisters, aunts-nieces, cousins, and unrelated individuals, Greenberg was able to demonstrate a highly significant positive relationship between the coefficient of relationship and the proportion of females that the guards allowed to pass. The discriminators used in this behavior are largely produced by the females themselves, and females learn the odors of their nestmates rather than learning their own odor and comparing that to foreign odors (Buckle and Greenberg, 1981; Kukuk et al., 1977). Males of *L. zephyrum* are also capable of distinguishing their relationship to females and seem to use this information in choosing mates (Smith, 1983).

The ability of *L. zephyrum* males and females to discriminate on the basis of relatedness has important implications for the evolution of social behavior in this species. For females, there is a clear advantage to maintaining nests of closely related individuals, so that work in the nest is performed for relatives rather than strangers. The density of nests at aggregations would be a strong selective factor favoring discrimination between nestmates and strangers. The likelihood that some incoming bees may be non-nestmates but closely related would favor those bees that could discriminate degrees of relationship, since the admission of closely related non-nestmates would enhance the fitness of the admitting nest and possibly that of the incoming female as well. Relatedness of nests in aggregations can be sufficiently high to indicate that kin selection is important in those aggregations (Crozier et al., 1987). Further, Smith (1987) has demonstrated that queens in colonies that are artificially made up of related individuals are less aggressive toward workers in those colonies than are queens in colonies of unrelated individuals. Interestingly, there were fewer worker deaths in colonies of related individuals, possibly due to the lower level of aggressive behavior. Also, there were fewer workers with enlarged ovaries, and more admission of workers into nests, in the related colonies. Thus, the related colonies functioned significantly better than the unrelated colonies, which may have been an important factor in the evolution of social behavior in *L. zephyrum*. Finally, the ability of males to discriminate the relatedness of mates would reduce potentially deleterious inbreeding. Thus, while these considerations of evolutionary factors are speculative, it is certainly plausible to propose that semiochemicals in this insect have been an important part of its social evolution.

A major gap in our understanding of discriminator odors in *L. zephyrum*, as for all social insects, is the nature of the chemicals themselves. We are tantalizingly close to identifying these bee-produced pheromones, but have yet to make the major breakthrough that would definitively identify these compounds. The most likely recognition pheromone used by females is a blend of four macrocyclic lactones and their monounsaturated homologues which are found in the Dufour's gland (Smith and Wenzel, 1988; Smith et al., 1985). These lactones are important components of the headspace around females, are more similar between pairs of nestmates than random pairs, and elicit behaviors similar to natural extracts, although females are more responsive to natural extracts than to synthetic mixtures (Smith et al., 1985; Wcislo, 1987). At any rate, it appears likely that a complex blend is involved in recognition and that ratios of components may be more important than compounds distinct to family groups.

Semiochemical-based kin recognition has also been demonstrated in the primitively eusocial *Polistes* wasps (reviewed by Gamboa et al., 1986a and Michener and Smith, 1987). As for *L. zephyrum*, the ability to discriminate kin is important in regulating colony structure and, ultimately, in the evolution of social behavior in this group. *Polistes*, at least in temperate climates, produce gynes in the late summer or autumn, which diapause in clusters through the winter. In the spring, these gynes, now called foundresses, disperse and found new nests, often close to the site of their natal nest. Foundresses will frequently cofound nests, with two or more foundresses joining at the nest in the spring. In those cases, there is a dominant queen who lays most of the eggs and subordinates who lay eggs relatively rarely. Kin selection theory suggests, and experimental evidence supports, that multiple founding is mutually advantageous, primarily when the foundresses are sisters. Thus, kin recognition between adults, as well as recognition of brood on the nest, is expected if this system is to operate in a manner consistent with the expectations of kin selection theory (Klahn, 1979; Klahn and Gamboa, 1983).

Since cofoundresses are, on average, closely related (Metcalf and Whitt, 1977), recent research has concentrated on investigating whether or not nestmates can recognize their kin and, if so, what mechanisms might be involved. In numerous studies, preferences for multiple founding with sisters have been demonstrated, as well as increased aggressive behavior between non-sisters (Bornais et al., 1983; Noonan, 1979; Post and Jeanne, 1982; Ross and Gamboa, 1981). Further, foreign nests and brood are recognized and sometimes destroyed (Klahn and Gamboa, 1983). This ability would be particularly useful in multiple foundress nests and during attempted nest usurpations by other queens. There appears to be a recognition threshold, however; in *Polistes fuscatus*, aunts, nieces, and a minority of cousins are accepted by gynes onto nests, but most cousins and unrelated individuals are rejected (Gamboa, 1988).

While the precise source of the recognition cues has not been isolated, it is clear that semiochemicals are involved, both environmentally acquired and genetically determined. Recognition cues are learned rapidly following adult emergence, from nests, nest fragments, and/or brood (Gamboa et al., 1986b; Klahn and Gamboa, 1983; Pfennig et al., 1983a, b; Shellman and Gamboa, 1982). The recognition semiochemicals found in nests or brood could originate environmentally, from the pulp that makes up the paper nest or from food, but also may originate from the adult wasps; for example, females often wipe their sternal glands onto nests (Jeanne, Downing, and Post, 1983). Thus, as for *L. zephyrum*, semiochemically-based kin recognition is an important part of the evolution and functioning of sociality in these insects.

Semiochemically-based kin recognition may reach its most complex level in the honey bee, *Apis mellifera* L. (recently reviewed by Breed, 1985; Breed and Bennett, 1987; Getz, 1989), both because of the high level of sociality in this insect as well as multiple mating of honey bee queens. Honey bee colonies are monogynous, but since the queen mates with up to 17 drones, colonies are mixtures of half and full sisters (Laidlaw and Page, 1984; Page, 1986; Page et al., 1984; Winston, 1987, and references cited therein). This genetic complexity results in a situation where each worker would be expected to perform activities favoring not only her own nest-mates, but her full sisters as well. There is now strong evidence that workers (1) recognize nestmates and prevent non-nestmates from entering the nest (Breed, 1983; Kalmus and Ribbands, 1952; Ribbands, 1954); (2) recognize full sisters versus nonsisters and half sisters, and are more aggressive toward half sisters and unrelated workers than toward full sisters (Breed et al., 1985; Evers and Seeley, 1986; Frumhoff and Schneider, 1987; Getz and Smith, 1986); (3) are able to recognize their own queen (Ambrose et al., 1979; Boch and Morse, 1974, 1979, 1981; Breed, 1981; Page and Erickson, 1986); and (4) prefer sister larvae to unrelated, cousin, and half-sister larvae for rearing new queens, although not under all conditions (Breed et al., 1984; Noonan, 1986; Page and Erickson, 1984; Visscher, 1986). However, some caution should be used in accepting the sister and half sister discrimination experiments, since they involved laboratory colonies with only two or three patrilines, and in an active colony, workers may be faced with discriminating between many different choices.

Nevertheless, it is clear that workers possess the ability to discriminate between nestmates and non-nestmates and to recognize kin, at least under relatively simple conditions. The semiochemical basis of these discriminatory capabilities is also clear, but the sensory mechanisms used by bees and the odors involved remain obscure. There are two sources of recognition cues, and, as for the other species studied, both environmentally acquired and genetically determined odors seem to be involved. Since recognition can be influenced by applying floral-like scents to colonies (Ambrose et al., 1979; Breed et al., 1988; Renner, 1960), some recognition component involving externally acquired odors is likely. However, cuticular hydrocarbon patterns and/or mandibular gland secretions have also been proposed as a genetically-based recognition system. Crewe (1982) demonstrated that mandibular gland secretions of queens of different honey bee races showed consistent patterns of 8, 9, and 10 carbon keto- and hydroxy-acids, and that queens could be grouped by race according to these patterns. Racial and age-based differences in cuticular waxes have also been demonstrated (Blomquist et al., 1980; Carlson and Bolten, 1984; McDaniel et al., 1984, 1987), with at least 20 compounds identified which could contribute to a wax signature (Page, cited in Getz, 1989). Getz and Smith (1987) have shown that workers can discriminate between mixtures of closely related alkanes, and workers can also distinguish subtle differences in the concentrations of compounds (Kramer, 1976). Further, at least some of the variability in hydrocarbon extracts of individual workers is genetically determined, and provides sufficiently reliable genetic information to function as kin recognition labels (Page, Metcalf, Metcalf, Erickson, and Lampnon, in prep.).

A fairly consistent pattern has emerged from these and recognition studies in other social insects. First, nestmate and kin recognition has been found in many contexts and in species where it would be expected based on kin recognition theory and the need for colonies to maintain their genetic integrity. Second, this recognition is based on semiochemicals, both environmentally acquired odors and pheromones produced by the bees themselves. Third, the sources of the insect-produced odors may differ between groups, with mandibular, sternal, and Dufours glands being proposed in various species, as well as cuticular hydrocarbon blends. Finally, the recognition odors are likely complex blends, with individuals distinguishing differences between ratios of compounds rather than the presence or absence of particular components. It is clear that semiochemicals have the major role in maintenance of recognition-based colony functions and have been a significant force in the evolution of insect societies.

10.2. The Control of Reproduction and Queen Number

Control over reproduction in social insect colonies is important in two ways. First, control over the timing and extent of reproduction is essential to maximize offspring fitness as well as a colony's own fitness in the case of perennial colonies. Secondly, there is conflict in the nest between queens and workers in the extent of worker egg laying, with queens attempting to suppress egg laying by workers and workers "cheating" by overcoming the queen's control and laying eggs. Our knowledge of semiochemicals in these processes is fragmentary at best, and other factors such as physical dominance behavior, nutrition, season, and colony size clearly are important in reproductive control. However, the few cases of a semiochemical role that have been found offer powerful examples of how pheromones can mediate colony activities.

The only example of semiochemical control over the timing of reproduction in which at least some of the chemicals involved have been identified is that of the honey bee *Apis mellifera* L. Honey bees reproduce by swarming, whereby the old queen and a majority of workers leave the nest in search of a new nest site. Left behind are developing queens in special cells; these queens emerge and either issue in smaller swarms called afterswarms or else battle to the death until only one new queen remains in the original nest (reviewed by Winston, 1987). The swarming process begins with queen rearing, however, and swarms generally do not issue until the rearing of new queens is well under way. The rearing of new queens is largely under the control of the old queen, and queen removal results in the initiation of queen rearing within hours or days (Fell and Morse, 1984; Punnett and Winston, 1983; Winston, 1979). In normal queenright colonies, the initiation of queen rearing results from a complex combination of factors, one of which may be a colony's population growing to a size large enough so that the transmission of the queen's inhibitory pheromones is slowed, and the workers then begin the activities associated with queen rearing (Winston and Taylor, 1980; Winston, 1987).

The earliest research (Butler, 1959, 1960; reviewed by Free, 1987 and Winston, 1987) showed that substances produced in the queen's mandibular glands prevent workers from rearing new queens, and that the amount of queen substances necessary to inhibit a colony from rearing queens increases with the number of workers in a colony. Later work identified two acids from the mandibular glands as having partial inhibitory activity, and recently it has been shown that a blend of five compounds from the mandibular glands of queens can fully explain the inhibitory activity of those glands. This blend consists of 9-keto-2 (E)-decenoic acid (9ODA), R-(−)- and S-(+)-9-hydroxy-2 (E)-decenoic acid (9HDA), and the aromatics methyl-p-hydroxybenzoate (HOB) and 4-hydroxy-3-methoxyphenylethanol (HVA) (Slessor et al., 1988). When applied to colonies from which the queen has been removed, this pheromone suppresses the initiation of new queen rearing in a dose-dependent fashion for four to six days; the inhibition breaks down after that point (Winston et al., 1989; Winston et al., 1990). There are clearly other queen pheromones involved, possibly produced in the queen's abdomen (Velthuis, 1970).

It is interesting that queens may secrete massive amounts of the mandibular pheromone daily in order to suppress swarming, possibly as much as 10 times the amount in her mandibular glands every day (Winston et al., 1990). It is not clear why such a large quantity of pheromone is needed to control the inhibition of queen rearing; amounts of mandibular pheromone as low as 10^{-7} of that found in the queen's glands are highly attractive to workers. The answer may be rooted in queen-worker conflict and the evolution of queen dominance over workers. Physical dominance interactions between workers and queens are almost completely absent in a normally functioning honey bee colony, and the queen's secretion of large amounts of mandibular pheromone may be the result of an evolutionary "arms race" in which the workers attempt to break down the queen's compounds as rapidly as possible to reduce or eliminate her dominance, and the queens in turn have responded by producing copious pheromone secretions to maintain that dominance. Thus, it may be appropriate to view queens as pheromone factories in addition to their egg laying capabilities, producing enough pheromone to overwhelm the workers' pheromone catabolism pathways. When colonies get too populous, however, pheromone transmission is slowed to a level where worker catabolic rate is higher than the queen's secretory and/or the colony's transmission rate; at that point, workers may initiate queen rearing.

In addition to controlling the timing of honey bee reproduction, pheromones may influence the extent of that reproduction as well. There is a strong correlation between the brood area and the number of afterswarms that issue from colonies (Winston, 1979; Winston et al., 1981), which could be mediated by brood-secreted pheromones. This proposal remains speculative, however, in the absence of any chemical information about the nature and concentration of brood pheromones in post-swarming colonies.

Reproduction by individual honey bee workers also is controlled by semiochemicals, through the inhibitory effect of queen and brood-pheromones on worker ovary development and egg laying (reviewed by Jay, 1972; Winston, 1987). Although only a very low level of worker egg laying occurs in queenright colonies (Page and Erickson, 1986; Visscher, 1989), and most of those eggs are eaten by other workers (Ratnieks and Visscher, 1989), between 7% and 45% of workers show some ovary development, and up to 70% in colonies about to swarm. When queens are lost and colonies fail to rear a new queen, ovaries of some workers enlarge further, and they begin to lay eggs within a few weeks, although these eggs generally can only develop into drones because they are haploid (males in Hymenoptera usually result from haploid, unfertilized eggs) (Winston, 1987). The queen alone, with no brood present, only partially inhibits worker ovary development; the addition of worker larvae or pupae to colonies with their queens provides more complete inhibition, likely due to a brood-produced volatile substance (Jay, 1972). The queen pheromone active in ovary inhibition has been thought to be 9ODA, but data concerning its activity are ambiguous; a recent study has demonstrated that the five-component queen mandibular pheromone (which includes 9ODA) has no effect on worker ovary development (Willis, Winston, and Slessor, 1990). The source of the queen's inhibitory pheromone could still be mandibular in origin, but it is more likely from some other queen gland.

There is also some evidence from studies of false queens that queen pheromones suppress egg laying in those workers with more developed ovaries. Sometimes, one worker in a queenless colony will develop a higher level of queenliness than the others and will be treated as a queen (Sakagami, 1958). These false queens produce mandibular gland substances identical to those of real queens and appear to suppress egg laying by other workers, although they may not inhibit worker ovary development (Crewe and Velthuis, 1980; Plettner, Slessor, Page, Winston, and Robinson, in prep.; Velthuis et al., 1965). It is not known whether these same compounds, when produced by queens, also prevent egg laying by workers in queenright colonies.

Queens of other social insects also produce pheromones which influence reproduction. In the vespine wasp *Vespa orientalis*, a 16-carbon lactone from exocrine head glands is secreted by older queens which stimulates workers to commence building the large cells used for rearing new queens (Ishay et al., 1965; Ikan et al., 1969; Ishay, 1973). The presence of queens also inhibits ovary development and egg laying by workers, although it is not known whether these functions are semiochemical-based (Ishay, 1975; Spradbery, 1973). In ants, queens suppress gyne formation in some species and worker ovary development in most species in which the workers are capable of developing functional ovaries (reviewed by Brian, 1979). Pheromones have been implicated in at least one species, *Myrmica rubra*, since dead queens can suppress gyne rearing, although the shape of the queen is also important (Brian, 1973). Vapors from queens of the ponerine ant *Odontomachus haematodes* delay the onset of oviposition in queen-deprived colonies, as do squashed queens, suggesting that a pheromone is involved in this primitive ant species as well (Colombel, 1970, 1972).

Another type of semiochemical influence on social insect reproduction has been found in the fire ant *Solenopsis invicta* Buren. In this insect, mature colonies have only one functional queen, although winged virgin queens are found in colonies seasonally. Dealation or wing shedding is considered to initiate the physiological and behavioral events leading to egg laying, and the processes of dealation and oogenesis are prevented by pheromones secreted by the mother queen (Fletcher and Blum, 1981). Also, young colonies can be polygynous, but as the colonies grow, the foundress queens are progressively executed until only one remains; this process is mediated by the level of queen pheromone in the colony relative to the worker population (Fletcher and Blum, 1983). Neither of these pheromones have been identified (but see Vargo, 1988, for an ant primer pheromone bioassay).

This regulation of adult female reproductives in *S. invicta* colonies suggests worker recognition of both pheromone blends and individual queen signals; Fletcher (1987) and Fletcher and Blum (1983) have elaborated on these results to propose a model of pheromonal effects on reproduction which may apply to many social insects. They suggest that (1) workers recognize queens by means of complex pheromone blends produced only by female reproductives; (2) the different constituents of the pheromonal complex are produced in different proportions by different queens, providing each with a unique recognition signature; (3) the quantity of pheromones circulating in a colony is maintained within some optimal range, and deviation from that range results in worker behaviors to restore that level; and (4) there is a positive correlation between the pheromone production of a queen and her egg production, and workers discriminate among queens by killing the poorest pheromone producers first.

Termites also exhibit semiochemical control over reproduction, but in a more complex fashion than the other social insects, since both male and female pheromones are involved, and development of other castes is affected as well (reviewed by Howse, 1984). Termite colonies generally have a resident king in addition to the queen and can have numerous castes, including larvae, soldiers, pseudergates, and nymphs. Replacement reproductives can develop from almost any of these stages, and pheromones can inhibit their development, at least in the lower termites (Light, 1942–43; Luscher, 1974, 1976). In *Kalotermes flavicollis*, both the king and queen are necessary for full inhibition, and using barrier experiments, Luscher (1952) was able to demonstrate that pheromones were involved; similar results were found for *Porotermes adamsoni* (Mensa-Bonsu, 1976). The identity of the active compounds is not known.

It is clear from these examples that semiochemical regulation of reproduction can have a profound impact on colony organization and functioning. It is also clear that further elucidation of how such control functions in colonies is not going to be an easy task. For example, honey bee pheromones have been studied for well over 30 years in many laboratories worldwide, yet we still have only an imperfect understanding of the role pheromones have in colony reproduction, or even of how they act on workers. There are competing hypotheses for the mode of pheromone activity; one, a pharmacological model, suggests that semiochemicals act directly on worker physiology. The second, a sensory model, suggests the action of semiochemicals is mediated by the nervous system of the receiver. Differentiating between these mechanisms has considerable evolutionary importance and will certainly be an area of much research activity in the future, particularly as our ability to analyze, identify, and bioassay complex pheromone blends improves.

10.3. Semiochemical Distribution

Social insects differ in many ways from their more solitary relatives, because of the obvious need to function as a colonial unit rather than as individuals. In that regard, distribution of semiochemicals throughout the nest is of major importance if any benefit is to be accrued from their production. Since many workers do not contact the queen, some system is needed to transfer her chemical signals. Also, pheromones must be rapidly catabolized by workers, so that changes in queen or colony condition can be quickly communicated without the "contamination" of earlier messages. The little we know about semiochemical transfer between individuals indicates that sophisticated distribution and catabolism systems have evolved to facilitate message transfer, but we have only scratched the surface of this complex subject.

In the honey bee, some transmission mechanism besides direct worker-queen contact must operate, since workers not directly in contact with the queen show the effects of her pheromones. For example, in one experiment, two groups of workers were separated from each other by a mesh screen through which they could touch each other but not pass. One group was given a caged queen, while the other remained queenless, but neither group showed any ovary development (Verheijen-Voogd, 1959). Three transmission modes have been proposed: food exchange between workers, volatile transmission, and transmission between the body surfaces of workers, particularly during antennal contacts; evidence opposes the first two modes (reviewed by Winston, 1987).

There is abundant evidence for the surface transport hypothesis, particularly because of worker behavior during and following queen contacts (Ferguson and Free, 1980; Seeley, 1979; Velthuis, 1972; Verheijen-Voogd, 1959). Workers in the retinue surrounding the queen touch her with their antennae and mouthparts in a flurry of activity, for one or more minutes at a time. Following these contacts, workers generally groom their antennae with their forelegs, presumably passing pheromones between the antennae, tongue, and legs. Most significantly, workers that have contacted queens in this manner move through the colony making reciprocated antennal contacts with other workers at high rates for approximately 30 minutes; a typical worker will exchange nestmate antennations with an average of 56 other workers during this time. Seeley proposed the name "messenger bees" for these workers, and he and the others cited above have concluded that pheromones are distributed over the queen's body by her grooming, picked up by direct contacts between workers and the queen, and then transported throughout the colony by these messenger bees. Queen pheromones recently have been quantitatively traced along this route, confirming this hypothesis, and also indicating that some of the queen's pheromones are deposited directly on to comb and then picked up by workers (Naumann et al., 1991).

In contrast, termites transmit the king and queen pheromones via anal trophallaxis. This was demonstrated by Luscher (1952), who fixed a queen in a screen so that only her abdomen protruded into a colony. Inhibition of replacement reproductives occurred even when her tergal and sternal glands and genital opening were covered with varnish, but not if the anus was blocked. Given the high level of anal trophallaxis in termites to exchange gut symbionts, this transmission method is not surprising.

The nest substrate can also harbor semiochemical signals, particularly for nest and kin recognition. As noted earlier, recognition cues in *Polistes fuscatus* are learned from nests, nest fragments, and/or brood, utilizing exogenous and/or endogenous materials deposited in or on the nest. Breed et al. (1988) have shown a similar ability in honey bees to recognize kin through wax-mediated odors, although its function in the complex chemical world of a real nest needs to be examined further. Since volatiles from wax comb differ between colonies (Hepburn, 1986), it is certainly plausible that recognition signals could be mediated through wax-contained odors.

Finally, we know almost nothing about the dose of semiochemicals required for particular functions, which is not surprising, considering how few chemicals have been identified for the functions discussed in this chapter. The half-life of queen-mediated functions can be minutes to weeks, depending on the particular function. For example, worker honey bees are aware of their queen's absence within minutes; the half-life of the compounds that inform workers of her presence is approximately 15 minutes (Juska, 1978; Juska et al., 1981). In contrast, it takes many weeks for workers to develop ovaries and begin laying eggs following queen removal (Winston, 1987). Seeley (1979) has estimated that an individual worker attending the queen must be picking up less than 0.1 ng of pheromone, but Winston et al. (1990) have estimated that the rate of pheromone removal must be three or four orders of magnitude greater to explain the inhibitory effects of those pheromones on queen rearing. More dose-dependent data and tracking of pheromone transmission and breakdown in colonies is needed for all social insects in which semiochemicals are involved in the control of colony activities.

Conclusions

We are on the verge of making real advances in semiochemical studies of social insects, similar to the great strides made in the 1960s and 1970s in our understanding of the role releaser pheromones play in social insects. The new revolution, which will involve recognition and primer semiochemicals, is being fueled by two forces, increased analytical capabilities of chemists and a heightened awareness of the complexities in social insect colony structure by biologists. Close cooperation between chemists and biologists will be required to bring this exciting new field to maturity, however, as it was for previous advances in insect semiochemistry (see Roitberg, this volume). The challenges are great, but on the near horizon is the opportunity to unravel a level of functioning in social insects about which we know very little, but which promises exciting findings and some great surprises.

Acknowledgements

This manuscript benefited greatly from comments by M. Breed, W. Getz, and G. Gamboa, for which I am grateful. I also would like to acknowledge funding which has supported by own research on insect semiochemicals, from Operating and Strategic grants of the Natural Sciences and Engineering Research Council of Canada, the British Columbia Science Council, the Wright Institute, and Simon Fraser University. In addition, I would like to acknowledge my collaborator in this work, Dr. Keith Slessor, who has shaped much of my thinking about how to study insect pheromones.

References

Ambrose, J. T., R. A. Morse, and R. Boch. 1979. Queen discrimination by honey bee swarms. *Ann. Entomol. Soc. Amer.* 72:673–675.

Barrows, E. M., W. J. Bell, and C. D. Michener. 1975. Individual odor differences and their social functions in insects. *Proc. Nat. Aca. Sci. (U.S.A.)* 72:2824–2828.

Blomquist, B. J., A. J. Chu, and S. Remaley. 1980. Biosynthesis of wax in the honeybee *Apis mellifera* L. *Insect Biochem.* 10:313–321.

Boch, R. and R. A. Morse. 1974. Discrimination of familiar and foreign queens by honeybee swarms. *Ann. Entomol. Soc. Am.* 67:709–711.

Boch, R. and R. A. Morse. 1979. Individual recognition of queens by honeybees. *Ann. Entomol. Soc. Am.* 72:51–53.

Boch, R. and R. A. Morse. 1981. Effects of artificial odors and pheromones on queen discrimination by honeybees. *Ann. Entomol. Soc. Am.* 74:66–67.

Bornais, K. N., C. M. Larch, G. J. Gamboa, and R. B. Daily. 1983. Nestmate discrimination among laboratory overwintered foundresses of the paper wasp *Polistes fuscatus*. *Can. Entomol.* 115:655–658.

Bradshaw, J. W. S. and P. E. Howse. 1984. Sociochemicals of ants. In *Chemical Ecology of Insects*, eds. W. J. Bell and R. T. Carde, pp. 429–465. Sinauer Associates, Sunderland, MA.

Breed, M. 1981. Individual recognition and learning of queen odours by worker honeybees. *Proc. Natl. Acad. Sci. USA* 78:2635–2637.

Breed, M. 1983. Nestmate recognition in honeybees. *Anim. Behav.* 31:86–91.

Breed, M. 1985. How honeybees recognize their nestmates: a re-evaluation from new evidence. *Bee World* 66:113–118.

Breed, M. and B. Bennett. 1987. Kin recognition in highly eusocial insects. In *Kin Recognition in Animals*, eds. D. J. C. Fletcher and C. D. Michener, pp. 243–286. John Wiley, New York.

Breed, M., L. Butler, and T. M. Stiller. 1985. Kin discrimination by worker honeybees in genetically mixed groups. *Proc. Natl. Acad. Sci. USA* 82:3058–3061.

Breed, M., H. H. W. Velthuis, and G. E. Robinson. 1984. Do worker honeybees discriminate among unrelated and related larval phenotypes? *Ann. Entomol. Soc. Am.* 77:737–739.

Breed, M., K. R. Williams, and J. H. Fewell. 1988. Comb wax mediates the acquisition of nestmate recognition cues in honeybees. *Proc. Natl. Acad. Sci. USA* 85:8766–8769.

Brian, M. V. 1973. Queen recognition by brood-rearing workers of the ant *Myrmica rubra* L. *Anim. Behav.* 21:691–698.

Brian, M. V. 1979. Caste differentiation and division of labor. In *Social Insects*, Vol. 1, ed. H. R. Hermann, pp. 121–222. Academic Press, New York.

Butler, C. G. 1959. Queen substance. *Bee World* 40:269–275.

Butler, C. G. 1960. The significance of queen substance in swarming and supersedure in honeybee (*Apis mellifera* L.) colonies. *Proc. Roy. Entomol. Soc. London (A)* 35:129–132.

Buckle, G. R. and L. Greenberg. 1981. Nestmate recognition in sweat bees (*Lasioglossum zephyrum*): does an individual recognize its own odour or only odours of its nestmates? *Anim. Behav.* 29:802–809.

Carlson, D. and A. B. Bolten. 1984. Identification of Africanized and European honeybees, using extracted hycrocarbons. *Bull. Entomol. Soc. Am.* 30:32–35.

Colombel, P. 1970. Recherches sur la biologie et l'ethologie d'*Odontomachus haemotodes* L. (Hym. Formicoidea, Poneridae). Biologie des reines. *Insectes Soc.* 17:199–204.

Colombel, P. 1972. Recherches sur la biologie et l'ethologie d'*Odontomachus haemotodes* L. (Hym. Formicoidea, Poneridae). Biologie des ouvrieres. *Insectes Soc.* 19:171–194.

Crewe, R. M. 1982. Compositional variability, the key to the social signals produced by honeybee mandibular glands. In *The Biology of the Social Insects*, eds. M. D. Breed, C. D. Michener, and H. E. Evans, pp. 318–322. Westview Press, Boulder, CO.

Crewe, R. M. and H. H. W. Velthuis. 1980. False queens: a consequence of mandibular gland signals in worker honeybees. *Naturwiss.* 67:467–469.

Crozier, R. H., B. H. Smith, and Y. C. Crozier. 1987. Relatedness and population structure of the primitively eusocial bee *Lasioglossum zephyrum* (Hymenoptera: Halictidae) in Kansas. *Evolution* 41:902–910.

Duffield, R. M., J. W. Wheeler, and G. C. Eickwort. 1984. Sociochemicals of bees. In *Chemical Ecology of Insects*, eds. W. J. Bell and R. T. Carde, pp. 387–428. Sinauer Associates, Sunderland, MA.

Evers, C. A. and T. D. Seeley. 1986. Kin discrimination and aggression in honeybee colonies with laying workers. *Anim. Behav.* 34:924–925.

Fell, R. D. and R. A. Morse. 1984. Emergency queen cell production in the honeybee colony. *Insectes Soc.* 31:221–237.

Ferguson, A. W. and J. B. Free. 1980. Queen pheromone transfer within honeybee colonies. *Physiol. Entomol.* 5:359–366.

Fletcher, D. J. C. 1987. Recognition of actual and potential female reproductives in colonies of social Hymenoptera. In *From Individual to Collective Behavior in Social Insects*, eds. J. M. Pasteels and J. Deneubourg. Birkhauser Verlag, Basel, pp. 365–378.

Fletcher, D. J. C. and M. S. Blum. 1981. Pheromonal control of dealation and oogenesis in virgin queen fire ants. *Science* 212:73–75.

Fletcher, D. J. C. and M. S. Blum. 1983. Regulation of queen number by workers in colonies of social insects. *Science* 219:312–314.

Free, J. B. 1987. *Pheromones of Social Bees*. Cornell University Press, Ithaca, NY.

Frumhoff, P. C. and S. Schneider. 1987. The social consequence of honeybee polyandry: kinship influences worker interactions within colonies. *Anim. Behav.* 35:255–262.

Gamboa, G. J. 1988. Sister, aunt-niece, and cousin recognition by social wasps. *Behav. Genetics* 18:409–423.

Gamboa, G. J., H. K. Reeve, and D. W. Pfennig. 1986a. The evolution and ontogeny of nestmate recognition in social wasps. *Annu. Rev. Entomol.* 31:431–454.

Gamboa, G. J., H. K. Reeve, I. D. Ferguson, and T. L. Wacker. 1986. Nestmate recognition in social wasps: the origin and acquistion of recognition odours. *Anim. Behav.* 34:685–695.

Getz, W. M. 1989. The honeybee as a model kin recognition system. In *Kin Recognition*, ed. P.G. Hepper. Cambridge Univ. Press, Cambridge, England (in press).

Getz, W. M. and K. B. Smith. 1986. Honeybee kin recognition: learning self and nestmate phenotypes. *Anim. Behav.* 34:1617–1626.

Getz, W. M. and K. B. Smith. 1987. Olfactory sensitivity and discrimination of mixtures in the honeybee *Apis mellifera*. *J. Comp. Physiol. A.* 160:239–245.

Greenberg, L. 1979. Genetic component of bee odor in kin recognition. *Science* 206:1095–1097.

Hamilton, W. D. 1964. The genetical evolution of social behavior, I and II. *J. Theoretical Biology* 7:1–52.

Hepburn, H. R. 1986. *Honeybees and Wax*. Springer-Verlag, Berlin.

Howse, P. E. 1984. Sociochemicals of termites. In *Chemical Ecology of Insects*, eds. W. J. Bell and R. T. Carde, pp. 475–520. Sinauer Associates, Sunderland, MA.

Ikan, R., R. Gottlieb, and E. D. Bergmann. 1969. The pheromone of the queen of the oriental hornet *Vespa orientalis*. *J. Insect Physiol.* 15:1709–1712.

Ishay, J. 1973. The influence of cooling and queen pheromone on cell building and nest architecture by *Vespa orientalis*. (Vespinae, Hymenoptera). *Insectes Soc.* 20:243–252.

Ishay, J. 1975. Caste determination by social wasps: cell size and building behaviour. *Anim. Behav.* 23:425–431.

Ishay, J. R. Ikan, and E. D. Bergmann. 1965. The presence of pheromones in the oriental hornet, *Vespa orientalis*. *J. Insect Physiol.* 11:1307–1309.

Jay, S. C. 1972. Ovary development of worker honeybees when separated from worker brood by various methods. *Can. J. Zool.* 50:661–664.

Jeanne, R. L., H. A. Downing, and D. C. Post. 1983. Morphology and function of sternal glands in polistine wasps. *Zoomorphol.* 103:149–164.

Juska, A. 1978. Temporal decline in attractiveness of honeybee queen tracks. *Nature* 276:261.

Juska, A., T. D. Seeley, and H. H. W. Velthuis. 1981. How honeybee queen attendants become ordinary workers. *J. Insect Physiol.* 27:515–519.

Kalmus, H. and C. R. Ribbands. 1952. The origin of odors by which honeybees distinguish their companions. *Proc. Royal Soc. (B)* 140:50–59.

Klahn, J. E. 1979. Philopatric and nonphilopatric foundress associations in the social wasp *Polistes fuscatus*. *Behav. Ecol. Sociobiol.* 5:417–424.

Klahn, J. E. and G. J. Gamboa. 1983. Social wasps: discrimination between kin and non-kin brood. *Science* 221:482–484.

Kramer, E. 1976. The orientation of walking honeybees in odour fields with small concentration gradients. *Physiol. Entomol.* 1:27–37.

Kukuk, P. F. and P. Decelles. 1986. Behavioral evidence of population structure in *Lasioglossum zephyrum* (Hymenoptera: Halictidae): female dispersion patterns. *Behav. Ecol. Sociobiol.* 19:233–239.

Kukuk, P. F., M. D. Breed, A. Sobti, and W. J. Bell. 1977. The contributions of kinship and conditioning to nest recognition and colony member recognition in a primitively eusocial bee, *Lasioglossum zephyrum*. *Behav. Ecol. Sociobiol.* 2:319–327.

Laidlaw, H. H. and R. E. Page. 1984. Polyandry in honeybees (*Apis mellifera* L.): sperm utilization and intracolony genetic relationships. *Genetics* 108:985–997.

Light, S. F. 1942–43. The determination of castes of social insects. *Q. Rev. Biol.* 17:312–326; 18:42–63.

Luscher, M. 1952. Die Produktion und Elimination von Ersatzgeschlechtstieren bei der Termite *Kalotermes flavicollis* Fabr. *Z. Vergl Physiol.* 34:123–141.

Luscher, M. 1974. Kasterund Kastendifferenzierung bei niederen Termiten. In *Sozialpolymorphismus bei Insekten*, pp. 694–739. Wiss. Verlag., Stuttgart.

Luscher, M. 1976. Evidence for an endocrine control of caste determination in higher termites. In *Phase and Caste Determination in Insects*, ed. M. Luscher, pp. 91–103. Pergamon Press, Oxford.

McDaniel, C. A. R. W. Howard, C. J. Blomquist, and A. M. Collins. 1984. Hydrocarbons of the cuticle, string apparatus, and sting shaft of *Apis mellifera* L. Identification and preliminary evaluation of chemotaxonomic characters. *Sociobiology* 8:287–298.

McDaniel, C. A., R. W. Howard, A. M. Collins, and W. A. Brown. 1987. Variation in the hydrocarbon composition of non-Africanized *Apis mellifera* L. sting apparatus. *Sociobiology* 13:133–143.

Mensa-Bonsu, A. 1976. The production and elimination of supplementary reproductives in *Porotermes adamsoni* (Froggatt) (Isoptera, Hodotermitidae). *Insectes Soc.* 23:133–153.

Metcalf, R. A. and G. S. Whitt. 1977. Intra-nest relatedness in the social wasp *Polistes metricus*. *Behav. Ecol. Sociobiol.* 2:339–351.

Michener, C. D. and B. H. Smith. 1987. Kin recognition in primitively eusocial insects. In *Kin Recognition in Animals*, eds. D. J. C. Fletcher and C. D. Michener, pp. 209–242. John Wiley, New York.

Nault, L. R. and P. L. Phelan. 1984. Alarm pheromones and sociality in pre-social insects. In *Chemical Ecology of Insects*, eds. W. J. Bell and R. T. Carde, pp. 237–256. Sinauer Associates, Sunderland, MA.

Naumann, K., M. L. Winston, K. N. Slessor, G. D. Prestwick, and F. X. Webster. 1991. The production and transmission of honey bee (*Apis mellifera* L.) mondibolar gland pheromone. *Behav. Ecol. Sociobiol.* (in press).

Noonan, K. M. 1979. Individual strategies of inclusive fitness maximizing in the social wasp, *Polistes fuscatus* (Hymenoptera: Vespidae). Ph.D. dissertation, Univ. of Michigan, Ann Arbor.

Noonan, K. C. 1986. Recognition of queen larvae by worker honeybees (*Apis mellifera* L.). *Ethology* 73:295–306.

Page, R. E. 1986. Sperm utilization in social insects. *Annu. Rev. Entomol.* 31:297–320.

Page, R. E. and E. H. Erickson. 1984. Selective rearing of queens by worker honeybees: kin or nestmate recognition. *Ann. Entomol. Soc. Am.* 77:578–580.

Page, R. E. and E. H. Erickson. 1986. Kin recognition and virgin acceptance by worker honeybees (*Apis mellifera* L.). *Anim. Behav.* 134:1061–1069.

Page, R. E., R. B. Kimsley, and H. H. Laidlaw. 1984. Migration and dispersal of spermatozoa in spermathecae of queen honeybees (*Apis mellifera* L.). *Experientia* 40:182–184.

Pfennig, D. W., H. K. Reeve, and J. S. Shellman. 1983a. Learned component of nestmate discrimination in workers of a social wasp, *Polistes fuscatus*. *Anim. Behav.* 31:412–416.

Pfennig, D. W., G. J. Gamboa, H. K. Reeve, J. Shellman-Reeve, and I. D. Ferguson. 1983b. The mechanism of nestmate discrimination in social wasps (*Polistes*, Hymenoptera: Vespidae). *Behav. Ecol. Sociobiol.* 13:299–305.

Post, D. C. and R. L. Jeanne. 1982. Recognition of former nestmates during colony founding in the social wasp *Polistes fuscatus. Behav. Ecol. Sociobiol.* 11:283–285.

Punnett, E. N. and M. L. Winston. 1983. Events following queen removal in colonies of European-derived honeybee races. *Insectes Soc.* 30:376–383.

Ratnieks, F. L. W. and P. K. Visscher. 1989. Worker policing in the honeybee. *Nature* 342:796–797.

Renner, M. 1960. Das Duftorgan der Honigbiene und die physiologische Bedeutung ihres Lockstoffes. *Z. Vergl. Physiol.* 43:411–468.

Ribbands, C. R. 1954. The defense of the honeybee community. *Proc. Roy. Soc. London (B)* 142:514–524.

Ross, N. M. and G. J. Gamboa. 1981. Nestmate discrimination in social wasps *Polistes metricus*, Hymenoptera: Vespidae). *Behav. Ecol. Sociobiol.* 9:163–165.

Sakagami, S. F. 1958. The false queen: fourth adjustive response in dequeened honeybee colonies. *Behaviour* 13:280–296.

Seeley, T. D. 1979. Queen substance dispersal by messenger workers in honeybee colonies. *Behav. Ecol. Sociobiol.* 5:391–415.

Shellman, J. S. and G. J. Gamboa. 1982. Nestmate discrimination in social wasps: the role of exposure to nest and nestmates (*Polistes fuscatus*, Hymenoptera, Vespidae). *Behav. Ecol. Sociobiol.* 11:51–53.

Slessor, K. N., L.-A. Kaminski, G. G. S. King, J. H. Borden, and M. L. Winston. 1988. The semiochemical basis of the retinue response to queen honeybees. *Nature* 332:354–356.

Smith, B. H. 1983. Recognition of female kin by male bees through olfactory signals. *Proc. Nat. Acad. Sci. USA* 80:4551–4553.

Smith, B. H. 1987. Effects of genealogical relationship and colony age on the dominance hierarchy in the primitively eusocial bee *Lasioglossum zephyrum*. *Anim. Behav.* 35:211–217.

Smith, B. H. and J. W. Wenzel. 1988. Pheromonal covariation and kinship in the social bee *Lasioglossum zephyrum* (Hymenoptera: Halictidae). *J. Chem. Ecol.* 14:87–94.

Smith, B. H., R. G. Carlson, and J. Frazier. 1985. Identification and bioassay of macrocyclic lactone sex pheromone of the halictine bee *Lasioglossum zephyrum*. *J. Chem. Ecol.* 11:1447–1456.

Spradbery, J. P. 1973. *Wasps*. Sidgwick and Jackson, London.

Thorne, B. L. 1982. Termite-termite interactions: workers as an agonistic caste. *Psyche* 89:133–150.

Vargo, E. L. 1988. A bioassay for a primer pheromone of queen fire ants (*Solenopsis invicta*) which inhibits the production of sexuals. *Insectes Soc.* 35:382–392.

Velthuis, H. H. W. 1970. Queen substances from the abdomen of the honey bee queen. *Z. Vergl. Physiol.* 70:210–222.

Velthuis, H. H. W. 1972. Observations on the transmission of queen substances in the honeybee colony by the attendants of the queen. *Behaviour* 41:105–129.

Velthuis, H. H. W., F. J. Verheijen, and A. J. Gottenbos. 1965. Laying worker honeybee: similarities to the queen. Nature 207:1314.

Verheijen-Voogd, C. 1959. How worker bees perceive the presence of their queen. *Z. Vergl. Physiol.* 41:527–582.

Visscher, P. K. 1986. Queen rearing by honeybees (*Apis mellifera*). *Behav. Ecol. Sociobiol.* 18:453–460.

Visscher, P. K. 1989. A quantitative study of worker reproduction in honeybee colonies. *Behv. Ecol. Sociobiol.* 25:247–254.

Wcislo, W. T. 1987. The role of learning in the mating biology of a sweat bee *Lasioglossum zephyrum* (Hymenoptera: Halictidae). *Behav. Ecol. Sociobiol.* 20:179–185.

Willis, L. G., M. L. Winston, and K. N. Slessor. 1990. The effect of queen honey bee (*Hymenoptera: Apidac*) mondibular pheromone on worker ovary development. *Canadian Entomol.* 122:1093–1099.

Winston, M. L. 1979. Intra-colony demography and reproductive rate of the Africanized honeybee in South America. *Behav. Ecol. Sociobiol.* 4:279–292.

Winston, M. L. 1987. *The Biology of the Honey Bee*. Harvard University Press, Cambridge, MA.

Winston, M. L. and O. R. Taylor. 1980. Factors preceding queen rearing in the Africanized honeybee (*Apis mellifera*) in South America. *Insectes Soc.* 27:289–304.

Winston, M. L., J. A. Dropkin, and O. R. Taylor. 1981. Demography and life history characteristics of two honeybee races (*Apis mellifera*). *Oecologia* 48:407–413.

Winston, M. L., K. N. Slessor, L. G. Willis, K. Naumann, H. A. Higo, M. H. Wyborn, and L.-A. Kaminski. 1989. The influence of queen mandibular pheromones on worker attraction to swarm clusters and inhibition of queen rearing in the honeybee (*Apis mellifera* L.). *Insectes Soc.* 36:15–27.

Winston, M. L., H. A. Higo, and K. N. Slessor. 1990. The effect of various dosages of queen mandibular pheromone on the inhibition of queen rearing in the honey bee (Hymenoptera: Apidae). *Ann. Entomol. Soc. Amer.* 83:234–238.

11

Evolutionary Perspectives and Insect Pest Control: An Attractive Blend for the Deployment of Semiochemicals in Management Programs

Jeremy N. McNeil

Université Laval

11.1. Introduction

The specificity and non-toxic nature of semiochemicals makes them an appealing option within an integrated pest management program, even though the use of behavior-modifying chemicals as management tools has not always lived up to initial expectations (Lewis, 1981; Cardé, 1990). With the benefit of 20/20 hindsight, this is not really surprising, as the projected goals often exceeded our understanding of the semiochemically-mediated systems that we wished to exploit. However, despite the fact that semiochemicals may not be "the panacea idealistic prophets believed them to be" (Boness et al., 1977), they have been successfully used in management programs for agricultural and forest pests (Jutsum and Gordon, 1989; Ridgway et al., 1990), clearly justifying continued efforts to develop and deploy insect behavior-modifying chemicals. An obvious question is, what can be done to increase the successful use of behavior-modifying chemicals? In the case of sex pheromones, the most intensively studied group to date, Tumlinson (1988) felt that research resulting in the elucidation of all components of a given species blend, together with a better comprehension of both the biosynthesis and the reception of these compounds, would greatly ameliorate their efficacy. In addition, I believe that a solid understanding of the behavioral ecology underlying the pheromone-mediated communication systems we wish to modify could greatly improve the degree of success (McNeil, 1991). Tumlinson (1988) also emphasised the need for more research on semiochemicals other than sex pheromones, such as epideictic pheromones, kairomones that modify the behavior of natural enemies, and host plant volatiles that influence oviposition.

In keeping with the objectives of this book, I wish to address two somewhat different perspectives of semiochemical use in pest management from those normally found in discussions on the subject. I first examine how an understanding of mate choice and male-male competition might help in our interpretation of trap catch data and, second, how the continued use of semiochemicals could affect certain biological characteristics of the target population. I have restricted my discussion to the use of lepidopteran sex pheromones as a monitoring tool (Wall, 1990, and references therein) or as a means of disrupting mating (Cardé, 1990, and references therein). However, there is little doubt that similar considerations could be applied to the use of sex pheromones for other insect order and to the deployment of other behavior-modifying chemicals in management programs. It should be noted from the outset that there have been very few studies undertaken to directly examine any semiochemicals from an evolutionary perspective, so I will attempt to show how such an approach might be applied to problems of potential practical importance.

11.2. Monitoring

Monitoring programs may be set up to (1) determine the distribution of an insect within a given geographic area (Elkinton and Cardé, 1981); (2) determine the period of adult activity in order to predict phenology and synchronize insecticide applications (Minks and de Jong, 1975; Riedl et al., 1976; Alford et al., 1979; Baker et al., 1980; Wall, 1988); (3) monitor long-term trends in density to detect when populations pass from endemic to epidemic levels (Sanders, 1981); (4) monitor for the level of insecticide resistance in a given population (Haynes et al., 1986, 1987); or (5) predict subsequent larval densities or resource losses (Allen et al., 1986), so that management decisions may be made proactively. In the majority of monitoring programs, it is not necessary to know the reproductive status of the males captured. However, if the objective is to predict the density of some future developmental stage in the life cycle, such information may help explain why, in some cases (e.g., Palaniswamy et al., 1990), there is a poor correlation between pheromone trap catch data and subsequent egg or larval densities.

In a number of Lepidoptera, male responsiveness to a pheromone source increases with age during the first few days following emergence (McNeil, 1991, and references therein), a pattern probably associated with sexual maturation. Kolodny-Hirsch and Schwalbe (1990) reported that under field conditions, some gypsy moth males mate repeatedly, while many others never succeeded in acquiring mates. Thus, while traps only capture mature males, they are not necessarily individuals that have successfully reproduced. Berg and Seabrook (1986b) used the presence or absence of a colored pigment in the reproductive tract of male spruce budworm, *C. fumiferana*, to identify virgin and mated males, respectively. They noted that a high proportion of males captured in pheromone traps throughout the adult flight period were unmated. In this species the pigmentation reappears some time after mating (Berg and Seabrook, 1986a), and it is possible, particularly at the end of the flight period, that males classified as virgins may actually have mated several days prior to being captured. Nevertheless, their data, together with similar information on the corn earworm (Henneberry and Clayton, 1984), suggest that a significant proportion of males captured in pheromone traps may be unmated.

There are a number of possible explanations, other than the absence of receptive females, as to why some males in the population do not acquire mates. There is clear evidence that, in many insects, female choice is a major factor determining male mating success (Thornhill and Alcock, 1983). Phelan and Baker (1986) provided the first empirical evidence of mate choice in Lepidoptera, with their work on the tobacco moth, *Ephestia elutella*, demonstrating that females clearly discriminate against small males. Thus, if female choice occurs, many males may be continuously rejected by receptive females.

Another factor that may result in certain individuals never acquiring a mate is intrasexual competition for receptive females. Marshall (1988) observed that while the average mass of *Parapediasia teterrella* and *Agriphila plumbifimbriella* males from field-collected mating pairs was significantly less than that of males captured on the wing during the mating period, the forewing length of mating and non-mating males was similar. There was little (*A. plumbifimbriella*) or no (*P. teterrella*) evidence of female choice under field conditions, and Marshall suggested that lighter males probably had a superior flight performance due to a lower wing loading. His conclusion was supported by the fact that heavier males were as successful as lighter ones in mate acquisition where the flight component of courtship was eliminated.

Sub-lethal pathogenic infects may also influence flight performance and thus affect the male's ability to compete with conspecifics. Spruce budworm males are often infected with *N. fumiferanae* (Wilson, 1987), but the duration of sustained flight to a pheromone source in a wind tunnel did not differ significantly between infected and healthy males (Sanders and Wilson, 1990). However, Gardiner (1990) found that oblique banded leafroller males, infected with a microspridian, have a reduced flight speed compared with healthy individuals. Under conditions of male-male competition, flight speed would be a more valid male attribute to consider as it, rather than flight duration, will likely have a greater influence on which male reaches the resource first. This would be of particular importance if, as reported in the oriental fruit moth (Baker, 1983), the first male to arrive has a higher probability of acquiring the female than one arriving shortly thereafter. The competitiveness of infected males could be further reduced if, as Sweeney and McLean (1987) reported in the western spruce budworm, *C. occidentalis*, receptivity to the female sex pheromone is reduced by pathogenic infections. The difference in reproductive success may be further accentuated, even when the infected male detected the pheromone and arrives first, if the female is able to discriminate between healthy and infected individuals.

There is very little information on the level of female choice or male-male competition in Lepidoptera under field conditions and at present one can only speculate to what extent any aspect of sexual selection may influence the probability that any given individual would be captured in a pheromone trap. However, it would be relatively easy to compare parameters, such as the body size or pathogen load, of males captured in pheromone traps with those obtained using other sampling methods (pupal collections, sweep netting, light traps, food baits). If significant differences were detected, then it would be worth undertaking detailed laboratory and field studies to determine if they are the result of female choice or male-male competition and whether this information actually improves our ability to interpret trap catch data.

11.3. Mating Disruption

The wide-scale application of sex pheromone may, through a number of different mechanisms (peripheral reception and central nervous system effects, competition between natural and synthetic sources, imbalance in sensory input), disrupt pheromone-mediated communication and successfully reduce pest densities (Cardé, 1990). However, repeated disruption might result in the selection of "resistant" populations (Cardé, 1976), similar to the situation observed in some species following the repeated use of chemical insecticides (Metcalf, 1980).

Obviously, for "resistance" to evolve, there must be (1) inter-individual variability in different parameters associated with signal emission and signal reception, (2) a genetic basis for these differences, and (3) covariance between the different characters and individual fitness, an area of insect pheromone biology that remains to be explored. There is clear documentation of inter-individual variation in the age at which females initiate pheromone release (Hirano and Muramoto, 1976; Kanno, 1979; Turgeon and McNeil, 1982; Howlader and Gerber, 1986), pheromone titer

(Mistrot Pope et al., 1982; Delisle and McNeil, 1987), emission rate (Mistrot Pope et al., 1982; Haynes et al., 1984) and the proportion of different pheromone components present in the blend (Miller and Roelofs, 1980; Haynes et al., 1984; Collins and Cardé, 1985; Roelofs et al., 1987a, b; Du et al., 1987; Witzgall and Frérot, 1989). There is also evidence that such characters are heritable (Collins and Cardé, 1985, 1989a; Roelofs et al., 1987a, b; Sreng et al., 1989; Collins et al., 1990; Han and Gatehouse, 1991). Similarly, inter-individual variability in male responsiveness to both pheromone concentration (Baker et al., 1981; Turgeon et al., 1983; Linn and Roelofs, 1985) and blend (Baker et al., 1981; Linn and Roelofs, 1985; Roelofs et al., 1987b) have been reported. The heritability of the male response to pheromone blend has been documented (Collins and Cardé, 1989b, 1990). The results of these different studies indicate that such traits could therefore be subject to selection.

For there to be a significant modification in the communication channel, concomitant change must occur in both emitter and receiver. Thus, the degree to which genes controlling female and male characters are the same or closely linked will be of major importance. The genetic basis of pheromone-mediated communication in Lepidoptera has lagged behind both the advances made in the biochemical and behavioral aspects of pheromone biology and genetic studies of other organisms (Glover and Roelofs, 1988), but interest in this area is increasing (See Lofstedt, 1990, and references therein). Work on the European corn borer, *Ostrinia nubilalis*, has shown that the ratio of (E) to (Z)-11-tetradecenyl acetate in the female pheromone (Klun and Maini, 1979; Roelofs et al., 1987b) and the olfactory response of receptor cells on the male antennae are both controlled by single autosomal genes (Roelofs et al., 1987b; Hansson et al., 1987). In contrast, the male behavioral response to the pheromone is determined by a single sex-linked gene on the Z chromosome (Roelofs et al., 1987b; Glover et al., 1990).

However, there are already questions concerning certain genetic relationships that require clarification. Klun and Huettel (1988) suggested that the autosomal genes controlling the pheromone blend and male antennal receptivity in the European corn borer are coupled in a complimentary manner, while Lofstedt et al. (1989) concluded that they are not closely linked and are probably inherited independently. A similar situation has arisen concerning the genetic relatedness of pheromone titer and blend composition in the pink bollworm, with different authors arriving at somewhat different opinions (Haynes et al., 1984; Collins and Cardé, 1985; Collins et al., 1990). Furthermore, in the pink bollworm, selecting females for a higher proportion of (Z,E)-7,11-hexadecadienyl acetate in the pheromone blend increased male responsiveness to high (Z,E) blends (Collins and Cardé, 1989b), while the selection of male responsiveness to a high (Z,E) blend modified neither blend composition nor pheromone titer in females (Collins and Cardé, 1990). In addition, there may be very marked species differences. For example, Hansson et al. (1989) found that, in contrast to the corn borer, the olfactory response of male sensillae to the female sex pheromone in two tortricid species was, at least in part, determined by a gene on the Z chromosome. It is obvious that a broader data base is necessary before we can realistically evaluate the possibility of selecting for "resistant" populations through simultaneous changes in both the emitter and the receiver.

Even if the male and female components of the pheromone communication were closely linked genetically, there are several lines of experimental evidence indicating that changes in pheromone ratios would not afford a major advantage to females when mating disruption techniques are deployed. The response window of both the pink bollworm and the red-banded leafroller males to different isomeric blends of their respective pheromones are considerably wider than the variability observed in female pheromone production (Lofstedt, 1990). Furthermore, under field conditions, the same male may respond to very different pheromone blends (Haynes and Baker, 1988), and the male response window may vary significantly with ambient temperature conditions (Linn et al., 1988). Thus, it is unlikely that marked changes in the blend ratio of the female sex pheromone will be observed in areas with continued use of pheromones for mating disruption.

If resistance was to occur, a more-probable scenario would be an increase in the concentration of the pheromone emitted by females. This could increase their probability of being distinguishable from the background pheromone source, yet requires no major change in the receiver. This may be occurring in populations of the pink bollworm where mating disruption through the application of the sex pheromone has been carried out on a commercial scale (Baker et al., 1990, and references therein). In an initial study examining the possibility of resistance, Haynes et al. (1984) compared both the release rate and isomeric ratio of the pheromone blend in females collected in fields where the pest had been controlled through the application of chemical insecticides with those from sites where mating disruption had been practiced for three to five years. They found no differences in either the mean pheromone emission rate or the ratio of (Z,Z) and (Z,E)-7,11-hexadecadienyl acetate, the two pheromone components. However, in a follow-up study, Haynes and Baker (1988) reported that during the period of 1982 to 1985, there was a significant increase in the mean pheromone release rate during the last two years, without any consistent pattern of change in the blend ratio. No parallel change in pheromone titer was observed over the same period in their laboratory colony. However, as the authors stated, their data do not permit them to conclude that these changes were due to selection resulting from the continued use of pheromone for the purpose of mating disruption, al-

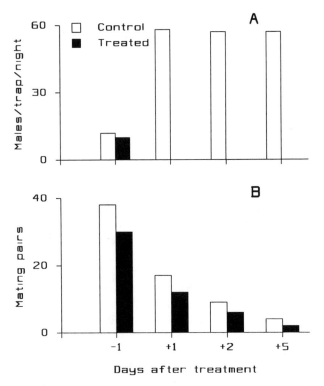

Figure 11.1. The number of cotton leafroller males captured in pheromone traps (A) and the number of mating pairs observed (B) in two plots, one treated with female sex pheromone for mating disruption and the other a control. Modified from Kehat et al. (1985).

though females emitting more pheromone may be less affected by the application of synthetic pheromone. This possibility is based on the observation that an increase in the concentration of lures increased male trap catch in fields where pheromone was used for mating disruption but not in controls (Doane and Brooks, 1981).

In a mating disruption study on the cotton leafworm Kehat et al. (1985) observed that the number of males captured in pheromone traps located in a plot treated with synthetic female pheromone was less than in the controls (Fig. 11.1A), although direct counts uncovered no marked differences in the number of mating pairs found in the treated and control plots (Fig. 11.1B). The authors attributed the absence of mating suppression to the low density of pheromone point sources used in the treated plot, an explanation that was substantiated in a subsequent study (Kehat et al.,

1986). However, even when the technical aspects of pheromone applications are resolved, there is rarely total inhibition of mating. If certain individuals are able to exploit other cues that may or may not be associated with the pheromone communication channel being disrupted by high concentrations of the female sex pheromones, they will increase their chances of locating a mate. Furthermore, if these traits are heritable, then a significant change in the genetic composition of the target population could result, and a form of behavioral resistance could arise.

In some moth species, female visual cues may play a role in mate location over some distance (Zagatti and Renou, 1984), while in others they are of importance only at very short distances (Charlton and Cardé, 1990). The extent to which such cues, or others such as sound (Conner, 1987), might serve to increase the probability of locating a mate when high concentrations of female pheromone are present in the environment has received little attention. The application of the synthetic female sex pheromone did not effectively disrupt mating in the fall cankerworm, *Alsophila pometaria*, even though it significantly reduced the ability of males to locate a pheromone source (Palaniswamy et al., 1986). Pheromone application technology, as seen in the case of the cotton leafworm (Kehat et al., 1985), may be one reason for the low level of mating disruption in this study. However, direct behavioral observations indicated that males of this species orient to vertical objects, as males were seen actively searching tree trunks in both treated and control plots. This idea was supported by the higher numbers of males that were captured in pheromone traps placed close (0.5 m) to trees than in those that were placed > 1 m away from the trunk. Furthermore, in the fall cankerworm (Palaniswamy et al., 1986), male orientation to and searching for vertical objects may be enhanced in the presence of the female sex pheromone, a behavior also reported for the gypsy moth (Charlton and Cardé, 1990). Thus, female sex pheromone applied for mating disruption could result in an increase in the probability of locating a mate through changes in certain male behaviors, especially at higher population densities.

Kipp et al. (1987) reported that in environments permeated with high titers of a two-component synthetic female sex pheromone, large spruce budworm females had a mating advantage over small ones. The reasons for the female size-related bias are presently unknown. The possibility that larger females emit higher pheromone concentrations or start calling earlier than smaller ones in the presence of conspecific pheromone (Palaniswamy and Seabrook, 1985) cannot be ruled out. However, males search host plants, and mating may occur prior to the female calling window (Kipp, personal communication), suggesting that visual, or tactile (Sanders, 1975), cues may play a role in the mating of this species. Thus, increased locomotory activity in the presence of pheromone (Palaniswamy

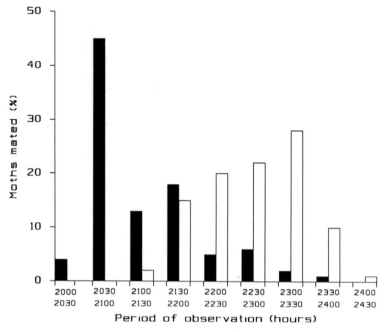

Figure 11.2. The proportion of *E. acrea* matings that occur in male leks (black bars) or as the result of male upwind flight to calling females (white bars) as a function of time. Modified from Willis and Birch (1982).

and Seabrook, 1978; Sanders, 1987) may result in larger females being more easily located by males than smaller individuals.

In the arctiid moth, *Estigmene acrea*, there is a dual pheromone-mediated mating system (Willis and Birch, 1982). Shortly after sundown, non-calling females are attracted into leks of males exposing their coremata, and at this time, mating may occur. Later in the scotophase, there is a second period of mating when males respond to calling females (Fig. 11.2). Wunderer et al. (1986) reported a similar system in two other arctiids. Very little is known about the dynamics of such dual systems, but the disruption of the female-mediated mating could select for greater reliance on male lekking as the means of reproduction. More recently, a dual system has been reported in the noctuid, *Trichoplusia ni*, with females flying upwind to males at dusk, several hours before calling females attract males (Landolt and Heath, 1990). In this species, selection of the male channel may be accentuated by the application of the female sex pheromone to the crop, as both the major component of the female sex pheromone and host plant's volatiles stimulate male sex pheromone release (Landolt and Heath, 1990). Thus, the use of the female sex pheromone for mating disruption of a species with a dual system may effectively reduce the female-mediated avenue of mate location but, by selecting for the male-mediated component, have little or no impact on overall mating success.

In the development of their model for intraspecific competition in the pheromone communication channel, Lundberg and Lofstedt (1987) assumed that males and females are randomly distributed within the habitat. However, this may be ecologically incorrect, as many insect species mate at specific sites within the habitat, termed conventional encounter sites (Parker, 1978). For example, the European corn borer in Iowa usually mates some distance from corn fields, in areas of dense grass where dew and free water are abundant (Showers et al., 1976; DeRozari et al., 1977). However, in southeastern Alberta, a more arid region where grasses are less abundant around fields, the majority of corn borer adults remain within the corn crop (Lee, 1988). Thus, if prevailing abiotic conditions modify mating behavior, is it possible that the continued use of pheromone in that part of the habitat containing the oviposition sites could result in mating activity occurring somewhere else? Any heritable behavioral traits that result in adults aggregating in sites outside the treated area, thereby increasing the probability of mating, would be selected for, especially if females are strongly attracted to host plant odors subsequent to mating (Landolt, 1989). The incidence of mating suppression of tethered codling moth females in a treated orchard was significantly lower for females placed on the tops of trees than for those at mid-canopy (Charmillot, 1990), indicating that even the selection of a different site on the host plant, where pheromone concentrations may be lower, could act to increase mating success. Females of certain species are able to detect conspecific pheromone (Palaniswamy and Seabrook, 1978; Noguchi and Tamaki, 1985), but the extent to which this modifies movement, and as a consequence distribution, under field conditions remains to be determined. Unfortunately, our knowledge of any factors affecting adult distribution and movement for most moth species is superficial at best, and considerable effort is required to rectify this situation.

11.4. Conclusions

It is evident that if we wish to optimize the deployment of pheromones in pest management, we must obtain a better understanding of the systems that we hope to manage. Significant progress is being made at the chemical, biochemical, and neurological levels (Payne et al., 1986;

Prestwich and Blomquist, 1987) but, in order to achieve our goals, considerably more attention must also be given to the behavioral, ecological, and genetic aspects of pheromone communication. To accomplish this, there must be a significant change in our philosophy with respect to future work on insect reproductive biology. The input of scientists taking basic and applied approaches is essential, but frequently, the level of communication between the two groups leaves much to be desired. Those interested in control often dismiss basic aspects as academically interesting but of no import to management, thereby failing to see how an understanding of the life history strategies may facilitate control. On the other hand, researchers studying the behavioral ecology and evolutionary biology often avoid pest species, in many cases an overt attempt to distance themselves from the perceived "spray and count" image, when in fact, insight gained from the study of basic questions could provide help to develop ecologically acceptable management programs. A good example of how the two approaches are complimentary is found in the research on European corn borer reproductive biology; not only did this work add to our basic understanding of encounter sites for mating, it resulted in a more effective deployment of pheromone traps for monitoring. In this chapter, I have suggested some areas where behavioral studies, examined from an evolutionary perspective, might help ameliorate the deployment of lepidopteran sex pheromones in management programs. However, if we are ever to meaningfully apply such an approach to any questions concerning the use of semiochemicals for the control of pest species, researchers in the basic and applied disciplines must open the lines of communication before real progress occurs.

Acknowledgements

I thank the editors, M. Isman and B. Roitberg, for the invitation to write this chapter. I am also grateful to T. Baker, R. Cardé, J. Delisle, L. Marshall, J. Millar, L. Royer, and L. Svard for their helpful comments and suggestions on an earlier version of this manuscript, and to P. Therrien for his assistance in the preparation of the figures.

References

Alford, D. V., P. W. Carden, E. B. Dennis, H. J. Gould, and J. D. R. Vernon. 1979. Monitoring codling and tortrix moths in United Kingdom apple orchards using pheromone traps. *Ann. Appl. Biol.* 91:165–178.

Allen, D. C., L. P. Abrahamson, D. A. Eggen, G. N. Lanier, S. R. Swier, R. S. Kelley, and M. Auger. 1986. Monitoring spruce budworm (Lepidoptera: Tortricidae) populations with pheromone-baited traps. *Environ. Entomol.* 15:152–165.

Baker, T. C. 1983. Variations in male oriental fruit moth courtship patterns due to male competition. *Experientia*. 39:112–114.

Baker, T. C., R. T. Cardé, and B. A. Croft. 1980. Relationship between pheromone trap capture and emergence of adult oriental fruit moths, *Grapholitha molesta* (Lepidoptera: Tortricidae). *Can. Entomol.* 112:11–15.

Baker, T. C., W. R. Meyer, and W. L. Roelofs. 1981. Sex pheromone dosage and blend specificity of response oriental fruit moths. *Entomol. Exp. Appl.* 30:269–279.

Baker, T. C., R. T. Staten, and H. M. Flint. 1990. Use of pink bollworm pheromone in the southwestern United States. In *Behavior-Modifying Chemicals for Insect Management: Applications of Pheromones and Other Attractants*, eds. R. L. Ridgway, R. M. Silverstein, and M. N. Inscoe. pp. 417–436. Marcel Dekker, New York.

Bergh, J. C. and W. D. Seabrook. 1986a. A simple technique for indexing the mating status of male spruce budworm, *Choristoneura fumiferana* (Lepidoptera: Tortricidae). *Can. Entomol.* 118:37–41.

Bergh, J. C. and W. D. Seabrook. 1986b. The mating status of spruce budworm males, *Choristoneura fumiferana* (Clem.) (Lepidoptera: Tortricidae), caught in pheromone-baited traps. *J. Entomol. Sci.* 21:254–262.

Boness, M., K. Eiter, and H. Disselnkötter. 1977. Studies of sex attractants of Lepidoptera and their use in crop protection, Pflanzenschutz-Nach. *Bayer* 30:213–236.

Cardé, R. T. 1976. Utilization of pheromones in the population management of moth pests. *Environ. Health Persp.* 14:133–144.

Cardé, R. T. 1990. Principles of mating disruption. In *Behavior-Modifying Chemicals for Insect Management: Applications of Pheromones and Other Attractants*, eds. R. L. Ridgway, R. M. Silverstein, and M. N. Inscoe, pp. 47–71. Marcel Dekker, New York.

Charlton, R. E. and R. T. Cardé. 1990. Orientation of male gypsy moths, *Lymantria dispar* (L.), to pheromone sources: the role of olfactory and visual cues. *Insect Behav.* 3:443–469.

Charmillot, P.-J. 1990. Mating disruption technique to control codling moth in western Switzerland. In *Behavior-Modifying Chemicals for Insect Management: Applications of Pheromones and Other Attractants*, eds. R. L. Ridgway, R. M. Silverstein, and M. N. Inscoe. pp. 165–182. Marcel Dekker, New York.

Collins, R. D. and R. T. Cardé. 1985. Variation in and heritability of aspects of pheromone production in the pink bollworm moth, *Pectinophora gossypiella* (Lepidoptera: Gelechiidae). *Ann. Entomol. Soc. Amer.* 78:229–234.

Collins, R. D. and R. T. Cardé. 1989a. Selection for altered pheromone-component ratios in the pink bollworm moth, *Pectinophora gossypiella* (Lepidoptera: Gelechiidae). *J. Insect Behav.* 2:609–621.

Collins, R. D. and R. T. Cardé. 1989b. Heritable variation in pheromone response of the pink bollworm, *Pectinophora gossypiella* (Lepidoptera: Gelechiidae). *J. Chem. Ecol.* 15:2647–2659.

Collins, R. D. and R. T. Cardé. 1990. Selection for increased pheromone response in the male pink bollworm, *Pectinophora gossypiella* (Lepidoptera: Gelechiidae). *Behav. Genet*. 20:325–331.

Collins, R. D., S. L. Rosenblum, and R. T. Cardé. 1990. Selection for increased pheromone titre in the pink bollworm moth, *Pectinophora gossypiella* (Lepidoptera: Gelechiidae). *Physiol. Entomol*. 15:141–147.

Conner, W. E. 1987. Ultrasound: its role in the courtship of the arctiid moth, *Cycnia tenera*. *Experientia*. 43:1029–1031.

Delisle, J. and J. N. McNeil. 1987. Calling behaviour and pheromone titre of the true armyworm *Pseudaletia unipuncta* (Haw.) (Lepidoptera: Noctuidae) under different temperature and photoperiodic conditions. *J. Insect Physiol*. 33:315–324.

DeRozari, M. B., W. B. Showers, and R. H. Shaw. 1977. Environment and the sexual activity of the European corn borer. *Environ. Entomol*. 6:657–665.

Doane, C. C. and T. W. Brooks. 1981. Research and development of pheromones for insect control on the pink bollworm, *Pectinophora gossypiella*. In *Management of Insect Pests with Semiochemicals: Concepts and Practice*, ed. E. R. Mitchell, pp. 285–303. Plenum Press, New York.

Du J.-W., C. Löfstedt, and J. Löfqvist. 1987. Repeatability of pheromone emissions from individual female ermine moths, *Yponomeuta padellus* and *Yponomeuta rorellus*. *J. Chem. Ecol*. 13:1431–1441.

Elkinton, J. S. and R. T. Cardé. 1981. The use of pheromone traps to monitor distribution and population trends of the gypsy moth. In *Management of Insect Pests with Semiochemicals: Concepts and Practice*, ed. E. R. Mitchell, pp. 41–55. Plenum Press, New York.

Gardiner, M. 1990. Influence of *Nosema fumiferanae* (Microspora) infection on flight of the obliquebanded leafroller (Lepidoptera: Tortricidae): a flight mill study. Master's thesis, Simon Fraser University, Burnaby, BC.

Glover, T., M. Campbell, P. Robbins, and W. Roelofs. 1990. Sex-linked control of sex pheromone behavioral responses in European corn borer moths (*Ostrinia nubilalis*) confirmed with TPI marker gene. *Arch. Insect Biochem. and Physiol*. 15:67–77.

Glover, T. J. and W. L. Roelofs. 1988. Genetics of lepidopteran sex pheromone systems. *ISI Atlas of Science* 1:279–282.

Han, E.-N. and A. G. Gatehouse. 1991. Genetics of precalling period in the oriental armyworm, *Mythimna separata* (Walker) (Lepidoptera: Noctuidae), and implication for migration. *Evolution* 45:1502–1510.

Hansson, B. S., C. Löfstedt, and W. L. Roelofs. 1987. Inheritance of olfactory response to sex pheromone components in *Ostrinia nubilalis*. *Naturwiss*. 74:497–499.

Hansson, B. S., C. Löfstedt, and S. P. Foster. 1989. Z-linked inheritance of male olfactory response to sex pheromone components in two species of tortricid moths, *Ctenopseustis obliquana* and *Stenopseustis* sp. *Entomol. Exp. Appl*. 53:137–145.

Haynes, K. F. and T. C. Baker. 1988. Potential for evolution of resistance to pheromones: worldwide and local variation in chemical communication system of pink bollworm moth, *Pectinophora gossypiella. J. Chem. Ecol.* 14:1547–1560.

Haynes, K. F., L. K. Gaston, M. Mistrot Pope, and T. C. Baker. 1984. Potential for evolution of resistance to pheromones: interindividual and interpopulational variation in chemical communication system of pink bollworm moth. *J. Chem. Ecol.* 10:1551–1565.

Haynes, K. F., T. A. Miller, R. T. Staten, W.-G. Li, and T. C. Baker. 1986. Monitoring insecticide resistance with insect pheromones. *Experientia* 42:1293–1295.

Haynes, K. F., T. A. Miller, R. T. Staten, W.-G. Li, and T. C. Baker. 1987. Pheromone trap for monitoring insecticide resistance in the pink bollworm moth (Lepidoptera: Gelechiidae): new tool for resistance management. *Environ. Entomol.* 16:84–89.

Henneberry, T. J. and T. E. Clayton. 1984. Time of emergence, mating, sperm movement, and transfer of ejaculatory duct secretory fluid by *Heliothis virescens* (F.) (Lepidoptera: Noctuidae) under reversed light-dark cycle laboratory conditions. *Ann. Entomol. Soc. Amer.* 77:301–305.

Hirano, C. and H. Muramoto. 1976. Effect of age on mating activity of the sweet potato leaf folder, *Brachmia macroscopa* (Lepidoptera: Gelechiidae). *Appl. Ent. Zool.* 11:154–159.

Howlader, M. A. and G. H. Gerber. 1986. Effects of age, egg development, and mating on calling behavior of the bertha armyworm, *Mamestra configurata* Walker (Lepidoptera: Noctuidae). *Can. Entomol.* 118:1221–1230.

Jutsum, A. R. and R. F. S. Gordon, eds. 1989. *Insect Pheromones in Plant Protection.* John Wiley, New York.

Kanno, H. 1979. Effects of age on calling behaviour of the rice stem borer, *Chilo suppressalis* (Walker) (Lepidoptera: Pyralidae). *Bull. Ent. Res.* 69:331–335.

Kehat, M., S. Gothilf, E. Dunkelblum, N. Bar Shavit, and D. Gordon. 1985. Night observations on the cotton leafworm, *Spodoptera littoralis*: reliability of pheromone traps for population assessment and efficacy of widely separated pheromone dispensers for mating disruption. *Phytoparasitica.* 13:215–220.

Kehat, M., E. Dunkelblum, S. Gothilf, N. Bar Shavit, D. Gordon, and M. Harel. 1986. Mating disruption of the Egyptian cotton leafworm, *Spodoptera littoralis* (Lepidoptera: Noctuidae), in cotton with a polymeric aerosol formulation containing (Z-E)-9,11-tetradecadienyl acetate. *J. Econ. Entomol.* 79:1641–1644.

Kipp, L. R., J. C. Bergh, and W. D. Seabrook. 1987. A spruce budworm mating bias in two-component pheromone environments. *Entomol. Exp. Appl.* 45:139–144.

Klun, J. A. and S. Maini. 1979. Genetic basis of an insect chemical communication system: the European corn borer. *Environ. Entomol.* 8:423–426.

Klun, J. A. and M. D. Huettel. 1988. Genetic regulation of sex pheromone production and response: interaction of sympatric pheromonal types of European corn borer, *Ostrinia nubilalis* (Lepidoptera: Pyralidae). *J. Chem. Ecol.* 14:2047–2061.

Kolodny-Hirsch, D. M. and C. P. Schwalbe. 1990. Use of disparlure in the management of the gypsy moth. In *Behavior-Modifying Chemicals for Insect Management: Applications of Pheromones and Other Attractants*, eds. R.L. Ridgway, R. M. Silverstein, and M. N. Inscoe, pp. 363–385. Marcel Dekker, New York.

Landolt, P. J. 1989. Attraction of the cabbage looper to host plants and host plant odor in the laboratory. *Entomol. Exp. Appl.* 53:117–124.

Landolt, P. J. and R. R. Heath. 1990. Sexual role reversal in mate-finding strategies of the cabbage looper moth. *Science* 249:1026–1028.

Lee, D. A. 1988. Moth density and oviposition patterns of the European corn borer, *Ostrinia nubilalis* (Lepidoptera: Pyralidae), in Alberta. *Environ. Entomol.* 17:220–224.

Lewis, W. J. 1981. Semiochemicals: their role with changing approaches to pest control. In *Semiochemicals: Their Role in Pest Control*, eds. D. A. Nordlund, R. L. Jones, and W. J. Lewis, pp. 3–12. John Wiley, New York.

Linn, C. E., M. G. Campbell, and W. L. Roelofs. 1988. Temperature modulation of behavioural thresholds controlling male moth sex pheromone response specificity. *Physiol. Entomol.* 13:59–67.

Linn, C. E. Jr. and W. L. Roelofs. 1985. Response specificity of male pink bollworm moths to different blends and dosages of sex pheromone. *J. Chem. Ecol.* 11:1153–1590.

Löfstedt, C. 1990. Population variation and genetic control of pheromone communication systems in moths. *Entomol. Exp. Appl.* 54:199–218.

Löfstedt, C., B. S. Hansson, W. Roelofs, and B. O. Bengtsson. 1989. No linkage between genes controlling female pheromone production and male pheromone response in the European corn borer, *Ostrinia nubilalis* Hübner (Lepidoptera: Pyralidae). *Genet. Soc. Amer.* 124:553–556.

Lundberg, S. and C. Löfstedt. 1987. Intra-specific competition in the sex communication channel: a selective force in the evolution of moth pheromones? *J. Theor. Biol.* 125:15–24.

Marshall. L. D. 1988. Small male advantage in mating of *Parapediasia teterrella* and *Agriphila plumbifimbriella* (Lepidoptera: Pyralidae). *Amer. Mid. Nat.* 119:412–419.

McNeil, J. N. 1991. Behavioral ecology of pheromone-mediated communication in moths and its importance in the use of pheromone traps. *Annu. Rev. Entomol.* 36:407–430.

Metcalf, R. L. 1980. Changing role of insecticides in crop protection. *Annu. Rev. Entomol.* 25:219–56.

Miller, J. R. and W. L. Roelofs. 1980. Individual variation in sex pheromone component ratios in two populations of the redbanded leafroller moth, *Argyrotaenia velutinana*. *Environ. Entomol.* 9:359–363.

Minks, A. K. and D. J. de Jong. 1975. Determination of spraying dates for *Adoxophyes orana* by sex pheromone traps and temperature recordings. *J. Econ. Entomol.* 68:729–732.

Mistrot Pope, M., L. K. Gaston, and T. C. Baker. 1982. Composition, quantification, and periodicity of sex pheromone gland volatiles from individual *Heliothis virescens* females. *J. Chem. Ecol.* 8:1043–1055.

Noguchi, H. and Y. Tamaki. 1985. Conspecific female sex pheromone delays calling behavior of *Adoxopyes* sp. and *Homona magnanima* (Lepidoptera: Tortricidae). *Jpn. J. Appl. Entomol. Zool.* 29:113–118.

Palaniswamy, P., E. W. Underhill, C. Gillott, and J. W. Wong. 1986. Synthetic sex pheromone components disrupt orientation, but not mating, in the fall cankerworm, *Alsophila pometaria* (Lepidoptera: Geometridae). *Environ. Entomol.* 15:943–950.

Palaniswamy, P., B. Galka, and B. Timlick. 1990. Phenology and infestation level of the European corn borer, *Ostrinia nubilalis* (Hubner) (Lepidoptera: Pyralidae), in southern Manitoba. *Can. Entomol.* 122:1211–1220.

Palaniswamy, P. and W. D. Seabrook. 1978. Behavioral responses of the female eastern spruce budworm, *Choristoneura fumiferana* (Lepidoptera: Tortricidae) to the sex pheromone of her own species. *J. Chem. Ecol.* 4:649–655.

Palaniswamy, P. and W. D. Seabrook. 1985. The alteration of calling behaviour by female *Choristoneura fumiferana* when exposed to the synthetic sex pheromone. *Entomol. Exp. Appl.* 37:13–16.

Parker, G. A. 1978. Evolution of competitive mate searching. *Annu. Rev. Entomol.* 23:173–196.

Payne, T. L., M. C. Birch, and C. E. J. Kennedy, eds. 1986. *Mechanisms in Insect Olfaction*. Clarendon Press, Oxford.

Prestwich, G. D. and G. J. Blomquist, eds. 1987. *Pheromone Biochemistry*. Academic Press, New York.

Phelan, P. L. and T. C. Baker. 1986. Male-size-related courtship success and intersexual selection in the tobacco moth, *Ephestia elutella*. *Experientia* 42:1291–1293.

Riedl, H., B. A. Croft, and A. J. Howitt. 1976. Forecasting codling moth phenology based on pheromone trap catches and physiological-time models. *Can. Entomol.* 108:449–460.

Ridgway, R. L., R. M. Silverstein, and M. N. Inscoe, eds. 1990. *Behavior-Modifying Chemicals for Insect Management: Applications of Pheromones and Other Attractants*. Marcel Dekker, New York.

Roelofs, W. L., J.-W. Du, C. Linn, T. J. Glover, and L. B. Bjostad. 1987a. The potential for genetic manipulation of the redbanded leafroller moth sex pheromone blend. In *Evolutionary Genetics of Invertebrate Behavior*, eds. M. D. Huettel and H. Oberlander, pp. 265–272. Plenum Press, New York.

Roelofs, W., T. Glover, X.-H. Tang, I. Sreng, P. Robbins, C. Eckenrode, C. Löfstedt, B.S. Hansson, and B. O. Bengtsson. 1987b. Sex pheromone production and perception in European corn borer moths is determined by both autosomal and sex-linked genes. *Proc. Natl. Acad. Sci.* 84:7585–7589.

Sanders, C. J. 1975. Factors affecting adult emergence and mating behaviour of the eastern spruce budworm, *Choristoneura fumiferana* (Lepidoptera: Tortricidae). *Can. Entomol.* 107:967–977.

Sanders, C. J. 1981. Sex attractant traps: their role in the management of spruce budworm. In *Management of Insect Pests with Semiochemicals: Concepts and Practice*, ed. E. R. Mitchell, pp. 75–91. Plenum Press, New York.

Sanders, C. J. 1987. Flight and copulation of female spruce budworm in pheromone-permeated air. *J. Chem. Ecol.* 13:1749–1758.

Sanders, C. J. and G. G. Wilson. 1990. Flight duration of male spruce budworm (*Choristoneura fumiferana* [Clem]) and attractiveness of female spruce budworm are unaffected by microsporidian infection or moth size. *Can. Entomol.* 122:419–422.

Showers, W. B., G. L. Reed, J. F. Robinson, and M. B. DeRozari. 1976. Flight and sexual activity of the European corn borer. *Environ. Entomol.* 5:1099–1104.

Sreng, I., T. Glover, and W. Roelofs. 1989. Canalization of the redbanded leafroller moth sex pheromone blend. *Arch. Insect Biochem. Physiol.* 10:73–82.

Sweeney, J. D. and J. A. McLean. 1987. Effect of sublethal infection levels of *Nosema* sp. on the pheromone-mediated behavior of the western spruce budworm, *Choristoneura occidentalis* Freeman (Lepidoptera: Tortricidae). *Can. Entomol.* 119:587–594.

Thornhill, R. and J. Alcock. 1983. *The Evolution of Insect Mating Systems*. Harvard Univ. Press, Cambridge.

Tumlinson, J. H. 1988. Contemporary frontiers in insect semiochemical research. *J. Chem. Ecol.* 14:2109–2130.

Turgeon, J. and J. McNeil. 1982. Calling behaviour of the armyworm, *Pseudaletia unipuncta*. *Entomol. Exp. Appl.* 31:402–408.

Turgeon, J. J., J. N. McNeil, and W. L. Roelofs. 1983. Responsiveness of *Pseudaletia unipuncta* males to the female sex pheromone. *Physiol. Entomol.* 8:339–344.

Wall, C. 1988. The application of sex-attractants for monitoring the pea moth, *Cydia nigricana* (F.) (Lepidoptera: Tortricidae). *J. Chem. Ecol.* 14:1857–1866.

Wall, C. 1990. Principles of monitoring. In *Behavior-Modifying Chemicals for Insect Management: Applications of Pheromones and Other Attractants*, eds. R. L. Ridgway, R. M. Silverstein, and M. N. Inscoe, pp. 9–23. Marcel Dekker, New York.

Willis, M. A. and M. C. Birch. 1982. Male lek formation and female calling in a population of the arctiid moth *Estigmene acrea*. *Science*. 218:168–170.

Wilson, G. G. 1987. Observations on the level of infection and intensity of *Nosema fumiferanae* (Microsporida) in two different field populations of the spruce budworm, *Choristoneura fumiferana*. *Inf. Report FPM-X-79*. Ontario, Canadian Forestry Service.

Witzgall, P. and B. Frérot. 1989. Pheromone emission by individual females of the carnation tortrix, *Cacoecimorpha pronubana*. *J. Chem. Ecol.* 15:707–717.

Wunderer, H., K. Hansen, T. W. Bell, D. Schneider, and J. Meinwald. 1986. Sex pheromones of two Asian moths (*Creatonotos transiens, C. gangis*; Lepidoptera: Arctiidae): behavior, morphology, chemistry and electrophysiology. *Exp. Biol.* 46:11–27.

Zagatti, P. and M. Renou. 1984. Les phéromones sexuelles des zygènes III. Le comportement de *Zygaena filipendulae* L. (Lepidoptera: Zygaenidae). *Ann. Soc. Entomol. Fr.* (*N.S.*) 4:439–454.

Taxonomic Index

ANIMAL KINGDOM

Arthropods

Insecta

Coleoptera

Anthonomus grandis (boll weevil) 115, 133
Brachinus sp. 107
Caryedes brasiliensis 169, 171, 194
Chrysolina sp. 97
Chrysomela 20-punctata 230
Cleridae 128
Deloyala guttata 198
Dendroctonus ponderosae 131
Dibolia spp. 227
Epicauta 115
Harrmonia axyridis 223
Hylobius pales (pales weevil) 115
Hypera sp. (clover weevils) 115
Hypera postica (alfalfa weevil) 115
Ips paraconfusus 164
Ips pini 164
Leptinotarsa decemlineata 159
Listroderes costirostris (vegetable weevil) 115
Myrmecaphodius excavaticollis 130, 133
Phratoria tibialis 170, 230
Phratoria vitellinae 169, 230
Pissodes approximatus 284, 285
Pissodes strobi 284, 285
Plagiodera versicolora 230
Popillia japonica (japanese beetle) 115
Scolytidae (scolytid beetles) 115
Triboleum castaneum 115

Diptera

Culex quinquefasciatus 197
Drosophila sp. 52, 71, 115, 190, 196, 246
Drosophila melanogaster 163, 191, 196, 199, 274
Drosophila pseudoobscura 187
Liriomyza sativae 188, 198
Mayetolia destructor 201
Musca domestica (housefly) 115, 186, 195, 200
Prodecidochares australis 201
Rhagoletis mendax 165
Rhagoletis pomonella 12, 165, 189, 190, 245, 247, 254

Hemiptera

Lygaeus equestris 226
Oncopeltus fasciatus 105, 167, 170, 220, 221, 223
Rhodnius prolixus 227

Homoptera

Acyrthosiphon nipponicus 223
aphididae 25
Aphis nerii 226
Echenopa binotata 187
Myzus persicae 197
Nilaparvata lugens 202
Shizaphis graminum 51

Hymenoptera

Andrena bees 130
Apanteles karyai 131

PLANT KINGDOM

Subject Index